Studies in Computational Intelligence

Volume 1066

The series "Studies in Computational Intelligence" (SCI) publishes new developments and advances in the various areas of computational intelligence—quickly and with a high quality. The intent is to cover the theory, applications, and design methods of computational intelligence, as embedded in the fields of engineering, computer science, physics and life sciences, as well as the methodologies behind them. The series contains monographs, lecture notes and edited volumes in computational intelligence spanning the areas of neural networks, connectionist systems, genetic algorithms, evolutionary computation, artificial intelligence, cellular automata, self-organizing systems, soft computing, fuzzy systems, and hybrid intelligent systems. Of particular value to both the contributors and the readership are the short publication timeframe and the world-wide distribution, which enable both wide and rapid dissemination of research output.

This series also publishes Open Access books. A recent example is the book Swan, Nivel, Kant, Hedges, Atkinson, Steunebrink: The Road to General Intelligence https://link.springer.com/book/10.1007/978-3-031-08020-3

Indexed by SCOPUS, DBLP, WTI Frankfurt eG, zbMATH, SCImago.

All books published in the series are submitted for consideration in Web of Science.

Khalid Raza
Editor

Nature-Inspired Intelligent Computing Techniques in Bioinformatics

Springer

Editor
Khalid Raza
Department of Computer Sciences
Jamia Millia Islamia
New Delhi, India

ISSN 1860-949X ISSN 1860-9503 (electronic)
Studies in Computational Intelligence
ISBN 978-981-19-6378-0 ISBN 978-981-19-6379-7 (eBook)
https://doi.org/10.1007/978-981-19-6379-7

This Springer imprint is published by the registered company Springer Nature Singapore Pte Ltd.
The registered company address is: 152 Beach Road, #21-01/04 Gateway East, Singapore 189721, Singapore

This volume is dedicated to my parents (Late Abdul Jabbar Qauis and Late Zulekha Khatoon), beloved wife (Farhat), my son (Arham), and little angel (Hoorain).

Preface

Nature-inspired computing (NIC) is an interdisciplinary field of study that combines computer science with other branches of sciences in order to develop new computational models based on naturally occurring phenomena. It refers to a series of computational techniques that provides the opportunity to develop new computational methods for studying, modeling, and applying naturally occurring phenomena to address complex real-world problems. There are over 150 nature-inspired algorithms, and their variants are already reported. Although the emergence of computational intelligence primarily consists of artificial intelligence, fuzzy logic, and evolutionary computing, however, it is observed that NIC intersects with computing intelligence, and hence, several cross-fertilizations of these two emerging fields have been observed and foreseen.

The last few decades witnessed growing applications of NIC methods in the field of Bioinformatics for analyzing biological data and addressing various disease-related complex problems. The advances in both computational intelligence and NIC techniques, together with the availability of large-scale clinical, experimental, and multi-omics data from various high-throughput laboratories demonstrate big hope to the researchers, bioinformaticians, genome researchers, and clinical bioinformaticians to devise and employ NIC-based intelligent approaches for various complex problems in bioinformatics.

The main aim of this book is to encapsulate and occupy recent advances and state-of-the-art applications of NIC techniques in the field of bioinformatics and computational biology, which would aid medical sciences in various clinical applications. This edited volume would be covering fundamental applications, scope, and future perspectives of NIC techniques in bioinformatics, genomic profiling, gene expression data classification, DNA computation, systems and network biology, solving personalized therapy complications, antimicrobial resistance in bacterial pathogens, and computer-aided drug design, discovery, and therapeutics. Besides, this volume also aims to cover the role of NIC techniques in various diseases and disorders, including cancer detection and diagnosis, breast cancer, lung disorder detection, disease biomarkers, and potential therapeutics identifications.

New Delhi, India Khalid Raza

Acknowledgments The editor and contributors are thankful to all the anonymous reviewers for their valuable expert comments and suggestions on the chapters. Thank goes to the series editor, editors, publication and production manager, and other editorial staff of Springer Nature for their overwhelming support during the production of this book. Special thank goes to Ms. Sahar Qazi (DST-Inspire Fellow, Department of Computer Science, Jamia Millia Islamia) for her consistent and overwhelming support.

About This Book

The book titled *Nature-Inspired Intelligent Computing Techniques in Bioinformatics* encapsulates the recent applications of nature-inspired computing techniques in the field of bioinformatics, especially diagnosis, prognosis, and therapeutics of cancers, genome profiling of colorectal cancer, detection of lung disorder, classification of gene expression data, genomic diagnosis of antimicrobial resistance (AMR), biomarkers and potential therapeutics identification, and drug design, development, and therapeutics. It covers various nature-inspired techniques and hybrid intelligent systems which are widely appreciated for the diagnosis, prognosis, and optimization of therapeutics of various cancers and other diseases. This edited volume has 16 chapters, divided into three parts.

First Part contains two introductory chapters that cover the scope and applications of nature-inspired computing in bioinformatics and an overview and future perspectives of leveraging healthcare systems with nature-inspired computing techniques:

Second Part comprises eight chapters that cover applications of nature-inspired computing (NIC) in cancer research which includes NIC in breast cancer research; genomic profiling of colorectal cancer using NIC techniques; classification of high-dimensional and complex gene expression data; optimized NIC algorithms for lung disorder detection; swarm intelligence-based NIC algorithms and their applications in cancer detection and diagnosis; scope and applications of artificial immune systems toward analysis and diagnosis of complex problems; NIC in BAT echolocation to BAT algorithm; and social, emotional and ethical (SEE) attributes, which configures our bioinformatics systems to activate the hidden forces to shape human decisions,

Third Part is a collection of six chapters that demonstrate the application of NIC techniques in drug design, development, and therapeutics. Chapters of this section demonstrate the applications of NIC and AI algorithms for solving personalized therapy complications; genomic diagnosis of antimicrobial resistance; consequential innovations in NIC techniques for biomarkers and potential therapeutics identification; NIC techniques in drug design, development, and therapeutics; illustrious implications of NIC methods in therapeutics and computer-aided drug design; and

NIC methods in bioinformatics for the drug discovery against promising molecular targets carbonic anhydrases and serine/threonine kinases for cancer treatment.

The contributing authors of this volume are experts in their fields, and they are from various reputed universities and institutions across the globe. This volume is a valuable and comprehensive resource for computational biologist, medical oncologists, researchers from computer science, data science, medical sciences, bioinformatics, medical informatics, and health informatics, along with graduate and research students of computational biology, and computational and medical oncology.

Contents

Editor and Contributors

About the Editor

Dr. Khalid Raza is an Assistant Professor at the Department of Computer Science, Jamia Millia Islamia (Central University), New Delhi. Dr. Raza has been honored with "ICCR Chair Visiting Professor" at the Faculty of Computer and Information Sciences, Ain Shams University, Cairo, Egypt. He has over 12 years of teaching and research experience in the field of Computational Intelligence and its various applications. He has contributed over 80 research articles in reputed Journals and Edited Books, including one sole-authored book, and several edited books. Dr. Raza is currently an Academic Editor of *PeerJ* Computer Science Journal and guest editor of Natural Products Communications. He has reviewed over 150 research articles for reputed Journals/Conferences in the last 5 years. Dr. Raza has delivered several keynote addresses, invited talks, public lectures, and seminars in National and International conferences, workshops, and chaired technical sessions at various conferences. He has also executed two Indian governments funded research projects. Dr. Raza is a member of MIR Lab (USA), CSI (India), and SCRS (India). Recently, Dr. Raza has been listed in the world's top 2% scientists in the study carried out by Stanford University, USA. His research interest includes Computational Intelligence methods and its applications in Bioinformatics, Oncology, Viro-informatics, and Health-informatics.

Contributors

Abuassi Mohammad Ahmad Department of Internal Medicine, Jordan Hospital, Amman, Jordan

Ahmad Shaban Department of Computer Science, Jamia Millia Islamia, New Delhi, India

Ajeeshkumar K. K. Tumor Biology, ICMR-National Institute of Pathology, New Delhi, India

Akermi Sarra Indian Science and Technology Foundation (ISTF), New Delhi, India

Alam Aftab Memorial Sloan Kettering Cancer Center, New York City, NY, USA

Ali Asim Central Council for Research in Unani Medicine (CCRUM), Ministry of AYUSH, Govt of India, New Delhi, India

Asif Aisha Department of Computer Science, Jamia Millia Islamia, New Delhi, India

Asim Mohammad Department of Clinical and Experimental Medicine, University of Surrey, Guildford, UK

Bano Nagmi Department of Computer Science, Jamia Millia Islamia, New Delhi, India

Biswas Ritushree Indian Science and Technology Foundation (ISTF), New Delhi, India

Dasegowda K. R. Department of Biotechnology, School of Applied Sciences, REVA University, Bangalore, Karnataka, India

Dey Abira Indian Science and Technology Foundation (ISTF), New Delhi, India

Dey Suparna St. Xavier's College (Autonomous), Kolkata, India

Idoipe Jean Bernard Insight Biosolutions (IBS), Rennes, France

Iqbal Naiyar Department of Computer Science and IT, Maulana Azad National Urdu University, Hyderabad, India

Islamovna Umurzakova Gavkharoi International Medical Faculty, Osh State University, Osh City, Kyrgyzstan

Jana Chandramohan Department of Chemistry, Michael Madhusudan Memorial College, Durgapur, West Bengal, India

Jayant Sunil Annotation Analytics Pvt. Ltd., Gurgaon, India

Johari Surabhi Institute of Management Studies (IMSUC), Ghaziabad, Uttar Pradesh, India

Kalmatov Roman International Medical Faculty, Osh State University, Osh City, Kyrgyzstan

Kareem Mohammed Abdul Department of Biochemistry, School of Sciences, Indira Gandhi National Open University, New Delhi, India

Karwasra Ritu Central Council for Research in Unani Medicine (CCRUM), Ministry of AYUSH, Govt of India, New Delhi, India

Kati Ahmet Experimental Medicine Research and Application Center, University of Health Sciences Turkey, Uskudar, Istanbul, Turkey;
Department of Biotechnology, Institution of Health Sciences, University of Health Sciences Turkey, Uskudar, Istanbul, Turkey

Kaur Punit Department of Biophysics, All India Institute of Medical Sciences, New Delhi, India

Khan Fatima Nazish Department of Biotechnology, New Delhi, India

Khanam Ayesha Department of Computer Science, Jamia Millia Islamia, New Delhi, India

Khanna Kushagra Llyod Institute of Management and Technology, Greater Noida, Uttar Pradesh, India

Lakshman Narayana V. Vignan's Nirula Institute of Technology and Science for Women, Guntur, Andhra Pradesh, India

Lakshmi Patibandla R. S. M. Department of IT, VFSTR, Guntur, Andhra Pradesh, India

Larzat Nathalie Insight Biosolutions (IBS), Rennes, France

Lee Nicholas Franciss Technical University of Munich, Munich, Germany

Lee Ruoya Insight Biosolutions (IBS), Rennes, France

Magarde Bhupendra Kumar International Medical Faculty, Osh State University, Osh City, Kyrgyzstan

Malhotra Rishabh Llyod Institute of Management and Technology, Greater Noida, Uttar Pradesh, India

Meyyazhagan Arun Cytogenetic Department, Biomedical Research Centre, Euroespes, Spain

Mitalipova Ainura International Medical Faculty, Osh State University, Osh City, Kyrgyzstan

Mohammed Mohammed Razeeth Shait Moores Cancer Center, University of California, La Jolla, CA, USA

Mouna Azaroual Tambov State Medical University, Tambov City, Russia

Mrad Yves Najm Gilbert and Rose-Marie Chagoury School of Medicine, Lebanese American University, Blat, Lebanon

Muhammad Waseem International Medical Faculty, Osh State University, Osh City, Kyrgyzstan

Nigam Anshul Amity University Mumbai, Somathne, Panvel, Mumbai, Maharashtra, India;
FBE Technologies Pvt. Ltd., Kanpur, India

Parween Zeenat Patliputra University, Patna, Bihar, India

Pavani V. Vignan's Nirula Institute of Technology and Science for Women, Guntur, Andhra Pradesh, India

Peerzada Mudasir Nabi Tumor Biology Department, National Institute of Pathology, Indian Council of Medical Research, New Delhi, India

Peri Sameera Department of Biotechnology, School of Applied Sciences, REVA University, Bangalore, Karnataka, India

Prabhavathi H. Department of Studies & Research in Biotechnology, Tumkur University, Tumakuru, Karnataka, India

Qazi Sahar Department of Computer Science, Jamia Millia Islamia, New Delhi, India

Qazi Waaris Department of Biotechnology, Jamia Millia Islamia, New Delhi, India

Qureshi Mohammad Irfan Department of Biotechnology, Jamia Millia Islamia, New Delhi, India

Radhakrishnan Akshar Department of Biotechnology, School of Applied Sciences, REVA University, Bangalore, Karnataka, India

Radhika P. Vignan's Nirula Institute of Technology and Science for Women, Guntur, Andhra Pradesh, India

Rafeeq Misbahuddin M. Department of Pharmacology, Faculty of Medicine, King Abdulaziz University, Rabigh, Jeddah, Kingdom of Saudi Arabia

Rahat Syed Ali Abbas International Medical Faculty, Osh State University, Osh City, Kyrgyzstan

Rambabu Majji Department of Biotechnology, School of Applied Sciences, REVA University, Bangalore, Karnataka, India

Rana Shweta Division of BMI, ICMR, New Delhi, India

Ranjan Mritunjay Kr. Department of Computer Science, Magadh University, Bodh Gaya, India;
SOCSE, Sandip University, Nashik, India

Raza Khalid Department of Computer Science, Jamia Millia Islamia, New Delhi, India

Rikhari Deeksha Department of Biotechnology, Motilal Nehru National Institute of Technology, Allahabad, India

Rizvi Masood Ahmad Department of Chemistry, University of Kashmir, Srinagar, India

Sahu Ankita Tumor Biology, ICMR-National Institute of Pathology, New Delhi, India

Sattar Arif Md. Department of Computer Science, Anugrah Memorial College, Gaya, India

Sayeed Salwa Department of Computer Science, Jamia Millia Islamia, New Delhi, India

Sethi Geetika ICMR-AIIMS Computational Genomics Centre, New Delhi, India

Sharma Ashwani Insight Biosolutions (IBS), Rennes, France

Sharma Nitin Shriram Murti College of Engineering & Technology (Pharmacy), Bareilly, India

Sharma Priyanka Department of Biophysics, All India Institute of Medical Sciences, New Delhi, India;
Division of BMI, ICMR, New Delhi, India

Sharma Shivkant Department of Pharmacology, Gurugram University, Gurugram, India

Sheikh Kayenat CSIR-Institute of Genomics and Integrative Biology, New Delhi, India;
Department of Computer Science, Jamia Millia Islamia, New Delhi, India

Sherbaevna Salieva Rana International Medical Faculty, Osh State University, Osh City, Kyrgyzstan

Siddiqui Mohd Faizan International Medical Faculty, Osh State University, Osh, Kyrgyzstan

Simak Jasbir Kaur Amity University Mumbai, Somathne, Panvel, Mumbai, Maharashtra, India

Singh Astha Tumor Biology, ICMR-National Institute of Pathology, New Delhi, India

Singh Gajendra Piramal Foundation for Education Leadership, Jhunjhunu, India

Singh Harpreet Division of BMI, ICMR, New Delhi, India

Singh Surender Department of Pharmacology, All India Institute of Medical Sciences, New Delhi, India

Sinha Subrata Centre for Biotechnology and Bioinformatics, Dibrugarh University, Assam, India

Srivastava Ankit Department of Biotechnology, Motilal Nehru National Institute of Technology, Allahabad, India

Srivastava Sameer Department of Biotechnology, Motilal Nehru National Institute of Technology, Allahabad, India

Tashmatova Nazgul International Medical Faculty, Osh State University, Osh City, Kyrgyzstan

Tripathi Manish Kumar Department of Biophysics, All India Institute of Medical Sciences, New Delhi, India

Vasudevan Karthick Department of Biotechnology, School of Applied Sciences, REVA University, Bangalore, Karnataka, India

Verma Saurabh Tumor Biology, ICMR-National Institute of Pathology, New Delhi, India

Villela Ricardo National Autonomous University of Honduras, Tegucigalpa, Honduras

Yadav Manoj Kumar Department of Biomedical Engineering, SRM University, Sonepat, Haryana, India;
Department of Bioinformatics, SRM University, Delhi-NCR, Haryana, India

Preliminaries

The Scope and Applications of Nature-Inspired Computing in Bioinformatics

Waaris Qazi, Sahar Qazi, Naiyar Iqbal, and Khalid Raza

Abbreviations

AAA	Artificial algae algorithm
AI	Artificial intelligence
AIS	Artificial immune system
CI	Computational intelligence
CSO	Cat swarm optimization
CSOA	Chicken swarm optimization algorithm
DE	Differential evolution
EA	Evolutionary algorithms
ESA	Elephant search algorithm
FSA	Fish swarm algorithm
GA	Genetic algorithm
GBC	Genetic bee colony
GP	Genetic programming
CGP	Cartesian genetic programming
GWO	Grey wolf optimization
MFO	Moth flame optimization
NIC	Nature-inspired computing

W. Qazi
Department of Biotechnology, Jamia Millia Islamia, New Delhi, India

S. Qazi · K. Raza (✉)
Department of Computer Science, Jamia Millia Islamia, New Delhi, India
e-mail: kraza@jmi.ac.in

N. Iqbal
Department of Computer Science and IT, Maulana Azad National Urdu University, Hyderabad, India

K. Raza (ed.), *Nature-Inspired Intelligent Computing Techniques in Bioinformatics*,
Studies in Computational Intelligence 1066,
https://doi.org/10.1007/978-981-19-6379-7_1

PSO Particle swarm optimization
SI Swarm intelligence
WOA Whale optimization algorithm
WSN Wireless sensor networks

1 Introduction

From the family of—"*natural computing*" (Brabazon et al. 2015), nature-inspired computing (NIC) adopts its functionality from the behavior of humans, animals, birds, and insects. A few examples to cite are—ant colonization, the swarm of bees, the echoing of bats, and discipline of fish. These natural phenomena raised interest in computational scientists to develop computational algorithms that can help in solving various problems faced in science and engineering (Darwish 2018). This new field of computer science has opened up a way for sub-fields such as—neural networking, swarm intelligence, artificial immune systems, and evolutionary algorithms. Nonetheless, simulating natural reactions and behaviors using the software can be tricky and pretty challenging. However, a typical NIC system is based on self-analyzing, learning, and adapting to the environment or problem nature, aiding in making better decisions (What is Nature-Inspired Computing?—Computer Science Degree Hub 2016).

Nature-inspired computing has been formed by combining artificial intelligence (AI), biological phenomena, and computational intelligence (CI). The NIC algorithms encapsulate natural features such as—(a) *diversity*, (b) *self-realization*, (c) *learning*, (d) *adaptation*, (e) *rationality and processing*, (f) *evolving*, and (g) *dynamic behavior*. Apart from being deployed on electronic hardware, these NIC algorithms can also be used on biological macromolecules—DNA, RNA, proteins, ligands, ions, etc., on computing devices (Computing Nature 2013). Figure 1 depicts how NIC came into existence, which simply combines—artificial intelligence (A), biological phenomenon (B), and computational intelligence (C). Henceforth, the formula: $A + B + C = NIC$.

Biologists are heavily dependent on computational methods and strategies to analyze humongous biological/medical data. The NIC encapsulates an ensemble of myriad studies of computer science, statistics, mathematics, and biological sciences where the essence is to adapt and develop robust competing techniques just like nature. It is a novel approach to optimization algorithms that is motivated by the dynamics of the biological evolution of our natural milieu. The main objective of this chapter is to disseminate information about NIC systems, their types, scope, applications in bioinformatics research, advantages and disadvantages along with future prospects.

The chapter briefly introduces the readers to the concept of NIC systems, Sect. 2 provides a detailed description of the two types of NIC models—traditional and new

Fig. 1 Precursors for the development of nature-inspired computing

models. Section 3 further elaborates on the scope and applications of the discussed NIC models in bioinformatics research. Section 4 lists out major limitations of NIC. Section 5 highlights the future prospects of NIC systems. This chapter summarizes the different models of NIC computation that are heavily deployed in bioinformatics and their scope in biological research.

2 Nature-Inspired Models of Computation

There are two categories of NIC models—(a) *traditional models* that encapsulate—cellular automata, neural networking, and evolutionary algorithms, while (b) *new models* that consist of—swarm intelligence, artificial immune systems, membrane computing, amorphous computing, etc. Figure 2 depicts the two predominant classes of NIC models used today below. Initially, the human brain and the Darwinian concept of natural selection and survival-of-the-fittest were used as inspirations for the development of NIC algorithms. However, with their success and effective solutions to complex problems, scientists further searched for bio-inspiration such as birds, human innate immune systems, plasma membranes semi-permeability function, and various microorganisms. These were clubbed as new models of NIC.

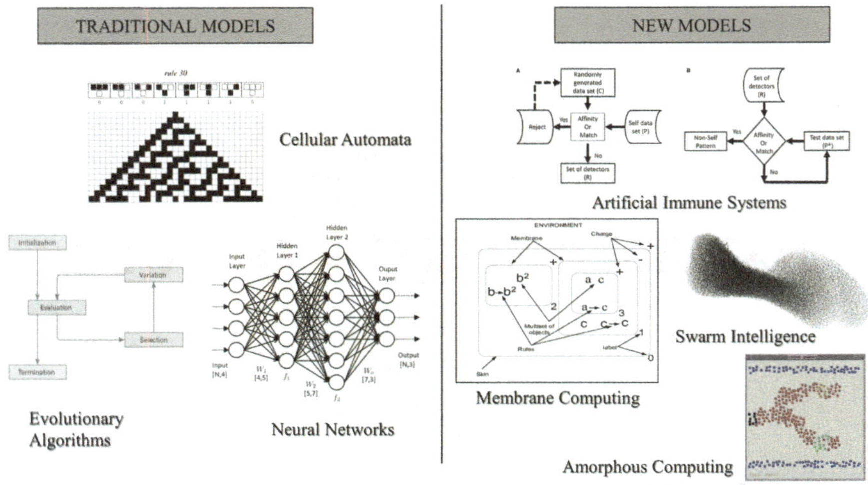

Fig. 2 Two predominant classes of nature-inspired computing models

2.1 *Traditional NIC Models*

(a) *Cellular automata:* It is a dynamic NIC model that encapsulates an array of cells where each cell can be in a fixed number of states. The space and time are discrete in this model. A cellular automaton (automata–plural) transforms the state of the cells as per the transformation rules that are predefined by the user. The transformation of states is in a synchronous manner. States of cells are computed by a transformation rule that depends on the cell's present state and the state of all the neighboring cells (Cosentino et al. 2013). The best example of cellular automata is *Conway's Game of Life.* This NIC model has also been deployed for biological phenomena such as growth, reproduction, and evolution (Akdur 2011).

(b) *Neural networks:* Taking inspiration from the highly complex processing units—the human brain, neural network models have a significant role in solving myriad biological and engineering problems today. It was first proposed back then in 1943 by McCullock and Pitts and has been considered the philosophical computational theory (McCulloch and Pitts 1943). This NIC model aims to employ the concepts of the human brain and connectionism to deduce effective solutions for complex problems. The NIC model can be considered as weighted directed graphs where the neurons are simply nodes and the connection between two neurons (called edges) has some weight. A set of instructions allows executing a command and updates the state of each neuron, thus producing a signal to the computing system (Piccinini and Bahar 2012).

(c) *Evolutionary algorithms:* Inspired by the Darwinian evolution, evolutionary algorithms have made NIC more robust and efficient. An evolutionary algorithm model consists of a population of individuals, a fitness measure and some operators based on which the population reproduces offspring, thus forming a new generation. The main operators used in an evolutionary algorithm model are—mutation and recombination which allows for variation in each offspring produced. At every step, the individuals from the population are tested based on their fitness measures. This is where the "*survival-of-the-fittest*" comes into play. The new generation of offspring is formed by the fit individuals from the previous population (parents) by using operators. This process of forming new populations keeps going on until it converges on a genetically robust population of individuals; i.e., all the individuals pass the fitness score with the least error/bias (Back et al. 1997). The commonly used evolutionary algorithms include—genetic algorithms (GA), genetic programming (GP) (Thierens 2010), and Cartesian genetic programming (CGP) (Manazir and Raza 2019, 2022).

2.2 New NIC Models

(a) *Swarm intelligence:* Also known as—"*Collective intelligence*" has been inspired by microorganisms such as bacteria and insects like—ants, termites, bees, spiders, birds, and fish. The way these organisms associate with one another concerning their surroundings is the main essence of the swarm intelligence model. Therefore, swarm intelligence can be described as a problem-solving approach that utilizes the association of individuals among themselves taking their surroundings into consideration. *Particle swarm optimization* (PSO) uses the same idea; however, it has a larger search space; thus, it gives more solutions to a given problem. The PSO model consists of a swarm of particles where each particle represents a solution. These particles have their velocity with which they tend to move toward the best possible position. This movement thus occurs in a larger search space, and particles ultimately converge toward the best possible positions. PSO approach has been used in myriad optimization problems, unsupervised machine learning, gaming problems, etc. Swarm intelligence is known to have characteristics of adaptability, scalability, self-organization, parallelism, and tolerance (Darwish 2018). Some of the algorithms that have been widely used to optimize and find the best solutions for complex biological problems are discussed as follows:

- *Chicken swarm optimization algorithm* (CSOA)—It is one of the approaches that utilize a chicken swarm to find Pareto optimal solutions by deploying an aggregate function to describe the social arrangement of chickens in search space. In simpler terms, the algorithms use chickens' movement to find optimal solutions, thus achieving optimization. It is heavily deployed in biological problem optimization (Zouache et al. 2019).
- *Artificial bee colony* (ABC)—It is another swarm intelligence meta-heuristic algorithm that is inspired by the foraging behavior of honeybees. It was first described by Karaboğa in 2005. Here, the computational colony comprises

three bee groups—employed bees which are equivalent to possible solutions for a given problem, onlookers, and scouts. The source food (nectar) represents possible solutions to the problem while the amount of nectar represents the robustness of the solution. The search takes place for nectar wherein the scouts and onlookers hunt for nectar positions, whereas the employed bee keeps on changing the position of the nectar and keeps the new position of nectar in her memory (forgetting the previous position). The cycle keeps on repeating until all the employed bees share the nectar positions with scouts and onlooker bees, and then they start to compute the amount of nectar from each position. The greater the nectar amount, the better the nectar position (Karaboga 2005). This method also has a new extension called—*genetic bee colony* (GBC)—that has been effectively used to classify gene microarrays (Alshamlan et al. 2015).

- *Fish swarm optimization* (FSO)—It is another intelligent swarm approach that is inspired by fish social behavior. It consists of searching, swarming, and mimicking behaviors. The search for food by stealthily avoiding the predators' eyes and overall socializing is the main behavior we observe in fish swarms. Very similarly, the algorithm behaves in such a way that it learns to hunt for possible best solutions for a given problem by identifying numerous new directions (larger search space). This algorithm is known for its rapid convergence with effective and robust solution ensembles (Tsai and Lin 2011).
- *Cat swarm optimization* (CSO)—First proposed by Chu and Tsai (2007), this algorithm is powered by the behavior of felines (Chu and Tsai 2007). This approach mimics a cat resting and tracing behaviors wherein every cat indicates a set of possible solutions with a position, fitness measure, and a flag. Every cat has its own speed with which it moves in search space, and the fitness measure decides how fit a cat is, the flag ultimately classifies the cats into either tracing or resting mode. This algorithm keeps on running until the best cat (best possible solution) is achieved. It has various extensions and hybrid models and is highly used in facial recognition, image segmentation, removing noises from CT scans, ECG signal classification, etc. (Ahmed et al. 2020).
- *Whale optimization algorithm* (WOA)—It is a meta-heuristic algorithm that copies the social etiquette of humpback whales and is inspired by the bubble-net hunting strategy (Mirjalili and Lewis 2016).
- *Elephant search algorithm* (ESA)—A meta-heuristic algorithm inspired by the social traits of elephants. The algorithm utilizes male elephants for global exploration of solutions while female elephants for local exploration, thus finding an optimal solution. The searching elephants terminate with optimal solutions irrespective of their search positions. This algorithm allows for identifying robust solutions from a larger search space (Tian et al. 2018).
- *Moth flame optimization* (MFO)—It takes inspiration from the navigation method employed by moths (transverse orientation) during the night. During nights, moths tend to take up a fixed angle respective to the moon for better

flying. This helps them to fly in a straight line for greater distances. Nonetheless, these moths often get trapped or diverted because of the artificial lights—streetlights, car lights, etc., which leads these moths to get trapped in a spiral path around these artificial sources of light. This strategy has been deployed to model for efficient optimization in engineering, biological complex process, biochemical process modeling, etc. (Mirjalili 2015).

- *Grey wolf optimization* (GWO)—A meta-heuristic optimization algorithm that aims to copy the attacking behavior of grey wolves and is generally a large-scale search algorithm. It considers four grey wolves for four different tasks—hunting, searching for prey, encircling prey, and attacking prey, thus similarly identifying the best possible solutions for an optimization problem (Mirjalili et al. 2014).
- *Artificial algae algorithm* (AAA)—A meta-heuristic NIC method that is inspired by photosynthetic algae. It is based on the evolutionary process, movement, and adaptation of the algae. This algorithm has been recognized to have the immense computational power to produce reliable optimization in real-life problems (Uymaz et al. 2015).

(b) *Artificial immune systems* (AIS): It is also known as—"*immunocomputing*" and is inspired by the natural immune systems of organisms in real life. Immune systems play a pivotal role in every organism—they help in identifying the foreign bodies, kill bacteria/viruses/any other microorganism, thus shielding the body, keeping a log memory of the immunity of the body, self-healing, and providing tolerance. They have a role in guarding organisms against any harmful external agent, thus with this as an inspiration, computational scientists developed artificial immune systems that have the potential of natural immune systems. These AIS models are heavily deployed today for different purposes such as virus identification and termination, problem rectification in time series data, error identification, pattern rectification, machine learning, bioinformatics, and optimization problems (Timmis et al. 2004).

(c) *Membrane computing*: As the name suggests, this NIC model has been adopted from the membranes of living cells. Membranes are compartmentalized and have a nested hierarchical framework. Inside membrane, surrounded cells are different organelles that perform essential different tasks for cells to multiply and grow further. In reality, these membranes allow for passive and active transport of important material for cells. Thus, artificial membranes have been developed by taking natural membranes as inspiration. In membrane computing, these artificial membranes start with an arrangement that encapsulates some objects (for multiplication and growth). This model begins by selecting rules for different objects in a random fashion. The result of membrane computing is accumulated from a predefined output position. Membrane computing has been deployed to understand various natural phenomena such as photosynthesis, signaling pathways of diseases, cell-mediated immunity, modeling of biological processes, and machine learning (Păun 2010).

(d) *Amorphous computing*: This NIC model has been inspired by the concept of morphogenesis which helps cells in a living organism associate by following the genetic rule of the nucleic acid DNA of the organism. This association provides a sync between cells of the organism. Thus, amorphous computing also helps to identify novel programming approaches to produce a sync and reliability between computational elements in a system (Stark 2013). Amorphous computing has been used in systems biology, synthetic biology and modeling biological processes. It is also deployed to solve various engineering problems (Grozinger et al. 2019). This model has helped to achieve morphological and cognitive characteristics of living beings in computers (Pfeifer and Bongard 2006).

3 Scope and Applications of NIC Models in Bioinformatics

These NIC models that have been discussed in detail in Sect. 2 are widely popular in engineering, medical, communication, R&D, etc. These models also have impacted biological research, especially since they have a crucial role in bioinformatics. These NIC models have provided a way in transforming bioinformatics into *translational bioinformatics* (Qazi and Raza 2021; Ahmad et al. 2021; Alam and Raza 2021). Today, nature-inspired computing is heavily used in synthetic biology, systems biology, molecular computing, epidemiology, drug designing and repurposing, cellular computing, quantum computing, simulation, etc. These NIC models are high in demand due to these reasons: (i) they are easy to use; (ii) no data arrangement is required; (iii) easily identify the nearest optimal solution; and (iv) has greater applicability and are flexible. In bioinformatics, major research areas are *sequence alignment, structure prediction of macromolecules, structure alignment, motif and domain identification, molecular docking, molecular dynamic simulation, disease prediction, risk susceptibility, functional analyses*, etc. NIC models have extended themself into various research-based applications in bioinformatics for reducing data dimensionality, ease of implementation and quick solution identifications. Figure 3 highlights the scopes and applications of NIC models in bioinformatics.

Different inspirations from nature have provided with umbrellawide options of computational approaches that have aided researchers to solve apex problems. Table 1 is a summary of all the natural inspirations and their artificial models that are popular in bioinformatics as applications today.

4 Limitations of Nature-Inspired Computing Models in Bioinformatics

Nature-inspired computing has proved its power for the past couple of decades now. However, with appreciation also comes a few disappointments. Although we have developed algorithms inspired by the human brain, however, the human brain is a

Fig. 3 Scope and applications of nature-inspired computing models in bioinformatics

complex organ. Its cognition, thinking and learning capacity and processing capabilities are yet to be understood and worked on. Therefore, the NIC systems that have been developed using the brain as an inspiration cannot be accepted for complete processing purposes. Neural networks and deep learning models that are currently being deployed almost in every sphere of biological research cannot be clubbed as—"*one-algorithm-solves-all*"! Moreover, due to constrained computational power and infrastructure, NIC models are still limited to only a few real-life applications. Most of these NIC system chips are based on the traditional von Neumann architecture that is made of semiconductor materials. To match up with the dynamic biological processes, it becomes important to develop sensors, memory containers, etc., that can work with external software to better analyze biological data (Markram et al. 2015). However, it is evident that NIC models have significantly helped in eliminating and reducing the curse of dimensionality (CoD). However, these algorithms still have a higher time complexity when compared to traditional algorithms namely—neural networks and evolutionary algorithms, etc. (Ghareh et al. 2019).

5 Future Prospects

Nature-inspired computing algorithms work perfectly either individually. However, these individual algorithms could further be combined for enhanced performance for complex optimization problems. Because of combining two or more NIC algorithms, the problem of speed and run time complexity may exceed tremendously. Therefore, NIC algorithms could be combined and worked with quantum computing and chaotic theory instead. A few studies have revealed how quantum computing has been integrated with NIC algorithms to improve the convergence rate and efficiency

Table 1 Summarized description of nature-inspired computing systems used in bioinformatics

S. No.	Nature-inspired computing approach	Natural inspiration	Bioinformatics applications	Reference
1	*Evolutionary algorithms*	Darwinian evolution	Multiple sequence alignments, structure prediction, motif identification, cheminformatics, evolutionary analysis, gene expression analysis, clinical decision support systems performance, molecular docking, molecular dynamic simulations, protein structure analysis	Willett (1995), Notredame and Higgins (1996), Hill et al. (2005), To and Vohradsky (2007), Wong et al. (2011), Wong (2015), Gondro and Kinghorn (2018)
2	*Biodegradability prediction*	Biodegradation	Biodegradability pathway prediction, chemical toxicity prediction, databases, high-throughput screening strategies	Arora and Bae (2014)
3	*Cellular automata*	Existence and survival	Classification of proteins, promotor prediction, protein structure prediction, etc	Xiao et al. (2011), Sree and Ramesh Babu (2014)
4	*Self-organization*	Ants, termites, bees, wasps	Developmental evolution, cartography of tissues, image processing, sequence classification, transcription factor binding site prediction	Garrity and Lilburn (2005), Wirth et al. (2011), Bozorgmehr (2014)

(continued)

Table 1 (continued)

S. No.	Nature-inspired computing approach	Natural inspiration	Bioinformatics applications	Reference
5	*Neural networks*	Brain	Quantification of biomedical data, development of decision-making platforms, medical image processing, and analysis, gene expression analysis, prediction of protein structures	Cao et al. (2018), Shahid et al. (2019), Tang et al. (2019), Torada et al. (2019)
6	*Artificial life*	Survival	Synthetic biology, systems biology, mimicking evolution of life, artificial synthesis of microorganisms, omics-based analysis	Guruprasad and Sekar (2006), Aguilar et al. (2014), Chicco et al. (2020)
7	*Artificial immune system*	Immune system	Synthetic immunology, therapeutics, identification of differentially expressed genes (DEGs), biomarker prediction	Roybal and Lim (2017), Zhang et al. (2020)
8	*Computer graphics*	Patterning of living beings (skins, feathers, shells, etc.)	Genome-wide association analysis (GWAS), interactive visualization of genomic, proteomic, transcriptomic data, high-throughput screening	Nobile et al. (2016), O'Donoghue (2021)
9	*Lindenmayer systems*	Plant structures	Predicting realistic models of organisms, especially plants	Chen and Colomb (2003)

(continued)

Table 1 (continued)

S. No.	Nature-inspired computing approach	Natural inspiration	Bioinformatics applications	Reference
10	*Membrane computing*	Intra-membrane molecular processes occurring in cells	Prediction of protein structures, functional analysis, drug discovery, disease biology,	Păun and Pérez-Jiménez (2006), Nugent and Jones (2012), Liang et al. (2019)
11	*Excitable media*	Forest fires, Axon behavior	Behavioral biology of cells, tissues, and organs—cardiac, neural, skeletal muscles	Bartocci et al. (2008), Herzog et al. (2018), Song and Qu (2020)
12	*Wireless sensor networks (WSNs)*	Sensory organs	Data signal compression, agricultural practices,	R.M.C. and Anand (2018), Li et al. (2022)
13	*Learning classifier systems*	Evolution	RNA sequence analysis, RNA structure prediction, DNA fragment assembly, protein functional analysis, drug discovery, analysis of single-cell transcriptomic data, structure stability prediction, immunology,	Tang et al. (2019), Li et al. (2020)

of the system. On the other hand, chaotic theory can be deployed to enhance the overall performance of the NIC model for obtaining better optimization (Darwish 2018).

6 Conclusion

Nature-inspired computing is a novel extension of artificial intelligence that is inspired by the behavior of humans, animals, birds, and insects. This new field of computer science has opened up a way for sub-fields such as—neural networking, swarm intelligence, artificial immune systems, and evolutionary algorithms. A typical NIC system is based on self-analyzing, learning, and adapting to the environment or problem nature, aiding in making better decisions. It has been widely appreciated and deployed in solving tricky and complex biological problems in healthcare and research. NIC systems are a part and parcel of bioinformatics and biological research. These are extensively used in biomedical and biological data analysis and interpretation, synthetic biology, systems biology, molecular computing, epidemiology, drug designing and repurposing, cellular computing, quantum computing, simulation, etc. Nature-inspired computation was inspired by living beings in the *past*, it is being heavily used to understand the nature of beings and surroundings in the *present*, and, most likely, it will be able to prophesize the risks of diseases, outbreaks of infections, genetic disorders, etc., in the *future*.

Acknowledgements SQ is supported by the DST-INSPIRE Fellowship provided by the Department of Science and Technology, Govt. of India.

References

Aguilar W, Santamaría-Bonfil G, Froese T, Gershenson C (2014) The past, present, and future of artificial life. Front Robot AI 1. https://doi.org/10.3389/frobt.2014.00008

Ahmad S, Qazi S, Raza K (2021) Translational bioinformatics methods for drug discovery and drug repurposing. Transl Bioinform Healthc Med 127–139. https://doi.org/10.1016/b978-0-323-89824-9.00010-0

Ahmed AM, Rashid TA, Saeed S AM (2020) Cat swarm optimization algorithm: a survey and performance evaluation. Comput Intell Neurosci 1–20. https://doi.org/10.1155/2020/4854895

Akdur G (2011) The use of biological cellular automaton models in medical, health and biological studies. Procedia Soc Behav Sci 28:825–831. https://doi.org/10.1016/j.sbspro.2011.11.151

Alam MT, Raza K (2021) Blockchain technology in healthcare: making digital healthcare reliable, more accurate, and revolutionary. Transl Bioinf Healthc Med 81–96. https://doi.org/10.1016/b978-0-323-89824-9.00007-0

Alshamlan HM, Badr GH, Alohali YA (2015) Genetic Bee Colony (GBC) algorithm: a new gene selection method for microarray cancer classification. Comput Biol Chem 56:49–60. https://doi.org/10.1016/j.compbiolchem.2015.03.001

Arora PK, Bae H (2014) Integration of bioinformatics to biodegradation. Biol Proced Online 16(1). https://doi.org/10.1186/1480-9222-16-8

Back T, Fogel DB, Michalewicz Z (1997) Handbook of evolutionary computation, 1st edn. IOP Publishing Ltd., GBR.988. ISBN 978-0-7503-0392-7. https://dl.acm.org/doi/book/https://doi.org/10.5555/548530

Bartocci E, Corradini F, Entcheva E, Grosu R, Smolka SA (2008) CellExcite: an efficient simulation environment for excitable cells. BMC Bioinform 9(S2). https://doi.org/10.1186/1471-2105-9-s2-s3

Bozorgmehr JE (2014) The role of self-organization in developmental evolution. Theory Biosci Theorie den Biowissenschaften 133(3–4):145–163. https://doi.org/10.1007/s12064-014-0200-4

Brabazon A, O'Neill M, McGarraghy S (2015) Natural computing algorithms. In: Natural computing series. Springer, Berlin, Heidelberg. https://doi.org/10.1007/978-3-662-43631-8

Cao C, Liu F, Tan H, Song D, Shu W, Li W, Zhou Y, Bo X, Xie Z (2018) deep learning and its applications in biomedicine. Genomics Proteomics Bioinf 16(1):17–32. https://doi.org/10.1016/j.gpb.2017.07.003

Chen Y-PP, Colomb RM (2003) Database technologies for L-system simulations in virtual plant applications on bioinformatics. Knowl Inf Syst 5(3):288–314. https://doi.org/10.1007/s10115-002-0087-0

Chicco D, Heider D, Facchiano A (2020) Editorial: artificial intelligence bioinformatics: development and application of tools for omics and inter-omics studies. Front Genet 11. https://doi.org/10.3389/fgene.2020.00309

Chu S-C, Tsai P-W (2007) Computational intelligence based on the behavior of cats. Int J Innov Comput Inf Control ICIC, Int C 3(1):163–173. http://www.ijicic.org/isme06-30-1.pdf

Computing Nature (2013) Google Books. https://books.google.se/books?id=SV1GAAAAQBAJ&printsec=frontcover&hl=sv&source=gbs_ge_summary_r&cad=0#v=onepage&q&f=false

Cosentino C, Vescio B, Amato F (2013) Cellular automata. Encyclopedia Syst Biol 381–385. https://doi.org/10.1007/978-1-4419-9863-7_989

Darwish A (2018) Bio-inspired computing: algorithms review, deep analysis, and the scope of applications. Future Computi Inf J 3(2):231–246. https://doi.org/10.1016/j.fcij.2018.06.001

Garrity GM, Lilburn TG (2005) Self-organizing and self-correcting classifications of biological data. Bioinformatics 21(10):2309–2314. https://doi.org/10.1093/bioinformatics/bti346

Ghareh F, Amini M, Arabnia H (2019) Applications of nature-inspired algorithms for dimension applications of nature-inspired algorithms for dimension reduction: enabling efficient data analytics. https://digitalcommons.fiu.edu/cgi/viewcontent.cgi?article=1018&context=cs_fac. Accessed on June 6 2022

Gondro C, Kinghorn BP (2018) A simple genetic algorithm for multiple sequence alignment. Genet Mol Res GMR 6(4). https://pubmed.ncbi.nlm.nih.gov/18058716/

Grozinger L, Amos M, Gorochowski TE, Carbonell P, Oyarzún DA, Stoof R, Fellermann H, Zuliani P, Tas H, Goñi-Moreno A (2019) Pathways to cellular supremacy in biocomputing. Nat Commun 10(1). https://doi.org/10.1038/s41467-019-13232-z

Guruprasad S, Sekar K (2006) Artificial life and living systems: Insight into artificial life and its implications in life science research. Bioinformation 1(4):139–140. https://doi.org/10.6026/97320630001139

Herzog S, Wörgötter F, Parlitz U (2018) Data-driven modeling and prediction of complex spatio-temporal dynamics in excitable media. Front Appl Math Stat 4. https://doi.org/10.3389/fams.2018.00060

Hill T, Lundgren A, Fredriksson R, Schiöth HB (2005) Genetic algorithm for large-scale maximum parsimony phylogenetic analysis of proteins. Biochim Biophys Acta (BBA) Gen Subj 1725(1):19–29. https://doi.org/10.1016/j.bbagen.2005.04.027

Karaboga D (2005) An idea based on honey bee swarm for numerical optimization. ScienceOpen (2020). Scienceopen.com. https://www.scienceopen.com/document?vid=5c9119f7-3147-42aa-8cd6-0f11c54f98c7

Li H, Tian S, Li Y, Fang Q, Tan R, Pan Y, Huang C, Xu Y, Gao X (2020) Modern deep learning in bioinformatics. J Mol Cell Biol 12(11):823–827. https://doi.org/10.1093/jmcb/mjaa030

Li C, Chen D, Xie C, Tang Y (2022) Algorithm for wireless sensor networks in ginseng field in precision agriculture. PLoS ONE 17(2):e0263401. https://doi.org/10.1371/journal.pone.0263401

Liang X, Zhu W, Lv Z, Zou Q (2019) Molecular computing and bioinformatics. Molecules 24(13):2358. https://doi.org/10.3390/molecules24132358

Manazir A, Raza K (2022) Comparative evaluation of genetic operators in cartesian genetic programming. In: Proceedings of of 21st international conference intelligent systems design and applications, December 13-15, 2021, vol 418. Lecture notes in networks and systems, Springer, Berlin, pp 765–774. https://doi.org/10.1007/978-3-030-96308-8_71

Manazir A, Raza K (2019) Recent developments in cartesian genetic programming and its variants. ACM Comput Surv 51(6):122. https://doi.org/10.1145/3275518

Markram H, Muller E, Ramaswamy S et al (2015) Reconstruction and simulation of neocortical microcircuitry. Cell 163(2):456–492. https://doi.org/10.1016/j.cell.2015.09.029

McCulloch WS, Pitts W (1943) A logical calculus of the ideas immanent in nervous activity. Bull Math Biophys 5(4):115–133. https://doi.org/10.1007/bf02478259

Mirjalili S (2015) Moth-flame optimization algorithm: a novel nature-inspired heuristic paradigm. Knowl-Based Syst 89:228–249. https://doi.org/10.1016/j.knosys.2015.07.006

Mirjalili S, Lewis A (2016) The whale optimization algorithm. Adv Eng Softw 95:51–67. https://doi.org/10.1016/j.advengsoft.2016.01.008

Mirjalili S, Mirjalili SM, Lewis A (2014) Grey wolf optimizer. Adv Eng Softw 69:46–61. https://doi.org/10.1016/j.advengsoft.2013.12.007

Nobile MS, Cazzaniga P, Tangherloni A, Besozzi D (2016) Graphics processing units in bioinformatics, computational biology and systems biology. Briefings Bioinf bbw058. https://doi.org/10.1093/bib/bbw058

Notredame C, Higgins DG (1996) SAGA: sequence alignment by genetic algorithm. Nucleic Acids Res 24(8):1515–1524. https://doi.org/10.1093/nar/24.8.1515

Nugent T, Jones DT (2012) Membrane protein structural bioinformatics. J Struct Biol 179(3):327–337. https://doi.org/10.1016/j.jsb.2011.10.008

O'Donoghue SI (2021) Grand challenges in bioinformatics data visualization. Front Bioinf 1. https://doi.org/10.3389/fbinf.2021.669186

Păun G (2010) A quick introduction to membrane computing. J Logic Algebraic Program 79(6):291–294. https://doi.org/10.1016/j.jlap.2010.04.002

Păun G, Pérez-Jiménez MJ (2006) Membrane computing: brief introduction, recent results and applications. Biosystems 85(1):11–22. https://doi.org/10.1016/j.biosystems.2006.02.001

Pfeifer R, Bongard JC (2006) How the body shapes the way we think—a new view on intelligence. https://www.semanticscholar.org/paper/How-the-body-shapes-the-way-we-think-a-new-view-on-Pfeifer-Bongard/2910099b7a7c555af9f14bfb2bc20e9475d0588f

Piccinini G, Bahar S (2012) Neural computation and the computational theory of cognition. Cogn Sci 37(3):453–488. https://doi.org/10.1111/cogs.12012

Qazi S, Raza K (2021) Translational bioinformatics in healthcare: past, present, and future. Transl Bioinf Healthc Med 1–12. https://doi.org/10.1016/b978-0-323-89824-9.00001-x

Manjunath CR, Anand S (2018). Secure multicasting using geographical information: A cluster based approach in wireless sensor networks. Int J Eng Res Technol 2(14). https://www.ijert.org/secure-multicasting-using-geographical-information-a-cluster-based-approach-in-wireless-sensor-networks

Roybal KT, Lim WA (2017) Synthetic immunology: hacking immune cells to expand their therapeutic capabilities. Annu Rev Immunol 35(1):229–253. https://doi.org/10.1146/annurev-immunol-051116-052302

Shahid N, Rappon T, Berta W (2019) Applications of artificial neural networks in health care organizational decision-making: a scoping review. PLoS ONE 14(2):e0212356. https://doi.org/10.1371/journal.pone.0212356

Song Z, Qu Z (2020) Delayed global feedback in the genesis and stability of spatiotemporal excitation patterns in paced biological excitable media. PLoS Comput Biol 16(10):e1007931. https://doi.org/10.1371/journal.pcbi.1007931

Sree P, Ramesh Babu I (2014) Cellular automata and its applications in bioinformatics: a review. Glob Perspect Artif Intell (GPAI) 2(2):16. https://arxiv.org/ftp/arxiv/papers/1404/1404.0453.pdf

Stark WR (2013) Amorphous computing: examples, mathematics and theory. Nat Comput 12(3):377–392. https://doi.org/10.1007/s11047-013-9370-0

Tang B, Pan Z, Yin K, Khateeb A (2019) Recent advances of deep learning in bioinformatics and computational biology. Front Genet 10. https://doi.org/10.3389/fgene.2019.00214
Thierens D (2010) The linkage tree genetic algorithm. In: Parallel problem solving from nature, vol PPSN XI, pp 264–273. https://doi.org/10.1007/978-3-642-15844-5_27
Tian Z, Fong S, Deb S, Tang R, Wong R (2018) Vitality-based elephant search algorithm. Oper Res Int Journal 18(3):841–863. https://doi.org/10.1007/s12351-018-0419-9
Timmis J, Knight T, de Castro L N, Hart E (2004) An overview of artificial immune systems. Nat Comput Ser 51–91. https://doi.org/10.1007/978-3-662-06369-9_4
To CC, Vohradsky J (2007) A parallel genetic algorithm for single class pattern classification and its application for gene expression profiling in Streptomyces coelicolor. BMC Genomics 8(1). https://doi.org/10.1186/1471-2164-8-49
Torada L, Lorenzon L, Beddis A, Isildak U, Pattini L, Mathieson S, Fumagalli M (2019) ImaGene: a convolutional neural network to quantify natural selection from genomic data. BMC Bioinf 20(S9). https://doi.org/10.1186/s12859-019-2927-x
Tsai H-C, Lin Y-H (2011) Modification of the fish swarm algorithm with particle swarm optimization formulation and communication behavior. Appl Soft Comput 11(8):5367–5374. https://doi.org/10.1016/j.asoc.2011.05.022
Uymaz SA, Tezel G, Yel E (2015) Artificial algae algorithm (AAA) for nonlinear global optimization. Appl Soft Comput 31:153–171. https://doi.org/10.1016/j.asoc.2015.03.003
What is Nature Inspired Computing? Computer Science Degree Hub. (2016, September). Computer Science Degree Hub. https://www.computersciencedegreehub.com/faq/what-is-nature-inspired-computing
Willett P (1995) Genetic algorithms in molecular recognition and design. Trends Biotechnol 13(12):516–521. https://doi.org/10.1016/s0167-7799(00)89015-0
Wirth H, Löffler M, von Bergen M, Binder H (2011) Expression cartography of human tissues using self organizing maps. BMC Bioinf 12(1). https://doi.org/10.1186/1471-2105-12-306
Wong K-C, Peng C, Wong M-H, Leung K-S (2011) Generalizing and learning protein-DNA binding sequence representations by an evolutionary algorithm. Soft Comput 15(8):1631–1642. https://doi.org/10.1007/s00500-011-0692-5
Wong K-C (2015) Evolutionary algorithms: concepts, designs, and applications in bioinformatics: evolutionary algorithms for bioinformatics. ArXiv.org. https://doi.org/10.48550/arXiv.1508.00468
Xiao X, Wang P, Chou K-C (2011) Cellular automata and its applications in protein bioinformatics. Curr Protein Pept Sci 12(6):508–519. https://doi.org/10.2174/138920311796957720
Zhang Y, Qu P, Ji Y, Zhang W et al. (2020) A system hierarchy for brain-inspired computing. Nature 586(7829):378–384. https://doi.org/10.1038/s41586-020-2782-y
Zouache D, Ould Arby Y, Nouioua F, Ben Abdelaziz F (2019) Multi-objective chicken swarm optimization: a novel algorithm for solving multi-objective optimization problems. Comput Ind Eng 129:377–391. https://doi.org/10.1016/j.cie.2019.01.055

Leveraging Healthcare System with Nature-Inspired Computing Techniques: An Overview and Future Perspective

Mohd Faizan Siddiqui, Aftab Alam, Roman Kalmatov, Azaroual Mouna, Ricardo Villela, Ainura Mitalipova, Yves Najm Mrad, Syed Ali Abbas Rahat, Bhupendra Kumar Magarde, Waseem Muhammad, Salieva Rana Sherbaevna, Nazgul Tashmatova, Umurzakova Gavkharoi Islamovna, Mohammad Ahmad Abuassi, and Zeenat Parween

1 Introduction

The healthcare profession has been grossly inflated by technologies. Medical informatics, without a doubt, is altering the medical industry's landscape (Wani and Barh 2021). With technological improvements, it is now possible to properly identify and treat disorders. Many lethal illnesses have been eradicated thanks to modern

M. F. Siddiqui (✉) · R. Kalmatov · A. Mitalipova · S. A. A. Rahat · B. K. Magarde · W. Muhammad · S. R. Sherbaevna · N. Tashmatova · U. G. Islamovna
International Medical Faculty, Osh State University, Osh City, Kyrgyzstan
e-mail: fzn.imf@gmail.com

A. Alam
Memorial Sloan Kettering Cancer Center, New York City, NY, USA

A. Mouna
Tambov State Medical University, Tambov City, Russia

R. Villela
National Autonomous University of Honduras, Tegucigalpa, Honduras

Y. N. Mrad
Gilbert and Rose-Marie Chagoury School of Medicine, Lebanese American University, Blat, Lebanon

M. A. Abuassi
Department of Internal Medicine, Jordan Hospital, Amman, Jordan

Z. Parween
Patliputra University, Patna, Bihar, India
e-mail: Pzeenat06@gmail.com

K. Raza (ed.), *Nature-Inspired Intelligent Computing Techniques in Bioinformatics*,
Studies in Computational Intelligence 1066,
https://doi.org/10.1007/978-981-19-6379-7_2

immunizations and other therapeutic methods. However, a variety of chronic disorders, such as cardiovascular, respiratory, malignant, digestive, diarrheal, depression, malaria, cancer, diabetes, and others, do not seem to be going away anytime soon. If crucial human illnesses are detected correctly, they can be efficiently treated (Siddiqui 2021). Nature-inspired computing (NIC) is inspired by many features, components, and species of nature, including structures, plants, insects, flowers, waterfalls, and animals, among others. Swarm intelligence (SI), genetic algorithms (GA), artificial bee colony (ABC), particle swarm optimization (PSO), ant colony optimization (ACO), water wave algorithm (WWA), honeybee algorithm (HBA), bat algorithm (BA), cuckoo algorithm (CA), fish swarm algorithm (FSA), ray optimization (RO) are some of the important NIC techniques that have taken ideas from physics and biology (Agarwal and Mehta 2014). The size of the average data created across different activities has expanded by orders of magnitude due to advances in information and communication technology, as well as the data processing and storage capacity of computer devices in the contemporary period (Gupta and Raza 2019). This holds true for both the number of instances in the data and the number of characteristics. The various learning tasks (supervised or unsupervised) can yield better results by selecting the optimum amount and most useful features from such vast datasets. In the past, NICs were also used to choose the informative features in a high-dimensional space by removing unnecessary and duplicate characteristics (Alshamlan et al. 2016).

Swarm intelligence (SI) is a novel optimization approach inspired by the biology of animals, birds, and fish. SI approaches drew their inspiration from 'collective intelligence.' The collaboration of large numbers of homogenous agents such as wasps, bees, cuckoos, fish, ants, and so on identifies collective intelligence. It is a branch of computer science that seeks and investigates effective computing solutions inspired by the actions of actual swarms or insects (Carvalho Filho et al. 2017). SI attempts to discover an optimal solution to an NP-hard issue by leveraging the collective power and mutual collaboration of several agents. The genetic algorithm is a type of evolutionary algorithm developed by John Holland in the 1960s. It is the process of producing new children by fusing the genes of the population. It is determined by four factors: population size, generations, mutation rate, and crossover rate. This method produces efficient and accurate results when used to tackle real-world situations. Eberhart and Kennedy created particle swarm optimization, a novel evolutionary approach, in 1995. PSO investigates probabilistic algorithms driven by herding, flocking, and the behavior of swarms of fish or flocks of birds to find a suitable feeding location. This method is employed in the solution of n-dimensional problems. Furthermore, the answer is represented as a point or a surface (Aličković and Subasi 2017).

ACO investigated probabilistic techniques inspired by ant stigmergy and foraging behavior. ACO is a metaheuristic for solving a wide range of optimization issues. It gets its inspiration from the way actual ants select the quickest pathways from their colony to food sources. The pheromone, a chemical molecule that influences the behavior of other people with the same genes, allows the ants to communicate. The water wave algorithm (WWA) is an innovative metaheuristic approach. This method explores multidimensional solution space by utilizing the concepts of propagation and refraction. Ray optimization is an optimization approach that is based on refraction. Variables are represented as particles, which are often referred to as agents (Smith and Bryant 1975). The use of Shell's law aids in the examination of the search space so that the agents can converge on the final state. Honeybee algorithm (HBA) is derived from honeybee foraging behavior. It incorporates aspects of both exploitative and exploratory search. The entire tale of the honeybee algorithm centered on scout bee foraging activities and the wrangling dance necessary for communication. Karaboga presented another optimization approach, artificial bee colony (ABC), in 2005. This method is based on bee foraging behavior and solves the problem of finding an object by taking the best path. It is a self-adapting and self-growing algorithm. Bats are enticing animals that use high-frequency sound to locate their prey (Babaoğlu 2013). Xin-she Yang introduced the bat algorithm (BA) in 2010 by mimicking bat behavior. It is a bio-inspired algorithm that is based on bat echolocation behavior. The method of looking for an object based on the reflection of sound is known as echolocation. This algorithm runs indefinitely until it meets the convergence criterion. Another notable optimization tool is the cuckoo algorithm (CA). CA's main concept is based on a bird called a cuckoo. The cuckoo lifestyle, particularly egg-laying and breeding, is inspiring. The fish swarm algorithm (FSA) is inspired by fishes' collective behavior. It features a great degree of flexibility, convergence, and fault tolerance. The FSA mimics fish behavior. Any object can be treated as a fake fish. The workings of the FSA are quite similar to those of the GA. However, unlike GA, it does not involve crossover and mutation operations; therefore, it has higher convergence and better outcomes since there are fewer factors to alter (Alshamlan et al. 2015) The capabilities of NIC-based tools over normal AI-based technologies have been summarized in Table 1.

2 Nature-Inspired Algorithm: An Overview

Nature plays a critical role in resolving issues in a systematic and very efficient manner. Only a small number of academics are seeking to develop computational methods that might help humans in dealing with global health challenges. Natural techniques like swarm intelligence and bio-inspired, as well as physics and chemistry, may help solve difficult problems and provide the best solution. The parallel, dynamic, and self-organizing nature of nature-inspired techniques makes them ideal for use in soft computing applications. Ants, bees, and bugs inspired the development of these algorithms. It is all here in this chapter: a comprehensive look at nature-inspired techniques in health care. Nature-inspired computing (NIC) and computational intelligence (CI) cross-fertilization will surely provide optimal solutions to present global health concerns and offer up new research and development avenues (Siddique and Adeli 2015). The main classification and significant applications of NIC have been illustrated in Fig. 1.

Table 1 Comparison of AI-based technologies and NIC capabilities

Shortcomings of AI-based technologies	Capabilities of NIC to resolve shortcomings
The accuracy of feature categorization is achieved by backward propagation techniques, that vary from over-fitting and under-fitting (Chan et al. 2020)	The randomization of parameters is used to discover the best or nearly optimum solution that optimizes classification accuracy (Hendrickson and Ward 1975)
Once networking has learned single-weight values, every further learning results in catastrophic forgetting (Le et al. 2019)	To maximize acquisition, it employs randomization to learn new parameters (Jaton et al. 1975)
Backward dissemination approaches, which suffer from over-fitting and under-fitting, are used to improve the accuracy of feature categorization (Vaikath et al. 2019)	The randomization of parameters is used to discover the best or nearly optimum solution that optimizes the accuracy of classification (Makar et al. 1975)
Forgetfulness occurs catastrophically when the network is exposed to new weights (Čartolovni et al. 2022)	New parameters are discovered via randomization (Bland et al. 1976)
Local optimization of back propagation may improve classification accuracy; however, global optimization fails (Esmaeilzadeh 2020)	Searching for the most optimum global variable via local search (Poole-Wilson and Langer 1975)

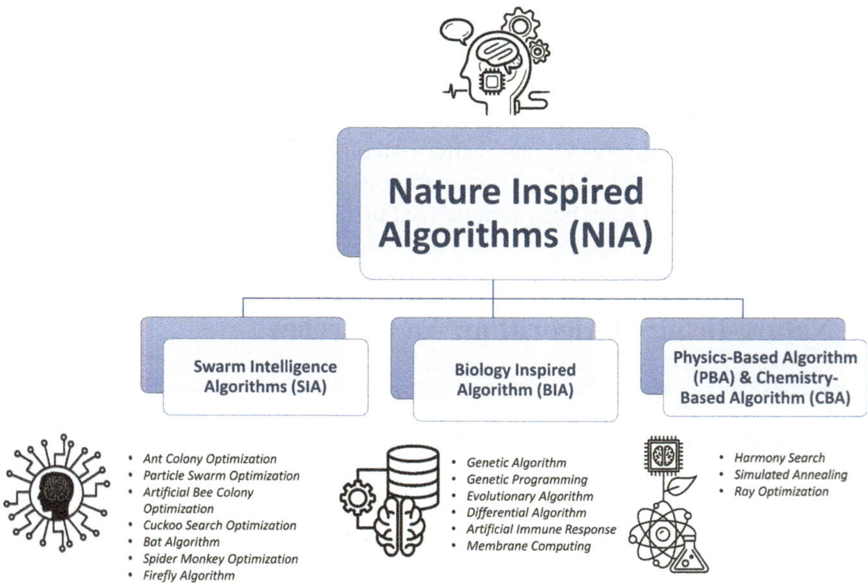

Fig. 1 Classification and important application based on nature-inspired algorithm

2.1 Swarm Intelligence

Decentralized and self-organized systems made up of relatively simple agents that interact with one another in their local environments and with the environment at large are the basis for swarm intelligence, a branch of artificial intelligence that is based on collective behavior of such systems. These algorithms have stimulated the growth of data analytics methods, which has been driven by the expansion of healthcare data availability. In the 1980s, the idea of swarm intelligence was initially put out. When it was first introduced, the scientific community began to pay more attention to it. This technique has an advantage over more traditional ones since it incorporates privacy-preserving technology by default, making it easier to compare scientific data from different sites. Research cooperation and information interchange, particularly in medicine, might benefit greatly from the use of swarm learning. Cancer diagnosis, brain tumor, diabetic retinopathy, heart disease, and medication creation and development all need the use of swarm intelligence (SI) and evolutionary algorithms (EA) in health care today. SI offers derivative-free optimization that is versatile, resilient, and inexpensive to deploy. SI with EAs are powerful global optimization strategies that are very beneficial for selecting features in medical systems. In the healthcare arena, SI and EA approaches and applications are the subject of ongoing study. Since processing healthcare data is essential for diagnosis, treatments, medicine, screening, and eventually lowering the mortality rate, there is a need to expand research advances in the areas of SI and EA approaches for processing healthcare data (Gupta et al. 2021). The SI's approach to an e-healthcare system that links and integrates patients or caregivers while protecting their privacy is really significant. This method allows patients or caregivers to converse with other patients experiencing similar challenges, to comprehend their emotions, and discuss many of their own concerns. During this procedure, however, confidential and sensitive information cannot be shared with anybody at any moment. Using common similarity metrics, the results of the proposed framework are compared to well-known privacy-preserving clustering techniques (Swathi and Sreedhar 2020). The significant SIA's applications in the healthcare domain are:

2.1.1 *Ant Colony Optimization:* Surgical case scheduling, health service data sharing.
2.1.2 *Particle Swarm Optimization:* Healthcare assistance, immediate hospitalization, emergency healthcare services.
2.1.3 *Artificial Bee Colony Optimization:* Smart healthcare system, healthcare data diagnosis, prediction of stroke, operating room scheduling problem.
2.1.4 *Cuckoo Search Optimization:* IoT-based healthcare application, big data analytics, heart disease, and diabetes prediction.
2.1.5 *Bat Algorithm*: Data classification, short-term COVID-19 forecasting.
2.1.6 *Spider Monkey Optimization:* COVID-19 prediction, attack detection in Internet of Medical Things (IoMT).
2.1.7 *Firefly Algorithms:* Sustainable and smart health care, healthcare data diagnosis.

2.2 *Biology-Inspired Algorithm*

Bio-inspired algorithms (BIA) are also a new area of study whose key benefit is the linking of subfields linked to connectionism, social behavior, and emergence in general. At its core, the study of life is used in conjunction with the study of technology to develop its capabilities. In several dimensions of BIA, biomedical image processing is conceptually analogous to biomedical signal processing. It involves the evaluation, enhancement, and display of pictures recorded by X-ray, ultrasound, magnetic resonance imaging (MRI), nuclear medicine, and visual imaging technologies. Optimization algorithms inspired by biology are advancing rapidly in many areas of scientific and technical study, including biomedical sciences (Subramanian et al. 2020). In order to generate classification rules, two of these methods use a similarity calculation between training and test data, while the other two use random population creation. Bio-inspired algorithms for supervised categorization of medical data were shown to be very effective. The use of devices and algorithms whose core notion is drawn from nature and which explore the design, analysis, and implementation of different bionics or bio-inspired algorithms is also a key benefit of these new instruments for the early identification of cognitive brain illnesses (Raza and Alam 2016). The significant BIA's application in the healthcare domain are:

2.2.1 *Genetic algorithms:* Secure cloud-based healthcare system, rehabilitation, optimization of wireless sensor networks, heart disease prediction.
2.2.2 *Genetic programming:* Improve interpretability in machine learning model, epidemiology curve prediction, detecting pulmonary tuberculosis.
2.2.3 *Evolutionary algorithm:* Healthcare data analysis, drug development.
2.2.4 *Differential algorithm:* Medical service optimization, differential diagnosis of disease, healthcare analysis.
2.2.5 *Artificial immune response:* Viral immune surveillance, artificial immunity prediction, regenerative medicine, digital mammogram database classification, innate immune responses regulation.
2.2.6 *Membrane computing:* IoT-based applications in health care, data acquisition, medical devices, smart health care.

2.3 *Physics-Based Algorithm and Chemistry-Based Algorithm*

An introduction to nature-inspired computing with a focus on real-world engineering challenges is provided via the use of physics and chemistry-based algorithms. The use of phenotypic and genotypic data, as well as medical imaging, has made it easier to diagnose and classify a wide range of disorders (Wang et al. 2021). The significant PBA's and CBA's applications in the healthcare domain are:

2.3.1 *Harmony search:* For patient admission, nurse rostering, and appointment scheduling.

2.3.2 *Simulated annealing:* Medical staff scheduling, healthcare management, telemedicine services, home healthcare supply chain.

2.3.3 *Ray optimization:* Radiography, dentistry, radiation dose optimization.

3 Combining AI Technologies and Nature-Inspired Computing Models for COVID-19

Life-threatening coronavirus may be transferred by coughing, spitting, or exhaling air from an infected individual. As a worldwide health and wellness issue, this pandemic necessitates a wide range of innovative solutions. COVID-19 may be controlled and treated using a variety of medicinal and technology approaches. One of the technologies employed in MRI classification is machine learning, whereas NIAs are used for image improvement (Alam et al. 2021). Machine learning and nature-inspired model for coronavirus (MLNI-COVID-19) have been utilized to analyze brain MRI images of COVID-19 patients during the epidemic. MRI picture categorization and optimization have been improved using this technique, allowing for more accurate diagnoses and treatment. This approach enhances overall performance, particularly in the ignored field of brain pictures owing to the lack of a dataset (Vaishya et al. 2020).

For AI-based technological interventions, the precision of automatic monitoring and recognition of the COVID-19 pandemic is now a difficulty. In addition, tracking down connections on a larger scale is prohibitively expensive since the number of contacts grows rapidly, necessitating greater outreach efforts. In light of the recent epidemic of COVID-19, new research avenues are being opened up for the use of NIC (Haleem et al. 2020). It is possible to use nature-inspired systems to improve existing screening and detection methods for COVID-19 instances because of their self-organization, simplicity of implementation, dynamics, and adaptability. Patients' reactions to therapy may now be predicted using a combination of CNN and the whale optimization algorithm. A whale optimization technique was used to identify the most important patient characteristics, and the results are encouraging. NIC reveals a broader range of real-world applications (Luo et al. 2019).

AI and NIC models have not yet been fully explored for COVID-19 pandemic prediction, according to a review of AI and nature-inspired computing models. To meet this need, the AI-NIC model for COVID-19 pandemic screening and detection was deployed. In order to shift low-dimensional feature space data to an appropriate clustering or low-dimensional mapping solution, swarm intelligence uses parameter tweaking (Qureshi et al. 2021). The combination of NIC and big data analytics is a promising research topic for the development of contact tracing algorithms. COVID-19 instances may be detected more accurately when NIC and big data on health are combined. For the COVID-19 pandemic, we have developed a contact tracing big data predictive analytics model that may be used to provide timely notifications or alerts on the regions to avoid (Bai et al. 2020).

4 Nature-Inspired Algorithms and Their Significant Use in Health Care

Using the principles of genetics as a guide, a nature-inspired algorithm (NIA) and GA attempt to develop practical answers to difficult situations. The first step is to produce a set of random people (solutions) with a variety of characteristics (chromosomes). Chromosomal crossover and mutations occur according to genetic rules and lead to the second generation of people with more varied characteristics. There are two primary techniques for diversifying individuals: crossover and mutation. Two chromosomes are selected for crossing in a zygote. The values up to the crossover point between the two chromosomes are then exchanged between the two chromosomes. Notably, GAs vary from optimization techniques based on derivatives. The first difference between GAs and classical derivative-based techniques is that GAs seek a population of points in the solution space, while classical methods only search for one point. Probabilistic transition rules and random number generators are used in GAs to pick the next population, whereas deterministic transition rules are used in derivative-based methods. Here are some examples of how GAs may be used in a range of medical fields (Tzanetos et al. 2020).

5 Healthcare Management

Health systems around the world rely heavily on the proper management of financial and human resources. The effective scheduling of patient admissions is an important element of hospital management that can improve patient service, satisfaction, and cost-effectiveness ratios. Patients at an ophthalmology facility were better served by a mathematical model that was improved using a GA. As a result of the new algorithm, the waiting list was reduced, the vacancies in hospital beds decreased, the preoperative waiting time for patients decreased, and the number of patients released from the hospital rose. There is another metaheuristic approach that may enhance patient scheduling and decrease time wastage and boost patient satisfaction: PSO (Travaglini et al. 2020).

Maintaining job skills and competence in clinical laboratories requires regular rotation of staff based on their abilities. In a clinical laboratory, staff rotation schedule has been improved by using GAs. Using GA-based software, one study found that it was possible to efficiently arrange staff rotation, assuring the preservation of methods and skills, as well as saving both time and money on the scheduling process (Tay et al. 2015).

6 Radiology

When it comes to diagnostic imaging, radiologists have to deal with a great volume of data that must be processed quickly. In order to help radiologists analyze images more quickly and accurately, computer-aided imaging detection and diagnosis (CADD) utilizes a variety of imaging modalities to identify normal and diseased patterns, segment them, and classify them. Magnetic resonance imaging (MRI) and computed tomography scanning (CT) are examples of these diagnostic procedures (Wei and Zhong 2019).

A scenario (such as a radiology picture of a human organ) is captured, analyzed, and interpreted using machine vision. An object's limits (form and size) must be established to have a better understanding of the entity. As a result, edge detection has become a vital feature of automated image processing systems. MRI, CT, and ultrasound are just a few of the imaging modalities that researchers have employed the GAs for. Breast cancer screening with mammography accounted for the vast majority of GA uses in radiology (Siddiqui et al. 2022).

It has been shown that the use of GAs for feature selection has helped researchers distinguish between benign and malignant breast cancers in mammograms and ultrasound pictures. Several studies have shown that GAs are effective for aligning MRI and CT scan pictures. MRI pictures and PET images were combined in another research to create colored breast cancer images. The development of a treatment strategy relies heavily on the accurate staging of the tumor (Zhang et al. 2021). Tumor staging necessitates the use of noninvasive imaging tests to precisely measure the size and volume of the tumor. Extraction of tongue cancer from head and neck magnetic resonance images has been developed by Zhou et al. An artificial neural network (ANN)-based symmetry-detection technique was used to minimize the amount of false-positive findings after the segmentation of pictures was done using a genetic algorithm (GA) (Gupta and Raza 2022). Using this method, tongue cancer may be extracted from an MRI with great accuracy and minimum user dependence (Fási et al. 2019).

6.1 Radiotherapy

With IMRT, a precise dosage of radiation is delivered to the brain, prostate, or the head and neck of the patient. IMRT planning requires selecting 5–10 angles for wavelet projection and estimating radiation dosage. The selection of gantry angles might be improved by using GA in a reasonable amount of time. Patients with cancers including pancreatic, rhabdomyosarcoma, and brain tumors have also been effectively treated with GA-based irradiation planning, as well as stereotactic radiosurgery and radiotherapy treatments (Ramakrishnan et al. 2020).

6.2 *Orthopedics*

Orthopedic surgery has benefited greatly from the advances made possible by biomedical engineering. Many debilitating hip joint illnesses have been alleviated with total hip arthroplasty (THA). In spite of this, it is possible for the therapy to fail if the femoral stem fails. Ishida et al. used a GA to design a femoral stem component with an optimal shape. Additionally, GAs have been used to determine the optimal design for tibial locking screws to limit the risk of screw breakage or loosening. Spinal pedicle screws used to stabilize fractured vertebrae were designed using ANNs and GAs in another study. An improved fatigue life, as well as perfect pullout and bending properties, were all made possible thanks to the hybrid algorithm (Akram and Raza 2018).

Scoliosis is a malformation of the spine's axis that manifests as a three-dimensional curve. Serial X-rays are used to track the evolution of the condition, which only affects a tiny fraction of people. Because X-rays may raise the risk of cancer, it is preferable to use non-invasive technologies to monitor the disease's progress. This approach was created by Jaremko et al. to predict the angle of spinal axis deformity from indicators of trunk surface deformity based on generalized additive models (GA-ANN). In more than two-thirds of cases, the hybrid technique was able to accurately assess the angle deformity (Chen et al. 2017).

7 Obstetrics and Gynecology

Obstetricians can decide the best time to intervene during labor based on the distinction between typical and protracted delivery. If the time it takes to attain complete cervical dilation is known, it may be used to estimate the expected delivery date and to distinguish between normal and extended labor. Three-parameter logistic models were developed by Hoh et al. utilizing the GA or Newton–Raphson (NR) methods to estimate cervical dilation time. The GA-based system beat the NR technique by forecasting the time to complete cervical dilatation more accurately. Cervical precancerous and cancerous alterations may be detected using a Pap smear, a kind of cytological examination (Pérez-Jiménez 2014). For the purposes of determining whether a cell is normal or aberrant, twenty characteristics of the cell are taken into consideration. These characteristics are then divided into seven groups. For the categorization of Pap smear findings, our hybrid model used GAs' feature-selection capacity to decrease the number of features required. The novel algorithm successfully classified the Pap smear findings, outperforming three other previously used methods (Schultz and Grieder 1987).

The use of GAs in prenatal diagnosis is another application that has come to fruition. Fetal macrosomia is one of the fetal characteristics that might make delivery more difficult. Second-trimester amniotic fluid from LGA and AGA newborns was analyzed by capillary electrophoresis in an effort to distinguish them from one other.

Data were analyzed using Bayesian statistics. In order to speed up the Bayesian calculation, a GA was employed to choose the most relevant wavelets (variables) from an electropherogram. When just two wavelets of albumin and a negatively charged unknown small molecule were used to distinguish between LGA and AGA, the system was 100% sensitive to the difference and 98% specific (Agbehadji et al. 2020).

It is possible to lessen the risks of low-birth-weight babies by knowing the fetal weight ahead of time. In order to estimate the fetal weight, Yu et al. used fuzzy logic to the support vector regression (FSVR). In order to pick the best characteristics for the FSVR system, gas was employed in an evolving FSVR. By having the lowest mean absolute percent error (6.6%) and the greatest correlation coefficient (0.902) between the estimated and actual fetal birth weight, it surpassed a back-propagation neural network (Kossmann 2000).

Cardiotocography may be used to monitor the fetal heart rate and contractions of the mother to evaluate the health of the fetus. A support vector machine (SVM) classifier was trained using cardiotocography recordings, and Ocak used a GA to choose the best features. The novel approach outperformed an ANN algorithm by 99.3% and 100% in terms of accurately classifying fetal health status as normal or abnormal. An autism spectrum disorder (ASD) is a neurological condition that first manifests in childhood and is marked by difficulties in social interaction, communication (both verbal and nonverbal), and repetitive behavior (Gupta and Raza 2020). GAs was utilized by Latkowski and Osowski to identify autism-related genes from the microarray gene expression data. RMI1, NRIP1, TOP1, ZFHX3, CEP350, NFYA, PSENEN, ANP32A, SEMA4C, and SP1 are often used genes. SVM and random forest classifiers were used to classify these genes as part of an ensemble of classifiers. The new technique has a sensitivity of 96% and a specificity of 83% for identifying autism (Leão et al. 2021).

The most prevalent kind of childhood leukemia, acute lymphoblastic leukemia (ALL), comprises several subgroups. Tumor cell gene expression data may be used to categorize malignancies. Lin et al. employed a GA to pick the most relevant genes for ALL classification out of the massive amount of data collected by microarray gene expression profiling. Six distinct subtypes of ALL were identified using silhouette statistics as a discriminant function. Compared to existing techniques, the suggested method achieved 100% classification accuracy and employed fewer distinguishing genes. There may be aneuploidy when one or more chromosomes are abnormally high or low in a cell's nucleus, depending on the species. Amniocentesis samples are tested for fetal aneuploidy using conventional chromosomal studies; however, the time necessary for these procedures demands the development of speedier diagnostic assays. To this goal, mass spectrometry was used to identify the amniotic fluid sample's proteome profile, and the resulting results were evaluated by a GA. The suggested approach has 100% sensitivity, 72%–96% specificity, 11%–50% positive predictive value, and 100% negative predictive value for aneuploidy detection (Allen et al. 2004).

8 Oncology

Screening tests may help discover cancer early, which might lead to better outcomes for patients if they get the correct treatment. Massive gene expression profiling might change the area of molecular diagnosis and prognosis thanks to the development of DNA microarray technology. It is necessary to identify significant predictive genes in order to deal with the statistical and analytical hurdles posed by the development of big datasets. GAs have been used to evaluate microarray data from numerous cancer cell lines because of its intrinsic capacity to explore and locate the best solution among big and complicated potential solutions with multiple simultaneous interactions. By staining breast cancer tissues with multiple antibodies specific for different markers, Dolled-Filhart et al. collected microarray data to uncover a minimal collection of biomarkers with maximal classification and prognostication values in breast cancer patients. There were three indicators that may be used to identify a group of patients who had a five-year survival rate of greater than 95% (Miles et al. 2021).

For cancer patients, the prognosis is a crucial and relevant component in determining the optimal treatment plan. More than 200 cancer patients were studied by Ergen et al., who examined four different data mining approaches to estimate the fate of cancer patients who were not in terminal stages when they were discharged from the hospital, patients' outcomes may be predicted with fewer explanatory factors using GA than any other technique (Ergen et al. 2020).

9 Cardiology

Different areas of cardiovascular medicine have made use of GAs. Most myocardial infarctions and strokes are characterized by the presence of atherosclerotic plaques. Plaque mechanical properties, such as elasticity, can help physicians better locate and map unstable or vulnerable plaques, thanks to this research. By using GAs for parameter estimation, it is possible to accurately measure the elasticity of tissues. For parameter estimation, this system is better than gradient-based methods, which are unable to handle inhomogeneous solution spaces with multiple local minima and require significant computational time (Johnson et al. 2018).

There has been a tremendous increase in the use of biomarker discovery and clinical proteomics in medical diagnosis and prognosis. Because of the high cost and complexity of advanced technologies like mass spectrometry and the computational and statistical approaches used to analyze the readouts, only a few relevant markers may be selected for the development of clinical assays. A recursive local floating improvement strategy was used to increase the GA's ability to forecast the likelihood of a catastrophic cardiac episode (MACE). A panel of seven proteins, including myeloperoxidase, was selected using this approach and was shown to be 77% accurate in predicting the likelihood of MACE. For years, doctors have relied on logistic regression models to help them make accurate medical diagnoses. Using a GA, researchers were able to pick the optimal variables for a logistic regression system to predict whether patients with chest discomfort had had a myocardial infarction (MI). The GA-based strategy outperformed previous methods in the selection of variables (Niazi et al. 2019).

10 Endocrinology

Insulin therapy's most prevalent side effect for those with type 1 diabetes is hypoglycemia (T1DM). Electroencephalogram patterns may be altered by hypoglycemia (EEGs). ANNs, GAs, and Levenberg–Marquardt (LM) training algorithms are used to identify hypoglycemia based on EEG data. Blood glucose levels and EEG waves were modeled using an ANN. The global search power of GA and the local search power of LM were coupled for the training of ANNs (Ghaheri et al. 2015). Analysis of EEG data using four factors from two EEG channels yielded a detection rate of 75% sensitivity and 60% specificity. Children with T1DM have been shown to have non-invasive bouts of nocturnal hypoglycemia using an inference method that uses a GA-based multiple regression with fuzzy inferences in another research. Hypoglycemia was diagnosed with a sensitivity of 75% and a specificity of more than 50% using heart rate and corrected QT interval (Kasendra and Wells 2020).

11 Pharmacotherapy

Pharmacovigilance, the study of the safety and side effects of medicines, is a crucial part of the review process for new treatments. On the other hand, algorithms may be able to give a more objective and less arbitrary evaluation of a drug's possible adverse effects than clinical judgment. A GA-based quantitative technique was also used to analyze adverse drug reactions. The new scoring system was able to predict the chance that a suspected medicine was the cause of an adverse drug reaction with an 84% sensitivity and a 71% specificity (Ji et al. 2022). After an organ transplant, Tacrolimus is given to supress the immune system so that the body doesn't reject the new organ. Because of the medication's high pharmacokinetic variability and narrow therapeutic window, maintaining a constant blood level is an extremely difficult task. An artificial neural network (ANN) was utilized to predict blood levels of tacrolimus in liver transplant patients. A GA was used to find the most clinically relevant set of candidate variables. Predicted results were tested against real data. When the ANN predicted tacrolimus blood levels, 84% of the datasets were within the clinically acceptable range of 3 ng/ml (Avila-Olias et al. 2013).

More than half of all drug development failures may be attributed to poor pharmacokinetics and inefficiency. The typical method of evaluating new investigational medicines' effectiveness and pharmacokinetics in animal models is time-consuming and expensive. Consequently, computational approaches have arisen to build QSPKR models for quick in silico screening of new prospective medications. QSPKR stands for quantitative structure–pharmacokinetic relation. A GA was used by Zandkarimi et al. to narrow down the more than 1480 alkaloid drug descriptions to a shortlist of the most promising candidates. An ANN was trained to build QSPKR prediction models using these properties, which were subsequently retrieved from well-known medications. Predicting the volume of distribution, clearance, and protein binding of alkaloid medicines with acceptable accuracy was achieved using a novel algorithm (Wang et al. 2019).

12 Neurology

There are white matter scars, sometimes known as plaques, that grow in patients with multiple sclerosis (i.e., MS), a devastating disease of the nervous system. These individuals have been diagnosed with the use of computer-aided diagnosis. Using brain MRIs and a GA, researchers were able to identify lesions caused by MS. There was an 87% similarity index between the lesions found by the GA and the one found by the radiology expert. Detecting aberrant brain electrical discharges during a seizure may be done with the use of an EEG. Several learning methods (LM, Quickprop, Delta-bar delta, and Momentum, and Conjugate gradient) were utilized to train an ANN for EEG-based categorization of epileptic vs healthy persons in order to build an automated system for identification of aberrant EEG signals (Julian et al. 2019). The ideal parameters and architecture of the ANN were discovered using a genetic algorithm (GA). Overall, the LM approach using the GA algorithm was shown to have a 96.5% success rate when it comes to training the ANN. Parkinson's disease has been linked to mitochondrial malfunction, according to a number of studies. Despite the small size of the mitochondrial genome, mtDNA mutations represent a significant contributor to hereditary illness. Genetic illness is often brought on by mutations in the mitochondrial genome (mtDNA), which provides crucial protein components to the oxidative phosphorylation enzyme complexes. Due to mitochondrial genetic differences, comparing mitochondrial mutations in healthy vs diseased states may not be as useful. Thus, Smigrodzki et al. created a GA to identify mitochondrial mutations in Parkinson's patients that are physiologically relevant. Based on mutational patterns in mitochondrial DNA, the suggested approach proved 100% accurate in diagnosing Parkinson's disease (Putrino et al. 2020).

13 Surgery

Artificial neural networks (ANNs) are strong mathematical algorithms that can accurately anticipate the behavior of systems. Non-small cell lung cancer patients who have surgery may be predicted thereafter using a GA-based ANN (GANN) (NSCLC). Optimizing against local minima was made easier with GA. The GANN model was able to properly and substantially predict the fate of NSCLC patients. The incorporation of tumor size into the algorithms also greatly increased the accuracy of predictions (Kim et al. 2020).

Geriatric cardiac surgery patients are becoming more common as the population ages. An accurate prognostication of postoperative morbidity and death would be instructive, preventing overestimation of risk and denial of surgery for individuals deserving it, which might happen with certain prediction algorithms, due to the high frequency of comorbid illnesses in the elderly. Short postoperative stay following cardiac surgery was shown to be associated with younger age, lack of preoperative beta-blocker usage, shorter cross-clamp duration, and the absence of congestive heart failure while using a GA (Desrosiers and Vallée-Bélisle 2017).

14 Infectious Disease

Human immunodeficiency virus (HIV) has made tuberculosis a possibly fatal infectious illness in both underdeveloped and rich countries (HIV). An ANN trained by a GA was designed using 38 examination factors and laboratory data to predict the diagnosis (tuberculosis vs. non-tuberculosis patients). In comparison with previous methods, the system was able to correctly classify around 95% of the data (Alam et al. 2022).

An essential aspect of HIV treatment, Highly Active Antiretroviral Therapy (i.e., HAART) is made of many antiretroviral drugs that work together to reduce the virus's ability to infect new cells and spread. It has been postulated that a systematic treatment break might be used to lessen the selection pressure on the virus and the generation of resistant particles since long-term HAART therapy requires patient compliance and may be linked with certain negative effects (Lu et al. 2018). As a result, Castiglione et al. developed a GA-based approach for selecting the ideal HAART treatment plan to manage HIV and aid in the restoration of the immune system. Virtual patients were utilized to test the impact of anti-HIV medications on a computerized immune system model. When combined with a full-length therapy, the novel organized interruption schedule may provide therapeutic benefits and protection against an opportunistic infection (Gao et al. 2021).

15 Rehabilitation Medicine

It is becoming more and more necessary to create and test new treatment equipment and procedures as the demand for physical rehabilitation grows. Many factors need to be tweaked and the procedures tested on people in order to fine-tune these new technologies (Franssen et al. 2019). It is possible to test the influence of new training adjustments on efficiency by using computer simulations of motions. Robotic physiotherapy of the lower leg was developed by Shi et al. Each patient's treatment plan was generated using a GA. A therapeutic robot for exercising the lower limbs was developed in another study. It was able to remember the physiotherapist's activities for each patient and replicate their behavior in the absence of a therapist with the use of an ANN and a GA (Shi et al. 2019).

16 Artificial Bee Algorithm

In the sphere of health care, artificial bee algorithm (i.e., ABC) algorithm is extensively utilized because of its benefits, such as the foraging behavior of honeybees and quick convergence speed. Also, the ABC method may be used in conjunction with another algorithm to get the desired outcome more quickly (Taherdangkoo 2014). The application of AI with the ABC algorithm has been a popular trend among academics in recent years. Here are a few examples of how the ABC algorithm has been put to use in the world of medicine.

16.1 Biomedical Sensors

Low-amplitude and low-frequency biomedical signals are influenced by a variety of disturbances, including interference from power lines and the movement of recording electrodes. Medical signals may benefit from better signal-to-noise ratios by using digital filters specifically designed for that purpose (Yan et al. 2018). An IIR filter for transcranial Doppler signals based on ABC in order to boost the SNR. In order to capture signals from the human body, biomedical sensors may include noise. If a human body organ is not working properly, this signal will be out of sync with the rest of the body. A universal filter for different noise signals is not possible because of the differing probabilities of noise (Karaboga et al. 2021).

16.2 Noise Filtering and Cancellation

Electroencephalogram (EEG)/event-related potential (ERP) is filtered using an adaptive noise canceller (ANC) with bounded range ABC (BR-ABC). Signal-to-noise ratio in dB, correlation between templates ERP, difference between resultant, and mean values are used to evaluate the algorithm's performance. ABC outperforms LMS and RLS in noisy environments, demonstrating its superiority. An EEG signal that has been impacted by noise may be filtered using adaptive noise cancellation techniques such as LMS, RLS, and the enhanced ANC based on the bounded range ABC (BR-ABC) algorithm. There are many advantages to the ABC technique, such as its ability to filter and remove artifacts from the EEG signal (Mohamed et al. 2021).

16.3 DNA Microarray

Analysis of genes may be done using DNA microarrays. At the gene level, numerous messenger ribonucleic acid (mRNA) molecules are synthesized, which is indicated by the gene (Avila-Olias et al. 2013). It is utilized in a variety of applications, including disease diagnosis, tumor identification, the selection of the appropriate medicine to prevent sickness, and the detection of mutations in other processes. In the process of finding several subtypes of a single illness, the categorization suffers (Zhang et al. 2017).

16.4 Confirmatory Diagnosis

In CT scans, the ABC algorithm may be used to partition the liver area for diagnostic purposes. ABC calculates the cluster centroid and the area corresponding to each cluster. It uses morphological surgeries to eliminate the rough edges of organs and tiny, thin sections of the liver covered by flesh. To improve things, we are using a region-growing technique. For Iris detection, ABC multilevel segmentation is suggested. If the number of thresholds is not kept under control, multilevel thresholding might become a laborious operation. G-best guided ABC algorithms, cuckoo search (CS), and PSO algorithms are utilized to choose the distribution of appropriate thresholds distribution delineating the iris components for improved iris recognition. EEG signals are classified using the ABC algorithm, which selects numerous characteristics (Weikert et al. 2020). Band power, autoregressive model, coherence, and phase-locking values are some of these characteristics. Artifact and background noise reduction, feature extraction, feature selection, and feature classification aspects may increase EEG classification accuracy. A surface Laplacian filter removes artifacts and background noise, while a support vector machine classifies sub-features (Roslan et al. 2020).

16.5 EEG Analysis

Epileptic disorders in the human brain may be studied using EEG signals and the properties of artificial bee colony (ABC) and radial basis function networks (RBFNNs). The RBFNNs were trained using a modified version of the ABC method for the categorization of EEG data. The spectators in this form of ABC are chosen by a binary tournament. These metrics are used to evaluate the method's performance in various combinations of healthy, seizure-free and seizure-prone segments, as well as healthy and seizure-prone segments. The test results reveal that RBFNNs trained with modified ABC's inverse-multi-quadric kernel is superior to RBFNNs trained with other kernels (Mohsin et al. 2021).

17 Ant Colony Optimization

Solutions to issues in the manufacturing and service industries may be found in transportation and storage logistics systems. Most contemporary warehouses include a number of characteristics that affect the effectiveness of order picking, including layout designs, storage assignments, and batching. Order-picking routes, which should be constantly altered in response to real-time warehouse management information, are among the variables regarded to have the greatest degree of flexibility (Takitou and Taneda 2019). Order picking, on the other hand, is one of the most common tasks because of its high operating expenses and time requirements. Order-picking mode takes up the lion's share of journey time. At least 55% of warehouse operations expenditures are attributed to order picking. As a result, warehouses have implemented effective methods for picking and replenishing routes, which have the ability to save costs and time. In warehouse operations, picking a routine strategy has a significant impact (Olaru et al. 2019).

17.1 Fast Medicine Dispensing System (FMDS)

Pharmacies and medication access management have moved from labor management to automated management with the introduction of sophisticated logistics technologies utilized in contemporary hospitals. The pharmacy automation system (PAS) combines automated control, computer network, medical Internet of Things, operations research, and machine manufacturing technologies. The PAS has accomplished pharmacy management digitization, networking, and automation. One of the automated drug distribution systems in the PAS is the FMDS (Bell et al. 2020). To replace conventional human-based medication administration and distribution, the system has achieved automation in pharmaceutical replenishing, storage, and dispensing. Administrating and distributing drugs is vital. The second line of defense, such as FMDS-assisted medicine administration and distribution, is required to reduce medication mistakes and increase patient safety. The pharmacist may employ automated dispensing, storage, and management systems to expedite pharmaceutical delivery and dispensing. Automation, integration, and intelligentization of pharmaceutical logistics increase the distribution and storage rates of medicine dispensing systems (Lertxundi et al. 2020).

17.2 Drug Management in Hospitals and Pharmacies: OPM

Pharmacies and medication access management have moved from labor management to automated management with the introduction of sophisticated logistics technologies utilized in contemporary hospitals. The pharmacy automation system

(PAS) combines automated control, computer network, medical Internet of Things, operations research, and machine manufacturing technologies. The PAS has accomplished pharmacy management digitization, networking, and automation. One of the automated drug distribution systems in the PAS is the FMDS. To replace conventional human-based medication administration and distribution, the system has achieved automation in pharmaceutical replenishing, storage, and dispensing. Administrating and distributing drugs is vital. The second line of defense, such as FMDS-assisted medicine administration and distribution, is required to reduce medication mistakes and increase patient safety. The pharmacist may employ automated dispensing, storage, and management systems to expedite pharmaceutical delivery and dispensing. Automation, integration, and intelligentization of pharmaceutical logistics increase the distribution and storage rates of medicine dispensing systems (Strobach et al. 2021). Medication-intensive storage and automated medicine dispensing are the key FMDS research areas. In certain nations, human and manipulator medication restocking is inefficient. To improve the conventional manual refilling and manipulator procedure, the order-picking mode (OPM) has been used to the automated refilling system by the manipulator. The OPM was created to determine the closest picking point in X- and Y-directions. The OPM was widely utilized in the FMDS, although the OPM's route selection mode cannot ensure that the next point picked is the closest, nor that the route value is precise. It is an ideal route selection method to choose the nearest place. This means that the OPM for refilling is still not completely tuned for practical usage (Liu et al. 2021).

18 Conclusion

Nature-inspired computing (NIC) was presented in this chapter, as well as a few of its important and significant medical applications. It is not uncommon for NICs and other metaheuristics based on biology to be utilized to tackle complicated issues by professionals in other disciplines of study. Optimization approaches might be very beneficial to medical professionals and researchers due to the inherent complexity of medicine. This predicament is a result of a lack of effective communication between computer scientists and doctors, as well as a lack of knowledge of sophisticated mathematical formulae among medical professionals. If this is the case, bringing together doctors, computer scientists, and engineers in combined journal clubs or attending doctors' ground rounds and case report presentations might alleviate the issue. It is also possible to find innovative answers to medical issues by enhancing multidisciplinary courses, as well as by involving engineering researchers in healthcare settings and hospitals, especially in the lower-middle-income countries and in the rural areas where it needs the most.

Acknowledgement M F Siddiqui, R Kalmatov, AM, SAAR, BKM, WM, SRS, NT and UGI are grateful to International Medical Faculty, Osh State University, Osh City, Kyrgyzstan, for providing the research infrastructure. M F Siddiqui is thankful Dr. Gregory Nicolas and Dr. Nancy Emmanuel of Medical Student Advisory Board (MSAB) and Faculty of Medicine of the University of Sao Paulo, Sao Paulo, Brazil for sharing their constant guidance and support throughout this study. M F Siddiqui would like to express his heartfelt gratefulness to Dr. Marian L. Waterman and Dr. John Lowengrub from the University of California, Irvine; for helping him during the preparation Oncology segment of this chapter. A Mouna is thankful to her teachers and Tambov State Medical University, Tambov City, Russia for their guidance and support. All authors are grateful to the Editor, Dr. Khalid Raza for this research collaboration.

References

Agarwal P, Mehta S (2014) Nature-inspired algorithms: state-of-art, problems and prospects. IJCA 100(14):14–21

Agbehadji IE, Awuzie BO, Ngowi AB, Millham RC (2020) Review of big data analytics, artificial intelligence and nature-inspired computing models towards accurate detection of COVID-19 pandemic cases and contact tracing. Int J Environ Res Public Health 17(15):E5330

Akram M, Raza A (2018) Towards the development of robot immune system: a combined approach involving innate immune cells and T-lymphocytes. Biosystems 172:52–67

Alam A, Khan A, Imam N, Siddiqui MF, Waseem M, Malik MZ et al (2021) Design of an epitope-based peptide vaccine against the SARS-CoV-2: a vaccine-informatics approach. Brief Bioinform 22(2):1309–1323

Alam A, Abubaker Bagabir H, Sultan A, Siddiqui MF, Imam N, Alkhanani MF et al (2022) An integrative network approach to identify common genes for the therapeutics in tuberculosis and its overlapping non-communicable diseases. Front Pharmacol 27(12):770762

Aličković E, Subasi A (2017) Breast cancer diagnosis using GA feature selection and Rotation Forest. Neural Comput Appl 28(4):753–763

Allen GM, Shortall BP, Gemci T, Corcoran TE, Chigier NA (2004) Computational simulations of airflow in an in vitro model of the pediatric upper airways. J Biomech Eng 126(5):604–613

Alshamlan H, Badr G, Alohali Y (2015) mRMR-ABC: a hybrid gene selection algorithm for cancer classification using microarray gene expression profiling. Biomed Res Int 2015:1–15

Alshamlan HM, Badr GH, Alohali YA (2016) ABC-SVM: artificial bee colony and SVM method for microarray gene selection and multi class cancer classification. IJMLC 6(3):184–190

Avila-Olias M, Pegoraro C, Battaglia G, Canton I (2013) Inspired by nature: fundamentals in nanotechnology design to overcome biological barriers. TherDeliv. 4(1):27–43

Babaoğlu İ (2013) Diagnosis of coronary artery disease using artificial bee colony and K-nearest neighbor algorithms. IJCCE 56–59

Bai HX, Hsieh B, Xiong Z, Halsey K, Choi JW, Tran TML et al (2020) Performance of radiologists in differentiating COVID-19 from non-COVID-19 viral pneumonia at chest CT. Radiology 296(2):E46-54

Bell JS, Reynolds L, Freeman C, Jackson JK (2020) Strategies to promote access to medications during the COVID-19 pandemic. Aust J Gen Pract. 49(8):530–532

Bland RD, Clarke TL, Harden LB (1976) Rapid infusion of sodium bicarbonate and albumin into high-risk premature infants soon after birth: a controlled, prospective trial. Am J Obstet Gynecol 124(3):263–267

Čartolovni A, Tomičić A, LazićMosler E (2022) Ethical, legal, and social considerations of AI-based medical decision-support tools: a scoping review. Int J Med Inform 161:104738

Chan HP, Samala RK, Hadjiiski LM, Zhou C (2020) Deep learning in medical image analysis. Adv Exp Med Biol 1213:3–21

Chen Y, Shi S, Yao X, Nakano T (2017) Touchable computing: computing-inspired bio-detection. IEEE Trans Nanobiosci 16(8):810–821 de Carvalho Filho AO, Silva AC, de Paiva AC, Nunes RA, Gattass M (2017) Computer-aided diagnosis system for lung nodules based on computed tomography using shape analysis, a genetic algorithm, and SVM. Med Biol Eng Comput 55(8):1129–1146

Desrosiers A, Vallée-Bélisle A (2017) Nature-inspired DNA switches: applications in medicine. Nanomed (Lond) 12(3):175–179

Ergen SA, Dincbas FO, Yücel B, Altınok P, Akyurek S, Korkmaz Kıraklı E et al (2020) Risk factors of radiation pneumonitis in patients with NSCLC treated with concomitant chemoradiotherapy—are we underestimating diabetes?—Turkish oncology group (TOG)/Lung cancer study group. Clin Respir J 14(9):871–879

Esmaeilzadeh P (2020) Use of AI-based tools for healthcare purposes: a survey study from consumers' perspectives. BMC Med Inform DecisMak 20(1):170

Fási L, Di Meo F, Kuo CY, StojkovicBuric S, Martins A, Kúsz N et al (2019) Antioxidant-inspired drug discovery: antitumor metabolite is formed in situ from a hydroxycinnamic acid derivative upon free-radical scavenging. J Med Chem 62(3):1657–1668

Franssen FM, Alter P, Bar N, Benedikter BJ, Iurato S, Maier D et al (2019) Personalized medicine for patients with COPD: where are we? Int J Chron Obstruct Pulmon Dis 14:1465–1484

Gao M, Zhang Q, Feng XH, Liu J (2021) Synthetic modified messenger RNA for therapeutic applications. Acta Biomater 1(131):1–15

Ghaheri A, Shoar S, Naderan M, Hoseini SS (2015) The applications of genetic algorithms in medicine. Oman Med J 30(6):406–416

Gupta TK, Raza K (2020) Optimizing deep feedforward neural network architecture: a Tabu search based approach. Neural Process Lett 51(3):2855–2870

Gupta R, Srivastava D, Sahu M, Tiwari S, Ambasta RK, Kumar P (2021) Artificial intelligence to deep learning: machine intelligence approach for drug discovery. Mol Divers 25(3):1315–1360

Gupta TK, Raza K (2012) Optimization of artificial neural network: a bat algorithm-based approach. In: Abraham A, Gandhi N, Hanne T, Hong TP, Nogueira Rios T, Ding W (eds) Intelligent systems design and applications [Internet]. Lecture notes in networks and systems, vol 418. Springer International Publishing, Cham, pp 286–295 [cited 2022 Jul 9]. Available from https://link.springer.com/https://doi.org/10.1007/978-3-030-96308-8_26

Gupta TK, Raza K (2019) Optimization of ANN architecture: a review on nature-inspired techniques. In: Machine learning in bio-signal analysis and diagnostic imaging [Internet]. Elsevier, pp 159–82 [cited 2022 Jul 9]. Available from https://linkinghub.elsevier.com/retrieve/pii/B9780128160862000072

Haleem A, Vaishya R, Javaid M, Khan IH (2020) Artificial Intelligence (AI) applications in orthopaedics: an innovative technology to embrace. J Clin Orthop Trauma 11:S80–S81

Hendrickson WA, Ward KB (1975) Atomic models for the polypeptide backbones of myohemerythrin and hemerythrin. Biochem Biophys Res Commun 66(4):1349–1356

Jaton JC, Huser H, Braun DG, Givol D, Pecht I, Schlessinger J (1975) Conformational changes induced in a homogeneous anti-type III pneumococcal antibody by oligosaccharides of increasing size. Biochemistry 14(24):5312–5315

Ji W, Li Y, Peng H, Zhao R, Zhang X (2022) Nature-inspired dynamic gene-loaded nanoassemblies for the treatment of brain diseases. Adv Drug Deliv Rev 180:114029

Johnson KW, Torres Soto J, Glicksberg BS, Shameer K, Miotto R, Ali M et al (2018) Artificial intelligence in cardiology. J Am Coll Cardiol 71(23):2668–2679

Julian MC, Rabia LA, Desai AA, Arsiwala A, Gerson JE, Paulson HL et al (2019) Nature-inspired design and evolution of anti-amyloid antibodies. J Biol Chem 294(21):8438–8451

Karaboga HA, Gunel A, Korkut SV, Demir I, Celik R (2021) Bayesian network as a decision tool for predicting ALS disease. Brain Sci 11(2):150

Kasendra M, Wells JM (2020) A window into your gut: biologically inspired engineering of mini-gut tubes in vitro. Dev Cell 55(5):522–524

Kim IG, Wu Y, Park SA, Cho H, Shin JW, Chung EJ (2020) Tissue-engineered graft for circumferential esophageal reconstruction in rats. J Vis Exp (156)

Kossmann D (2000) The state of the art in distributed query processing. ACM Comput Surv 32(4):422–469
Le EPV, Wang Y, Huang Y, Hickman S, Gilbert FJ (2019) Artificial intelligence in breast imaging. Clin Radiol 74(5):357–366
Leão T, Wang M, Moss N, da Silva R, Sanders J, Nurk S et al (2021) A Multi-omics characterization of the natural product potential of tropical filamentous marine cyanobacteria. Mar Drugs 19(1):20
Lertxundi U, Hernández R, Medrano J, Orive G (2020) Drug pollution and pharmacotherapy in psychiatry: a "platypus" in the room. Eur Psychiatry 63(1):e33
Liu K, Huang H, Zhang L, Huang Y, Sun S, Chen X et al (2021) Effects of a physician- and pharmacist-managed clinic on pain management in cancer patients in China. Basic Clin Pharmacol Toxicol 129(1):36–43
Lu DY, Wu HY, Yarla NS, Xu B, Ding J, Lu TR (2018) HAART in HIV/AIDS treatments: future trends. Infect Disord Drug Targets 18(1):15–22
Luo H, Tang Q ling, Shang Y xi, Liang SB, Yang M, Robinson N et al (2020) Can Chinese medicine be used for prevention of corona virus disease 2019 (COVID-19)? A review of historical classics, research evidence and current prevention programs. Chin J Integr Med 26(4):243–50
Makar AB, McMartin KE, Palese M, Tephly TR (1975) Formate assay in body fluids: application in methanol poisoning. Biochem Med 13(2):117–126
Miles D, Ciruelos E, Schneeweiss A, Puglisi F, Peretz-Yablonski T, Campone M et al (2021) Final results from the PERUSE study of first-linepertuzumab plus trastuzumab plus a taxane for HER2-positive locally recurrent or metastatic breast cancer, with a multivariable approach to guide prognostication. Ann Oncol 32(10):1245–1255
Mohamed N, Kim HS, Kang KM, Mohamed M, Kim SH, Kim JG (2021) Heart and lung sound measurement using an esophageal stethoscope with adaptive noise cancellation. Sensors (Basel) 21(20):6757
Mohsin AH, Zaidan AA, Zaidan BB, Mohammed KI, Albahri OS, Albahri AS et al (2021) PSO-Blockchain-based image steganography: towards a new method to secure updating and sharing COVID-19 data in decentralised hospitals intelligence architecture. Multimed Tools Appl. 80(9):14137–14161
Niazi MKK, Parwani AV, Gurcan MN (2019) Digital pathology and artificial intelligence. Lancet Oncol 20(5):e253–e261
Olaru G, Wilhelm O, Nordin S, Witthöft M, Köteles F (2019) Modern health worries: deriving two measurement invariant short scales for cross-cultural research with ant colony optimization. PLoS ONE 14(2):e0211819
Pérez-Jiménez MJ (2014) The P versus NP problem from the membrane computing view. Eur Rev 22(1):18–33
Poole-Wilson PA, Langer GA (1975) Effect of pH on ionic exchange and function in rat and rabbit myocardium. Am J Physiol 229(3):570–581
Putrino D, Ripp J, Herrera JE, Cortes M, Kellner C, Rizk D et al (2020) Multisensory, nature-inspired recharge rooms yield short-term reductions in perceived stress among frontline healthcare workers. Front Psychol 11:560833
Qureshi KN, Alhudhaif A, Qureshi MA, Jeon G (2021) Nature-inspired solution for coronavirus disease detection and its impact on existing healthcare systems. Comput Electr Eng 95:107411
Ramakrishnan K, Prattipati M, Samuel P, Sankaranarayanan M (2020) Transcriptional control of gene expression in Pichia pastoris by manipulation of terminators. Appl Microbiol Biotechnol 104(18):7841–7851
Raza K, Alam M (2016) Recurrent neural network based hybrid model for reconstructing gene regulatory network. Comput Biol Chem 64:322–334
Roslan MF, Al-Shetwi AQ, Hannan MA, Ker PJ, Zuhdi AWM (2020) Particle swarm optimization algorithm-based PI inverter controller for a grid-connected PV system. PLoS ONE 15(12):e0243581
Schultz KT, Grieder F (1987) Structure and function of the immune system. ToxicolPathol 15(3):262–264

Shi D, Zhang W, Zhang W, Ding X (2019) A review on lower limb rehabilitation exoskeleton robots. Chin J Mech Eng. 32(1):74

Siddique N, Adeli H (2015) Nature inspired computing: an overview and some future directions. Cognit Comput 7(6):706–714

Siddiqui MF (2021) IoMT Potential Impact in COVID-19: combating a pandemic with innovation. In: Raza K (eds) Computational intelligence methods in COVID-19: surveillance, prevention, prediction and diagnosis [Internet]. Studies in computational intelligence, vol 923. Springer Singapore, Singapore, pp 349–361 [cited 2022 Apr 25]. Available from https://link.springer.com/https://doi.org/10.1007/978-981-15-8534-0_18

Siddiqui MF, Mouna A, Nicolas G, Rahat SAA, Mitalipova A, Emmanuel N et al (2022) Computational intelligence: a step forward in cancer biomarker discovery and therapeutic target prediction. In: Raza K (ed) computational intelligence in oncology [Internet]. Studies in Computational Intelligence, vol 1016. Springer Singapore, Singapore [cited 2022 Apr 25]. pp 233–250. Available from https://link.springer.com/https://doi.org/10.1007/978-981-16-9221-5_14

Smith RJ, Bryant RG (1975) Metal substitutions incarbonic anhydrase: a halide ion probe study. BiochemBiophys Res Commun. 66(4):1281–1286

Strobach D, Blassmann U, Gundl S, Krebs S, Querbach C, Schuhmacher C et al (2021) Assessing the quality of drug information provided by hospital pharmacies using a fictitious enquiry and simulated real-life conditions. Eur J Hosp Pharm 28(Suppl 2):e79-84

Subramanian M, Wojtusciszyn A, Favre L, Boughorbel S, Shan J, Letaief KB et al (2020) Precision medicine in the era of artificial intelligence: implications in chronic disease management. J Transl Med 18(1):472

Swathi M, Sreedhar KC (2020) A cloud-based privacy-preserving e-healthcare system using particle swarm optimization. In: Raju KS, Govardhan A, Rani BP, Sridevi R, Murty MR (eds) Proceedings of the third international conference on computational intelligence and informatics [Internet]. Advances in Intelligent Systems and Computing, vol 1090. Springer Singapore, Singapore, pp 133–143 [cited 2022 Jul 9]. Available from http://link.springer.com/https://doi.org/10.1007/978-981-15-1480-7_11

Taherdangkoo M (2014) Skull removal in MR images using a modified artificial bee colony optimization algorithm. Technol Health Care 22(5):775–784

Takitou S, Taneda A (2019) Ant colony optimization for predicting RNA folding pathways. Comput Biol Chem 83:107118

Tay D, Poh CL, Kitney RI (2015) A novel neural-inspired learning algorithm with application to clinical risk prediction. J Biomed Inform 54:305–314

Travaglini KJ, Nabhan AN, Penland L, Sinha R, Gillich A, Sit RV et al (2020) A molecular cell atlas of the human lung from single-cell RNA sequencing. Nature 587(7835):619–625

Tzanetos A, Fister I, Dounias G (2020) A comprehensive database of Nature-Inspired Algorithms. Data Brief 31:105792

Vaikath NN, Hmila I, Gupta V, Erskine D, Ingelsson M, El-Agnaf OMA (2019) Antibodies against alpha-synuclein: tools and therapies. J Neurochem 150(5):612–625

Vaishya R, Javaid M, Khan IH, Haleem A (2020) Artificial intelligence (AI) applications for COVID-19 pandemic. Diab Metab Syndr 14(4):337–339

Wang N, Świtalska M, Wang L, Shaban E, Hossain MI, El Sayed IET et al (2019) Structural Modifications of nature-inspired indoloquinolines: a mini review of their potential antiproliferative activity. Molecules 24(11):E2121

Wang Z, Qin C, Wan B, Song WW (2021) A comparative study of common nature-inspired algorithms for continuous function optimization. Entropy (Basel) 23(7):874

Wani N, Barh D (2021) Raza K (2021) Modular network inference between miRNA–mRNA expression profiles using weighted co-expression network analysis. J Integr Bioinf 18(4):20210029

Wei T, Zhong J (2019) A Preliminary study of knowledge transfer in multi-classification using gene expression programming. Front Neurosci 13:1396

Weikert D, Mai S, Mostaghim S (2020) Particle swarm contour search algorithm. Entropy (Basel) 22(4):E407

Yan J, Peng Y, Shen D, Yan X, Deng Q (2018) An artificial bee colony-based green routing mechanism in WBANs for sensor-based e-healthcare systems. Sens (Basel) 18(10):E3268

Zhang Y, Wang Z, Wang Z, Liu X, Yuan X (2017) A DNA-based encryption method based on two biological axioms of dna chip and polymerase chain reaction (PCR) amplification techniques. Chemistry 23(54):13387–13403

Zhang Y, Qazi S, Raza K (2021) Differential expression analysis in ovarian cancer: a functional genomics and systems biology approach. Saudi J Biol Sci 28(7):4069–4081

Nature-Inspired Computing in Cancer Research

Nature-Inspired Computing in Breast Cancer Research: Overview, Perspective, and Challenges of the State-of-the-Art Techniques

Ankita Sahu, K. K. Ajeeshkumar, Mudasir Nabi Peerzada, Manoj Kumar Yadav, and Saurabh Verma

1 Introduction

In the past couple of decades, various computational tools have been effectively practiced to solve many real-life multidisciplinary problems. Nature-inspired computational techniques (NICs) employ different performance metrics to address the problem inspired by observing the natural phenomenon in different environmental settings (Sharma et al. 2019). Nowadays, NIC is gaining interest and it adopted methodologies of developing computational algorithms for solving complex problems (Hariprasath et al. 2021).

Nature-inspired computing is an emerging advanced *computer-based* technique that uses unique behavioral features of different organisms such as insects, birds, animals, and human beings (Gupta and Raza 2019). These algorithms utilize the distinguish features such as autonomous, distributed, emergent, adaptive, and self-organized behavior of different organisms for finding the appropriate solutions to different real-life problems of various domains like education, research, business, healthcare, finance, and entertainment sectors (Zang et al. 2010).

Metaheuristics a high-level procedure, derives from the composition of two Greek words, namely *Meta* meaning one level above, and *heuristics* meaning to serve to find out. Although some high performance nature-inspired metaheuristic algorithms cannot guarantee the optimality of their solutions, they can offer a reasonable amount of time by exploring search spaces and often find good solutions with less computational effort. These algorithms came into practice for the first time probably in the 1940s (Turing 1948; Copeland 2004). Later on, the genetic algorithm (GA) is

A. Sahu · K. K. Ajeeshkumar · M. N. Peerzada · S. Verma (✉)
Tumor Biology, ICMR-National Institute of Pathology, New Delhi, India
e-mail: svarmasv1@rediffmail.com

M. K. Yadav
Department of Bioinformatics, SRM University, Delhi-NCR, Haryana, India

K. Raza (ed.), *Nature-Inspired Intelligent Computing Techniques in Bioinformatics*,
Studies in Computational Intelligence 1066,
https://doi.org/10.1007/978-981-19-6379-7_3

invented by John Holland in1975 inspired by evolutionary concepts like mutation and selection patterns (Holland 1992). The golden decade of the 1990s is known for evolutionary and metaheuristic-based computations (Yang and Karamanoglu 2020). The ant colony optimization (ACO) algorithm is one of the NIC techniques that search the optimal path through graphs by taking the motivation from the behavior of real ants, for the first time introduced by Dorigo (1992). The particle swarm optimization (PSO), a bio-inspired algorithm, was introduced by James Kennedy and Russell C. Eberhart in 1995 (Kennedy and Eberhart 1995). This algorithm is also an optimization search algorithm that utilizes the biological phenomenon of the flock of birds searching for their food.

NIC algorithms are extensively used for the early prediction of various human disorders and found to be effective (Kaur and Sharma 2017). NIC algorithms such as GA, artificial bee colonies (ABC), and PSO are extensively used in the early diagnosis of diseases such as cancer, diabetes, and cardiac diseases summarized in Table 1. Moreover, among ACO, PSO, and ABC algorithms, ABC is found to be superior for diagnosing breast cancer (Kaur and Sharma 2017). It is observed that the hybrid use of NIC algorithms produces better results as compared to their individual uses. The conventional machine-based approaches are widely used for the identification of breast cancer diagnosis but there is a lot of space for further improvement for advanced predictions and diagnosis (Derangula and Edara 2021). A high level of prediction accuracy can be achieved by utilizing NIC-based techniques along with the traditional machine-based approaches.

Table 1 Applications of NIC techniques

Algorithms	Applications	References
Genetic algorithm (GA)	Bioinformatics multiple sequence alignment, cancer biomarkers, microarray, building phylogenetic trees, gene expression profile analysis	Aličković and Subasi (2017), Alshammari et al. (2022), Mishra et al. (2020), Sathya et al. (2022)
Ant colony optimization (ACO)	Maintained the local consistency and global connectivity factors, disease diagnosis	Alshammari et al. (2022), Sayed et al. (2016)
Artificial bee colony (ABC)	Coronary artery disease, diabetes diagnosis training neural networks	Daelemans et al. (2003), He et al. (2014), Shah (2021),
Swarms intelligence (SI)	Cancer biomarkers, neurodegenerative disease, gene expression profile analysis, bioengineering	Best et al. (2017), Sayed et al. (2016), Schranz et al. (2020)
Gray wolf optimizer (GWO)	Parkinson's disease, cancer detection, neurodegenerative disease, heart problem	Mirjalili et al. (2014), Zang et al. (2010)

2 Nature-Inspired Computing Solutions for Breast Cancer

The NIC approaches are gaining attention in breast cancer research due to their efficacy and precision in prediction. Several experiments are performed to find out the role of various stand-alone as well as hybridized NIC algorithms to find out the breast cancer disease. The three swarm algorithms, namely ACO, firefly, and PSO algorithms, are applied to the Wisconsin breast cancer dataset to detect and classify the dataset, which records the measurements for breast cancer cases whether the cancer is benign or malignant using their available image features. These different swarm algorithms are employed to optimize the support vector machine (SVM) which helps in constructing better classifiers having the ability to differentiate between malignant and benign cancer images. When compared to conventional experimental methods, these NIC-based approaches are showing better diagnosis and prognosis (Hariprasath et al. 2021). Precise prediction remains one of the major challenges in breast cancer. Bio-inspired computing is found to be efficient to predict different disease models including breast cancer. GA techniques are showing promising results in terms of feature selection. Apart from this, there are other data mining machine learning algorithms, namely logistic regression, decision trees, random forest, radial basis function networks (RFRBFN), Bayesian network, multilayer perceptron (MLP), SVM, and rotation forest (RTF) to categorize the data profile. Nowadays, most of the data mining work is going on using hybridized algorithms. An accuracy of 99.48% is achieved in breast cancer detection when GA feature selection methods are employed in combination with rotation forest algorithms (Aličković and Subasi 2017). However, the ABC algorithm is a popular nature-inspired algorithm and is found to be effective in the prediction and diagnosis of breast cancer (Kaur and Sharma 2017; Shah et al. 2019), slow convergence due to poor exploration and exploitation processes remains a matter of concern. In order to overcome the convergence problem, the two honeybee-based algorithms, Gbest guided artificial bee colony (GGABC) and global artificial bee colony (GABC) algorithms, are improvised for solving complex optimization tasks. 3G-ABC is a hybrid algorithm constructed by using the foraging behavior of the global best from GGABC and guided the best honeybee from GABC. The 3G-ABC algorithm is showing outstanding performance in the prediction of different stages of breast cancer compared to any stand-alone algorithms. It provides accurate and critical information on breast cancer prediction which will be helpful for health authorities to efficiently evaluate and find a solution regarding the risk factors of breast cancer in an early stage (Shah et al. 2019). Swarm intelligence algorithms also enable accurate predictions, thereby medical practitioners can proceed with a better and more timely diagnosis of breast cancer. Four swarm algorithms such as gray wolf optimizer (GWO), moth flame optimization (MFO), PSO, and firefly algorithm optimization (FA) successfully used to solve the segmentation problem of biomedical imaging. Thermography-based breast cancer images are selected, and the nature-inspired swarm algorithm schemes have been tested to observe their efficacy and precision to sub-categorize the breast cancer images into normal and/ or abnormal. The SVM technique with its kernel functions was employed to identify breast cancer detection. The experimental outcomes clearly indicate the superiority of the SVM-linear kernel function in classifying the normal and abnormal tissues compared to all proposed swarm algorithms. Later on, linear kernel function aided with four swarm algorithms: GWO, MFO, PSO, and FA is applied on results for comparing them. The test result shows that FA swarm algorithm is the best one as it is classifying different stages of breast cancer with the highest accuracy, sensitivity, precision, and specificity (Sayed et al. 2016). The modified firefly algorithm is capable of handling a large amount of data in a relatively short execution time and generates a query plan which is a global optimal solution. The MFA is applied to breast cancer data available in the BioPortal dataset. The result indicates a significantly improve performance of the proposed work with the well-known nature-inspired query optimization algorithms (Ramalingam 2022).

Prediction of different phases of breast cancer such as benign and malignant breast cancer is identified by using the artificial immune recognition system (Sayed et al. 2016) using the Wisconsin dataset. A different algorithm, called clonal selection algorithm, hierarchical learning vector quantization, and multipass self-organizing map, is efficient in the supervised classification of breast cancer (Lovenia et al. 2019).

3 Nature-Inspired Algorithms

Several authors have employed different nature-inspired computing techniques for the early diagnosis of human disorders. These techniques can be used in health care, agriculture, engineering, and different disease diagnosis, viz. neurodegenerative diseases, Alzheimer's disease, heart problems, diabetes, skin problems, and various type of cancers including breast cancer, etc., depicted in Table 1 (Sharma et al. 2019).

3.1 Genetic Algorithm

Genetic algorithm (GA) is for the first time developed by integrating biological evolution concepts in computer simulations in the early 1950s. GA is inspired by Charles Darwin's theory of natural selection where only the fittest individuals among the entire population are selected to produce offspring for the next generation. So, GA is basically a computer science technique, inspired by the theory of natural selection, used to generate high-quality solutions to optimization and search problems (Sharma et al. 2021; Zang et al. 2010). The genetic algorithm seems to be a feasible approach to handling large-scale distributed systems that operate with sequential populations, and the distinguished periodic solution is received after a sequence of iterative processes until satisfactory outcomes are reached (Aličković and Subasi 2017; Sharma et al. 2021). This algorithm is used to identify the appropriate biomarkers for predicting breast cancer from blood analysis data and routine consultation of the case group which helps in the process of early detection of the disease (Mishra et al. 2020). The feature selection criteria of GA is providing high-level classification accuracy in the diagnosis of breast cancer (Sathya et al. 2022).

3.2 *Ant Colony Optimization*

Similar to GA, ant colony optimization (ACO) is a mixed procedure with many variants and applications. ACO is a population-based metaheuristic optimization technique, proposed by Marco Dorigo in 1992 (Dorigo 1992). It is inspired by the social behavior of real ants. The real ants live in ant colonies and find the shortest path between their colony to the source of food by using the experience of other members. The ACO algorithm mimics the same nature-inspired behavior to solve the combinatorial optimization problems using some sorts of graphs (Zang et al. 2010; Mirjalili et al. 2014). In the natural world, ants walk randomly until they find a sign of a food source and return to their colony while laying down pheromone on the ground in order to mark the optimal path and more ants follow the path. There are three well-known variations of ant colony optimization algorithms, namely ant system (AS), ant colony system (ACS), and MAX–MIN ant system (MMAS). The AS algorithm is the first optimization algorithm developed using real ant behavior. The ACS algorithm is based on the exploitation of ant's natural phenomenon of pheromones. If an ant finds the nearby food, then it secreted pheromones in large quantities to attract other ants of that colony. A similar concept is used in developing the ACS algorithm to solve the optimization problems. The MMAS algorithm controls the maximum and minimum pheromone amount on each trail and allows only the best tour to add for searching next possible step. The mentioned three algorithms are extensions of the ant system algorithm in order to improve its search performance and reduce its complexity. A multi-colony ant algorithm is also proposed based on MMAS to optimize distributed database queries and to exploit more strongly the finest solutions found during the search. This algorithm operates scalability and modularity. The experimental outcomes for breast cancer dataset show that an enhanced algorithm can discover clusters of data with arbitrary shapes in a dataset and is better than the clustering effect of prior techniques (Favaretto et al. 2009; Dey et al. 2016; Dhahri et al. 2019).

3.3 *Artificial Bee Colony Optimization*

The artificial bee colony (ABC) is a new, simple, and efficient nature-inspired optimization algorithm, popular for solving gene selection problems. The overall process of the proposed ABC approach is summarized in Fig. 1. The ABC approach is developed based on mimicking the collective behavior of bees for finding their food sources (Mirjalili et al. 2014). Many researchers are using the ABC algorithm for solving optimization problems of different domains. In the ABC model, the artificial bee colonies consist of three types of bees: scout, onlooker, and employed bees, on the basis of their functionality. First, the scout bees randomly search the environment that is responsible for looking for new food resources, later on, the information about good food sources is shared with the onlooker bees. The food source of higher quality (fitness) will have more chances to be selected by the onlooker bees (Xu et al. 2013). Thereafter, the nectar of food sources is exploited by employed bees and onlooker bees (He et al. 2014). The advantage of using this method is easy to use, tardiness, higher flexibility, and fewer control parameters than the other algorithms such as GA and PSO. ABC is one of the attractive nature-inspired algorithms which has been successfully applied in breast cancer data in terms of accuracy, precision, and recall. This algorithm provides higher classification performance when applied to breast and lung cancer dataset samples (Shah 2021).

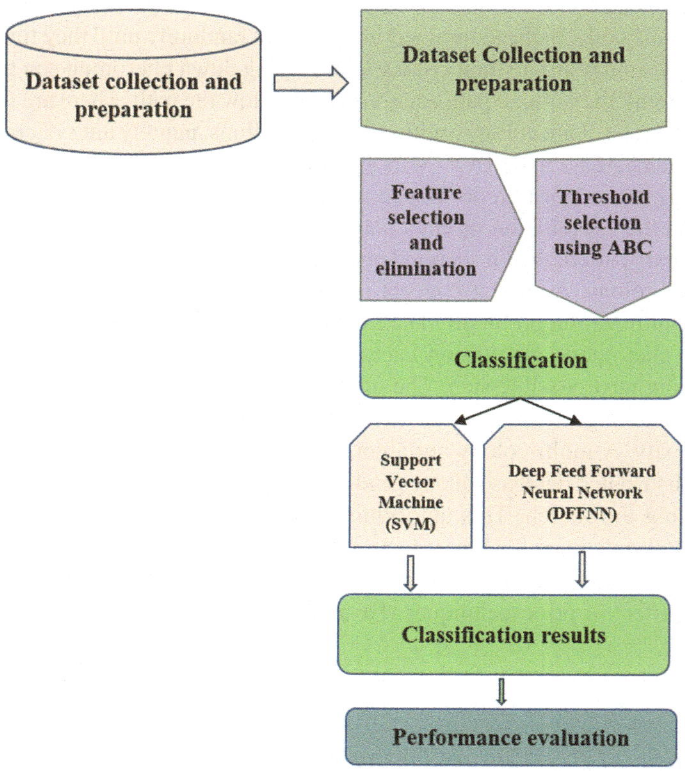

Fig. 1 Flowchart of ABC process

3.4 Swarm Optimization Algorithms

The swarm algorithm is a nature-inspired algorithm that works on the basis of the social behavior of birds and animals (Schranz et al. 2020). This is a population-based metaheuristic algorithm, proposed for the first time in 1993. It takes inspiration mostly from the social behavior of natural colonies, ant foraging, bird flocking, herds, and fish schooling. The major advantage of this swarm algorithm is easy to implement and preserves the information about the search space over the course of repetitive iterations. This algorithm usually uses fewer parameters, and fewer operators such as crossover, mutation, elitism, and so on, to adjust to NIC approaches. Here, we present the *four swarm*-based *algorithms*: PSO, GWO, moth flame optimization (MFO), and firefly algorithm optimization (FAO) as these algorithms are successfully employed to tackle classification, detection, and medical image segmentation problems.

Nowadays, different computing techniques are used to detect breast cancer by using images obtained from conventional imaging methods. There are a number of different factors responsible for the accuracy of computer-aided diagnosis (CAD) systems. The typical stages in medical imaging-based CAD systems are represented in Fig. 2. The different swarm algorithms are used for clustering in bio-images. The next step involves the extraction of the regions of interest (ROI) from the cluster of images. The ROI is basically the disease region, selected on the basis of clustering and morphology operators. In the ROI extraction phases, the channel shows the different colors that produce the clustered image for selecting the best cluster image for disease diagnosis (Dey et al. 2016). Furthermore, the SVM classifier is applied to detect breast cancer.

Particle Swarm Optimization The PSO algorithm is inspired by the intelligent behavior of a swarm of birds or a school of fish in searching for their food. This is a population-based stochastic algorithm, developed by Dr. Russell C. Eberhart and Dr. James Kennedy in 1995. The PSO algorithm can best be understood by studying the behavior of birds in a swarm. Suppose all birds of the swarm are hungry and they are in search of a nearby hidden food source. These hungry birds can be compared to tasks in a computation system hungry for data resources. An individual bird is equivalent to a particle or agent which moves around randomly to look for the food source in other words looking for the best solution. In addition, each particle keeps

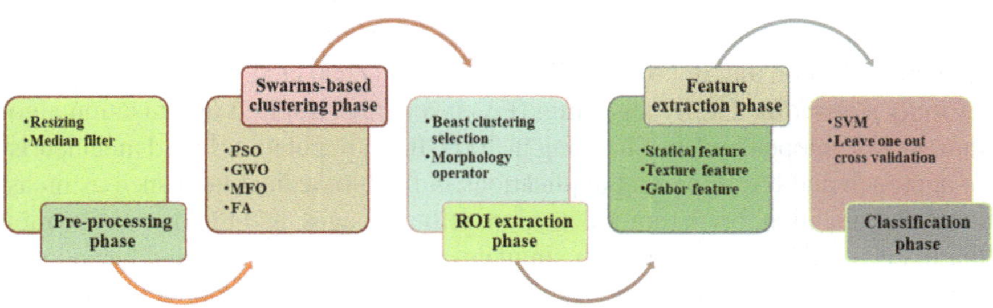

Fig. 2 Stages in the detection of breast cancer using swarm-based segmentation

track of its coordinates associated with the best local solution (fitness) which is called the personal best (pbest) value. Another best value is a global best, called gbest which is the best value. It gives the shortest path from the source to the destination. The algorithm aims to accelerate each particle toward pbest and gbest locations. The PSO algorithm performs better in less time as it needs only a few parameters to adjust when compared with GA (Hammouche et al. 2010).

Gray Wolf Optimizer Gray wolf optimizer (GWO) is based new population-based stochastic optimization approach inspired by the social hierarchy and hunting behavior of gray wolves (*Canis lupus*) in nature (Mirjalili et al. 2014). The pack of wolves follows dominance hierarchy system, where ranking is given according to strength and power. The alpha (α) male and female wolves are leaders of pack and are decision-makers. The decisions are followed by other members of pack. The second rank is given to beta (β) wolf which assists alpha males in taking decisions, and also taking care of discipline within the pack. The delta (δ) wolf is below the beta and possesses similar qualities. They are strong enough but lack leadership qualities. The omega (ω) wolves are giving the lowest rank in hierarchy, and they have to take orders from all other higher-order wolves and act accordingly. The GWO algorithm is used as a machine learning in image processing and breast cancer detection (Kamel et al. 2019; Vosooghifard and Ebrahimpour 2015). In the other study, GWO method combined with the decision tree is being used in the detection and cancer classification based on gene expression data to identify the different types of cancer (Tahoun et al. 2020; Vosooghifard and Ebrahimpour 2015).

Moth Flame Optimization Moth flame optimization (MFO) is a population-based metaheuristic algorithm proposed in 2015 by Mirjalili. The main concept of the moth flame optimization approach is inspired by a special navigation method of moths during the night for solving feature selection problems (Mirjalili 2015). Moths fly in the night by maintaining a fixed angle regarding the moon which is called transverse orientation. Moths express a similar kind of motion when artificial light sources are used in spite of natural moonlight. When the source light is at a long distance like moonlight, moths move in a straight line using a similar angle. The MFO algorithm is not the best option for solving complex optimization problems as it sinks into the local optimum due to the rapid dropping of population diversity and poor exploration. This screening technique is one of the earliest detection strategies for breast cancer (Alshammari et al. 2022). In the ROI phase, in moth flame optimization, the channel of the produced cluster image is chosen as the best cluster image for breast cancer detection (Sayed et al. 2016). This technique can effectively reduce the trouble of parameter selection and improve tracking accuracy (Zhao et al. 2022).

Firefly Algorithm Firefly algorithm (FA) is a nature-inspired optimization algorithm, first developed by Xin-She Yang in 2009. It is a population-based metaheuristics approach and has important applications in the biomedical field such as breast cancer classification, brain tumor detection, retinal images, and tumor classification (Sharma et al. 2021; Yang 2009). It simulates the flashing patterns and behavior of fireflies. The attraction of one firefly to others is directly proportional to the intensity of brightness. If more than one fireflies are there with a differential brightness pattern, then the firefly moves from less brightness to a higher one. FA is the best available swarm version as it results in higher flexibility in getting an optimal solution, effectiveness, sensitivity, precision, and specificity. Due to these features, this technique is the best choice for solving health care as well as biomedical problems (Nayak et al. 2020). FA brings a new strategy into the *optimization* field which is broadly used for the early detection and diagnosis of breast cancer. MFA is evaluated using the breast cancer image data available in the BioPortal dataset. The applied MFA shows significant improvement in terms of its performance and query execution time when compared with the existing nature-inspired query optimization algorithms (Ramalingam 2022; Sayed et al. 2016).

4 Choice of Fitness Functions and Parameter Setting

The different choices of fitness functions and parameter settings for three algorithms: ACO, ABC, and FA are discussed below.

4.1 Parameter Tuning and Fitness Functions for ACO

The optimal performance of the ACO algorithm depends upon the intelligent selection of suitable fitness function and their parameter tunings. The parameter setting techniques are categorized into two groups: (1) tuning and (2) control. Parameter tuning is performed prior to the optimization algorithm. However, parameter control is performed during the process of optimization. The parameter tuning can provide a clue for parameter control where the optimization is not well understood in advance. Collectively, both the parameter settings are essential for the successful function of nature-inspired swarm algorithms. In the case of parameter tuning by brute force, all possible combinations, all values, and all parameters are tested, or at the time of need only selected combinations of parameters are evaluated depending upon time and resource flexibility (Bansal et al. 2014; El-Gallad et al. 2002). However, in the case of parameter tuning by experimental design, the number of runs is reduced by appropriately designed experiments. In this case, each parameter is considered a factor, and its related possible values are described as levels of that particular factor. In the case of ant colony optimization, an ant develops a partial solution once it moves to another feasible city (Dorigo et al. 1996). The probability rule that each ant k in moves from city to city (i choose moves to city j) is calculated by the following equation:

$$P_{ij} = \frac{\tau i j^{\alpha} \cdot \eta i j^{\beta}}{\sum_{h \in N^k} (\tau j h^{\alpha} \cdot \eta i h^{\beta})}$$

$N^k$is the possible cities for an ant k from the city i. However, this set does not contain the visited cities in the partial solution of ant k. τ_{ij} is the pheromone that corresponds to the direction between the city (i and j). η_{ij} is a heuristic function and ranks alternatives in search algorithms which are calculated by inverting the distance between city i and city j (Dorigo and Gambardella 1997). It determines a value reflecting the attractiveness of the path between city i and city j. The pheromone trail τ_{ij} is rationalized with two mechanisms: pheromone deposit and pheromone trail evaporation (Dorigo and Stützle 2019; Stützle et al. 2012). During this updated procedure, all n ants deposit pheromones on iteration, and the pheromone trails evaporate by the equations:

$$\tau_{ij} = (1 - \rho) * \tau_{ij} + \sum_{k=1}^{n} \Delta\tau^k ij$$

$$\Delta\tau_{ij}^k = \begin{cases} Q/L_k \\ 0 \end{cases}$$

where ρ is the evaporation rate, Q is constant, and L_k is the length (the number of visited cities) of the partial solution of ant k (Dorigo et al. 1996). ACO algorithms are dependent on parameters. It is therefore worthwhile to formulate parameter adaptation techniques for these algorithms (Favaretto et al. 2009). The parameters that necessitate tuning and control in ACO include the total number of ants (n), pheromone weight (α), heuristic weight (β), initial pheromone (τ_0), evaporation rate (ρ), the total number of ants in the colony (n), and value denoted by Q. Dorigo et al. have experimented with parameter tuning for ant colony optimization to solve the traveling salesman problem (TSP) and have set the guidelines on the boundary of the basic parameters in ACO as follows α: the relative importance of pheromone trail, $\alpha \geq 0$; β: the relative importance of the heuristic information (visibility), $\beta \geq 0$; ρ: evaporation rate, $0 \leq \rho < 1$; Q: a constant quality function (Dorigo et al. 1996). It is experimentally concluded that good parameter values for ACO are include ant number m = problem size (for TSP), $\alpha = 1$, $\beta = 5$, $\rho = 0.5$, and $Q = 100$ (Dorigo and Stützle 2019). The utility of different parameter values in ACO indicated that two points are significant: (1) the performance of ACO is affected by its parameter values because parameters can unreservedly determine the strengthening and diversification of the search process, and (2) there are no parameter values which can be used in ant colony optimization to solve all optimization problems efficiently and effectively (Pilat and White 2002). The differences in parameter values are due to problem size, problem instance, and type of optimization problem. On the other hand, the fitness function (also called the evaluation function) evaluates how close a given solution is to achieving the desired problem. Problems taking place due to improperly premeditated fitness functions lead to difficulty in algorithm convergence. The fitness function should be concise and easily computable. ACO is an optimization technique that imitates an ant colony behavior (Stoykova and Spasov 2019).

4.2 Parameter Tuning and Fitness Function for ABC

The artificial bee colony (ABC) algorithm begins by the distribution of *n* scout bees to different flower patches randomly. Out of *n* visited sites, *m* sites include neighborhood site search. Out of the *m* sites, *e* elite sites are selected on the basis of their fitness values. The number of extra bees recruited for *e* elite sites (n_e) is greater than the number of bees recruited for the remaining *m-e* sites (n_m). The extra bees search within the neighborhood of the selected scout bee bounded by patch size, *ngh*. If p_i represents the position of a selected bee, recruited bees can be placed in the range: $p_i \pm ngh$. After each iteration, one best bee is selected to maintain the total number of bees at *n*. The remaining *n-m* bees that correspond to the unselected sites search randomly in the search space. The behavior of the ABC is influenced by parameters including the number of scout bees *n*, the number of selected sites *m*, the number of elite sites *e*, the number of recruited bees for elite sites n_e, the number of recruited bees for the remaining selected sites nm, and the initial patch size *ngh* (Phan et al. 2020).

A superior tuning of controlling parameters is treated as important for most of the algorithms in order to get the desired solutions. ABC algorithm necessitates determining a set of control parameters of the algorithm, viz. population size, mutation, amount of reproduction, etc. Like many of the NIC algorithms, the ABC algorithm also has to set the control parameters before running the algorithm. Values of these parameters affect the algorithm outcome (Akay and Karaboga 2009). The big colony size is not essential in this algorithm for solving the optimization problems. Control parameter limit is used to move the average chart dictating the occurrence of scout bees that are responsible for giving the diversity in the population. Three control parameters of the ABC algorithm employs are the number of food sources that in turn is matching the number of employed, the number of solutions in the population (SN), the maximum cycle number (MCN), and the value of limit in aggregate. The number of employed bees is similar to the number of solutions available in the population. In the first step, the ABC generates a randomly distributed initial food source positions of the number of solutions in the population (SN). Each solution x_i ($i = 1, 2, 3, 4, \ldots,$ SN) is a *D*-dimensional vector. *D* denotes the number of optimization parameters. In the employed bees' phase, each employed bee detects a new food source υ_i in the neighborhood of its existing source x_i by the following equation. The employed bee memorizes the good one by means of a greedy selection mechanism.

$$\upsilon_{ij} = x_{ij} + \Phi_{ij}\left(x_{ij} - x_{kj}\right)$$

where $k \in \{1, 2, 3, 4 \ldots, \text{SN}\}$ and $j \in \{1, 2, 3, 4, \ldots, D\}$ represent the randomly selected indexes. Although k is determined randomly, it has to be different from *i*. The range of $\varphi_{i,j}$ lies between $[-1, 1]$. It controls the production of a neighbor food source around $x_{i,j}$. The employed bees share information regarding the quality of the source after completing their searches. An onlooker bee chooses a food source depending on the fitness of a food source shared by employed bees. The following equation is used for calculating the chances of food sources to be selected by onlooker bees.

$$p_i = \frac{\text{fit}_i}{\sum_{n=1}^{SN} \text{fit}_n}$$

After onlooker bees select a food supply, every one of them finds a replacement solution within the neighborhood of the food source chosen by (1), and a greedy choice is applied between the new solution and therefore the food source chosen so as to boost the answer at hand. If a food source is not enhanced by means of "limit," it's expelled from the population and the employed bee of that food source turns out to be a scout. The scout bee finds a new random food source position by:

$$x_{ij}^{j} = x_{\min}^{j} + \text{rand}[0, 1]\left(x_{\max}^{j} - x_{\min}^{j}\right)$$

where $x^j{}_{\min}$ represents the lower limit and $x^j{}_{\max}$ represents the upper bounds of parameter j. These steps are repeated through a predetermined maximum cycle number, or until a termination criterion is satisfied.

Several parametric settings including swarm size and limits are required for the ABC algorithm for functioning. Any error in choosing these parameter settings for a particular process could lead to premature or no convergence. Certainly, the fitness function for this algorithm indicates the relative significance of a design. A good or higher fitness value for the algorithm indicates a better design. There are several ways to define the fitness function and one among them is the cost function, which defines it as follows:

$$F_i = (1 + \varepsilon) f_{\max} - f_i$$

where $f_{\max}$ indicates the largest recorded cost function value, f_i represents the cost function value for the ith design, and ε represents the small value to prevent any numerical difficulties (Arora 2017). The fitness function may decline the training error but with a small value. It is important to note that the ABC algorithm has technical weaknesses such as slow convergence, poor exploration–exploitation properties, and trapping in local solutions (Shah et al. 2016).

4.3 *Parameter Tuning and Fitness Function for Firefly Algorithm*

The firefly algorithm is one of the new classified swarm intelligent, metaheuristic, and nature-inspired techniques formulated by Xin-She Yang in 2008 by animating the characteristic behaviors of fireflies (Yang and Karamanoglu 2020). Yang formulated this method under the assumption that all fireflies are unisexual with attracted to each other firefly. The degree of attraction is proportionate to the brightness level of individuals. Therefore, the brighter flies move toward the low bright flies as they get attracted to them. Brighter fireflies move randomly if they don't find any fireflies brighter than them. The function parameter is linked with the bright light characteristic of the firefly population and is important in the formation of the firefly algorithm. Keeping under consideration the brightness of the light, the light intensity is inversely proportional to the square of the area and it defines the fitting function related to the distance between two fireflies in the algorithm. The fireflies are treated to move either systematically or randomly for the optimization of the fitting function in the case of the firefly algorithm. This optimization ensures that all the fireflies get attracted to the brighter ones to get the desired results. Here, the three parameters attractiveness, absorption, and randomness are taken under consideration to optimize the algorithm. The parameter attractiveness is dependent on the light intensity value between two fireflies. If this parameter becomes zero, then the randomization parameter is taken under consideration which is set by using the Gaussian distribution principle. The absorption parameter affects the attractiveness parameter as it goes from zero to infinity. The firefly appears at a random moment when the absorption parameter approaches infinity. Three basic rules are established for this algorithm (Yang 2009). The fireflies are unisex which means any firefly can be attracted to each other firefly. Both the brightness and attractiveness of fireflies are proportional to each other, and their values go on decreasing with the increase in distance between two fireflies. The brightness of a firefly is ascertained by the objective function in this algorithm (Sánchez et al. 2020). The variation of attractiveness represented by β with the distance between fireflies represented by r is defined by the equation given below:

$$\beta = \beta_0 \mathrm{e}^{-\gamma r^2}$$

where at $r = 0$ is the attractiveness denoted by β_0. However, the movement of firefly movement i to another more attractive (brighter) one j is defined by the following equation:

$$x_i^{t+1} = x_i^t + \beta_0 \mathrm{e}^{-\gamma r^2} \left(x_j^t - x_i^t \right) + \alpha_t \varepsilon_i^t$$

where x_i represents the position of a firefly in the iteration t, $\beta_0 \mathrm{e}^{-\gamma r^2} \left(x_j^t - x_i^t \right)$ indicates the attraction between firefly j and firefly i, and quantity $\varepsilon_i{}^t$ is a vector of random numbers with a randomization parameter described by α. This parameter is the initial randomness scaling factor and is described as follows:

$$\alpha_t = \alpha_t \delta^t$$

The parameters value for fireflies (n) is taken as 10; maximum number of iterations (t) 30 $\alpha = 0.015$, $\beta = 1$, $\delta = 0.95$ where for δ value between 0 and 1 are taken (Yang and He 2013).

5 Challenges of NIC Techniques in Breast Cancer Research

There are still many challenging issues concerning nature-inspired computing algorithms, especially from theoretical perspectives of breast cancer. Though researchers know the basic mechanisms of how such algorithms can work in clinical practice. Predicting the survival rate of breast cancer is a major challenging issue in the NIC technique although there are many instances in which nature-inspired computing can perform healthcare tasks as well or with better results. NIC-based diagnosis and treatment recommendations are sometimes challenging to embed in clinical workflows and biomedical science. In the present scenario, there are many limitations in NIC systems for breast cancer screening, such as the lack of public datasets on a large scale, the dependence on region of interest (ROI) annotation, high image quality necessities, and binary classification problems. These swarm algorithms were used to optimize the support vector machine (SVM) which robust deep feed-forward neural network (DFNN) to represent complex, nonlinear features of breast cancer data in a better way. The dataset is challenging to learn from since except for one case, the optimizers produced low accuracy. Understanding and addressing the challenges in breast cancer will lead to a better clinical outcome Chaudhury et al. have reported that early screening of breast cancer can help patients get the treatment they need. swarm algorithms such as ACO, PSO, GWO, MFO, and FA optimization were trained to classify the malignant and benign images from the Wisconsin breast cancer dataset (Diagnostic *Dataset*) (Hariprasath et al. 2021). The mortality rate of a breast cancer patient can be reduced significantly with an accurate disease diagnosis.

6 Future Research Direction

NIC is a fascinating new area that promises real-world applications like threat detection, medical imaging, motion tracking, robotics, aerodynamic design optimization, source localization, and many more. Natural-inspired computing has made a major contribution to computational biology, especially in a number of classical areas, and also in dynamic fields of the living systems including evolution, ecology, development, and behavior biology. The existing natural-inspired computing techniques, viz. GA, ACO, ABC, FA, and their modified applications will play an important role in future research direction in the field of health care, education, finance, agriculture, and bioengineering (Derangula and Edara 2021; Gupta and Raza 2019; Phan et al. 2020). Hence, the discussed NIC approaches are to motivate future research to increase the success percentage of the proposed methods and also increase the success rate of biomedical activities by using NIC-based computer-aided diagnosis (CAD) system to assist radiologists in classifying different diseases including breast tumors either benign or malignant by taking input from breast cancer datasets.

7 Conclusion

This chapter presents a systematic review of the role and performance of nature-inspired computing algorithms on medical image data. The NIC approaches are found to be very effective and efficient in the diagnosis of breast cancer. Meta-heuristic NIC algorithms act as a strong tool that can *solve* various *problems* related to the biomedical field. It provides good accuracy to classify diseases by providing a comprehensive list of prevailing platforms and projects and supplies a useful defensive approach. All algorithms are population-based, consisting of multiple agents such as ants, bees and fireflies, particles. These techniques are providing valuable insight and opening up new avenues in classifying, and identifying various diseases including breast cancer, with comparable accuracy to human competitive results. NIC techniques are inspired by the nature, provide a high-quality optimal solution quite quickly in practice, and can profoundly augment the prevailing technique for biomedical imaging applications and drug development.

Acknowledgements AS thanks to the Indian Council of Medical Research (ICMR), Government of India for financial support in the form of the ICMR-Research Associate fellowship (Grant No. 45/17/2019-PHA/BMS).

References

Akay B, Karaboga D (2009) Parameter tuning for the artificial bee colony algorithm. In: Nguyen NT, Kowalczyk R, Chen S-M (eds) Computational collective intelligence. Semantic web, social networks and multiagent systems. Springer, pp 608–619. https://doi.org/10.1007/978-3-642-04441-0_53

Aličković E, Subasi A (2017) Breast cancer diagnosis using GA feature selection and rotation forest. Neural Comput Appl 28(4):753–763. https://doi.org/10.1007/s00521-015-2103-9

Alshammari MM, Almuhanna A, Alhiyafi J (2022) Mammography image-based diagnosis of breast cancer using machine learning: a pilot study. Sensors 22(1):203. https://doi.org/10.3390/s22010203

Arora JS (2017) Chapter 17—nature-inspired search methods. In: Arora JS (ed) Introduction to optimum design, 4th edn. Academic Press, pp 739–769. https://doi.org/10.1016/B978-0-12-800806-5.00017-2

Bansal JC, Sharma H, Jadon SS, Clerc M (2014) Spider monkey optimization algorithm for numerical optimization. Memetic Computing 6(1):31–47. https://doi.org/10.1007/s12293-013-0128-0

Best MG, Sol N, In 't Veld SGJG, Vancura A, Muller M, Niemeijer A-LN, Fejes AV, Tjon Kon Fat L-A, Huis In 't Veld AE, Leurs C, Le Large TY, Meijer LL, Kooi IE, Rustenburg F, Schellen P, Verschueren H, Post E, Wedekind LE, Bracht J, Wurdinger T (2017) Swarm intelligence-enhanced detection of non-small-cell lung cancer using tumor-educated platelets. *Cancer* Cell 32(2):238–252.e9. https://doi.org/10.1016/j.ccell.2017.07.004

Copeland BJ (2004) The essential turing. Oxford University Press. https://doi.org/10.1093/oso/9780198250791.001.0001

Daelemans W, Hoste V, De Meulder F, Naudts B (2003) Combined optimization of feature selection and algorithm parameters in machine learning of language. In: Lavrač N, Gamberger D, Blockeel

H, Todorovski L (eds) Machine learning: ECML 2003. Springer, pp 84–95. https://doi.org/10.1007/978-3-540-39857-8_10
Derangula A, Edara S (2021) Identification of optimized features using nature-inspired meta-heuristics based optimizations in breast cancer detection. https://doi.org/10.1016/J.MATPR.2020.12.129
Dey N, Bhateja V, Hassanien AE (2016) Medical imaging in clinical applications: algorithmic and computer-based approaches. Springer
Dhahri H, Al Maghayreh E, Mahmood A, Elkilani W, Faisal Nagi M (2019) Automated breast cancer diagnosis based on machine learning algorithms. J Healthc Eng 2019:e4253641. https://doi.org/10.1155/2019/4253641
Dorigo M (1992) Optimization, learning and natural algorithms. Undefined. https://www.semanticscholar.org/paper/Optimization%2C-Learning-and-Natural-Algorithms-Dorigo/2b735a5cd94b0b5868e071255bd187a901cb975a
Dorigo M, Gambardella LM (1997) Ant colonies for the travelling salesman problem. Biosystems 43(2):73–81. https://doi.org/10.1016/S0303-2647(97)01708-5
Dorigo M, Maniezzo V, Colorni A (1996) Ant system: optimization by a colony of cooperating agents. IEEE Trans Syst Man Cybernet Part B (Cybernetics) 26(1):29–41. https://doi.org/10.1109/3477.484436
Dorigo M, Stützle T (2019) Ant colony optimization: overview and recent advances. In: Gendreau M, Potvin J-Y (eds) Handbook of metaheuristics. Springer International Publishing, pp 311–351. https://doi.org/10.1007/978-3-319-91086-4_10
El-Gallad A, El-Hawary M, Sallam A, Kalas A (2002) Enhancing the particle swarm optimizer via proper parameters selection. In: IEEE CCECE2002. Canadian conference on electrical and computer engineering. Conference proceedings (Cat. No.02CH37373) vol 2, pp 792–797. https://doi.org/10.1109/CCECE.2002.1013043
Favaretto D, Moretti E, Pellegrini P (2009) On the explorative behavior of MAX–MIN ant system. In: Stützle T, Birattari M, Hoos HH (eds) Engineering stochastic local search algorithms. Designing, implementing and analyzing effective heuristics. Springer, pp 115–119 https://doi.org/10.1007/978-3-642-03751-1_10
Gupta TK, Raza K (2019) Optimization of ANN architecture: a review on nature-inspired techniques. In: Machine learning in bio-signal analysis and diagnostic imaging. Elsevier, pp 159–182. https://doi.org/10.1016/B978-0-12-816086-2.00007-2
Hammouche K, Diaf M, Siarry P (2010) A comparative study of various meta-heuristic techniques applied to the multilevel thresholding problem. Eng Appl Artif Intell 23
Hariprasath K, Tamilselvi S, Saravana Kumar NM, Kaviyavarshini N, Balamurugan S (2021) Performance analysis of nature-inspired algorithms in breast cancer diagnosis. In: Nature-inspired algorithms applications. Wiley, pp 267–294 https://doi.org/10.1002/9781119681984.ch10
He Z, Ma C, Wang X, Li L, Wang Y, Zhao Y, Guo H (2014) A modified artificial bee colony algorithm based on search space division and disruptive selection strategy. Math Probl Eng 1–14. https://doi.org/10.1155/2014/432654
Holland JH (1992) Adaptation in natural and artificial systems: an introductory analysis with applications to biology, control, and artificial intelligence. A Bradford Book
Kamel SR, YaghoubZadeh R, Kheirabadi M (2019) Improving the performance of support-vector machine by selecting the best features by Gray Wolf algorithm to increase the accuracy of diagnosis of breast cancer. J Big Data 6(1):90. https://doi.org/10.1186/s40537-019-0247-7
Kaur P, Sharma M (2017) A survey on using nature inspired computing for fatal disease diagnosis. Int J Inf Syst Model and Des (IJISMD) 8(2):70–91. https://doi.org/10.4018/IJISMD.2017040105
Kennedy J, Eberhart R (1995) Particle swarm optimization. In: Proceedings of ICNN'95—international conference on neural networks, vol 4, pp 1942–1948. https://doi.org/10.1109/ICNN.1995.488968
Lovenia JDL, Freeda S, Jemima DD, Christopher N (2019) Breast cancer classification using nature-inspired algorithm. Int J Rec Technol Eng 8(2):1024–1027. https://doi.org/10.35940/ijrte.B1172.0982S1119

Mirjalili S (2015) Moth-flame optimization algorithm: a novel nature-inspired heuristic paradigm. Knowl-Based Syst 89:228–249. https://doi.org/10.1016/j.knosys.2015.07.006

Mirjalili S, Mirjalili SM, Lewis A (2014) Grey wolf optimizer. Adv Eng Softw 69:46–61. https://doi.org/10.1016/j.advengsoft.2013.12.007

Mishra AK, Roy P, Bandyopadhyay S (2020) Genetic algorithm based selection of appropriate biomarkers for improved breast cancer prediction. In: Bi Y, Bhatia R, Kapoor S (eds) Intelligent systems and applications. Springer International Publishing, pp 724–732. https://doi.org/10.1007/978-3-030-29513-4_54

Nayak J, Naik B, Dinesh P, Vakula K, Dash PB (2020) Firefly algorithm in biomedical and health care: advances issues and challenges. SN Comput Sci 1(6):311. https://doi.org/10.1007/s42979-020-00320-x

Phan HD, Ellis K, Barca JC, Dorin A (2020) A survey of dynamic parameter setting methods for nature-inspired swarm intelligence algorithms. Neural Comput Appl 32(2):567–588. https://doi.org/10.1007/s00521-019-04229-2

Pilat ML, White T (2002) Using genetic algorithms to optimize ACS-TSP. In: Dorigo M, Di Caro G, Sampels M (eds) Ant algorithms. Springer, pp 282–287. https://doi.org/10.1007/3-540-45724-0_28

Ramalingam G (2022) Modified firefly algorithm for optimizing biomedical breast cancer queries. https://doi.org/10.21203/rs.3.rs-171211/v1

Sánchez D, Melin P, Castillo O (2020) Chapter 12—modular granular neural network optimization using the firefly algorithm applied to time series prediction. In: Yang X-S (ed) Nature-inspired computation and swarm intelligence. Academic Press, pp 199–216. https://doi.org/10.1016/B978-0-12-819714-1.00023-3

Sathya M, Jeyaselvi M, Joshi S, Pandey E, Pareek PK, Jamal SS, Kumar V, Atiglah HK (2022) Cancer categorization using genetic algorithm to identify biomarker genes. J Healthc Eng 2022:e5821938. https://doi.org/10.1155/2022/5821938

Sayed GI, Soliman M, Hassanien AE (2016) Bio-inspired swarm techniques for thermogram breast cancer detection. In: Dey N, Bhateja V, Hassanien AE (eds) Medical imaging in clinical applications: algorithmic and computer-based approaches. Springer International Publishing, pp 487–506. https://doi.org/10.1007/978-3-319-33793-7_21

Schranz M, Umlauft M, Sende M, Elmenreich W (2020) Swarm robotic behaviors and current applications. Front Robot AI 7. https://www.frontiersin.org/article/https://doi.org/10.3389/frobt.2020.00036

Shah H (2021) Using new artificial bee colony as probabilistic neural network for breast cancer data classification. Front Eng Built Environ 1(2):133–145. https://doi.org/10.1108/FEBE-03-2021-0015

Shah H, Chiroma H, Herawan T, Ghazali R, Tairan N (2019) An efficient bio-inspired bees colony for breast cancer prediction. In: Abawajy JH, Othman M, Ghazali R, Deris MM, Mahdin H, Herawan T (eds) Proceedings of the international conference on data engineering 2015 (DaEng-2015). Springer, pp 597–608. https://doi.org/10.1007/978-981-13-1799-6_61

Shah H, Ghazali R, Herawan T, Rahman SU, Khan N (2016) Chapter 14—swarm based-artificial neural system for human health data classification. In: Al-Jumeily D, Hussain A, Mallucci C, Oliver C (eds) Applied computing in medicine and health. Morgan Kaufmann, pp 287–309. https://doi.org/10.1016/B978-0-12-803468-2.00014-X

Sharma M, Singh G, Singh R (2021) Clinical decision support system query optimizer using hybrid firefly and controlled genetic algorithm. J King Saud Univ Comput Inf Sci 33(7):798–809. https://doi.org/10.1016/j.jksuci.2018.06.007

Sharma S, Singh G, Singh D (2019) Role and performance of different traditional classification and nature-inspired computing techniques in major research areas. EAI Endorsed Trans Scalable Inform Syst 6(21). https://eudl.eu/doi/https://doi.org/10.4108/eai.13-7-2018.158419

Stoykova S, Spasov V (2019) Choice of fitness functions and parameter settings in genetic algorithms for analysis of induction motors. IOP Conf Ser Mater Sci Eng 618(1):012024. https://doi.org/10.1088/1757-899X/618/1/012024

Stützle T, López-Ibáñez M, Pellegrini P, Maur M, Montes de Oca M, Birattari M, Dorigo M (2012) Parameter adaptation in ant colony optimization. In: Hamadi Y, Monfroy E, Saubion F (eds) Autonomous search. Springer, pp 191–215. https://doi.org/10.1007/978-3-642-21434-9_8

Tahoun M, Almazroi AA, Alqarni MA, Gaber T, Mahmoud EE, Eltoukhy MM (2020) A grey wolf-based method for mammographic mass classification. Appl Sci 10(23):8422. https://doi.org/10.3390/app10238422

Turing A (1948) 10 intelligent machinery. Oxford University Press, pp 395–432. https://doi.org/10.1093/oso/9780198250791.003.0016

Vosooghifard M, Ebrahimpour H (2015) Applying grey wolf optimizer-based decision tree classifier for cancer classification on gene expression data. In: 2015 5th international conference on computer and knowledge engineering (ICCKE), 147–151. https://doi.org/10.1109/ICCKE.2015.7365818

Xu Y, Fan P, Yuan L (2013) A simple and efficient artificial bee colony algorithm. Math Probl Eng 2013:e526315. https://doi.org/10.1155/2013/526315

Yang X-S (2009) Firefly algorithms for multimodal optimization. In: Watanabe O, Zeugmann T (eds) Stochastic algorithms: foundations and applications. Springer, pp 169–178

Yang X-S, He X (2013) Firefly algorithm: recent advances and applications. Int J Swarm Intell 1(1):36–50. https://doi.org/10.1504/IJSI.2013.055801

Yang X-S, Karamanoglu M (2020) Chapter 1—nature-inspired computation and swarm intelligence: a state-of-the-art overview. In: Yang X-S (ed) Nature-inspired computation and swarm intelligence. Academic Press, pp 3–18. https://doi.org/10.1016/B978-0-12-819714-1.00010-5

Zang H, Zhang S, Hapeshi K (2010) A review of nature-inspired algorithms. J Bionic Eng 7:S232–S237. https://doi.org/10.1016/S1672-6529(09)60240-7

Zhao X, Fang Y, Liu L, Xu M, Li Q (2022) A covariance-based Moth–flame optimization algorithm with Cauchy mutation for solving numerical optimization problems. Appl Soft Comput 119:108538. https://doi.org/10.1016/j.asoc.2022.108538

Advances in Genomic Profiling of Colorectal Cancer Using Nature-Inspired Computing Techniques

Deeksha Rikhari, Ankit Srivastava, and Sameer Srivastava

Abbreviations

ABC	Artificial bee colony
AI	Artificial intelligence
CI	Computational intelligence
CSA	Cuckoo search algorithm
CRC	Colorectal cancer
GA	Genetic algorithm
IWO	Invasive weed optimization
ML	Machine learning
MMR	Mismatch repair system
NGS	Next-generation sequencing
NIC	Nature-inspired computational
NN	Neural network
PSO	Particle swarm optimization
SVM	Support vector machine

D. Rikhari · A. Srivastava · S. Srivastava (✉)
Department of Biotechnology, Motilal Nehru National Institute of Technology, Allahabad 211004, India
e-mail: sameers@mnnit.ac.in

K. Raza (ed.), *Nature-Inspired Intelligent Computing Techniques in Bioinformatics*,
Studies in Computational Intelligence 1066,
https://doi.org/10.1007/978-981-19-6379-7_4

1 Introduction

Colorectal cancer (CRC) is a heterogeneous disease resulting due to accumulation of various genetic and epigenetic aberrations that affect the overall process of cell proliferation and development (Mundade et al. 2014). CRCs are a leading cause of cancer-related mortality worldwide and account for more than 1.9 million cases worldwide, with a total of 935,000 deaths reported in 2020 (Sung et al. 2021). The progression of CRC occurs through various stages, starting from the formation of adenomatous polyps that eventually leads to malignant carcinomas due to stepwise alterations in various tumor suppressor, mismatch repair genes, and oncogenes (Srivastava et al. 2022a). Mutation in several tumor suppressor genes like APC (Adenomatous polyposis coli), SMAD2/4, TP53 (Tumor Protein 53), and oncogenes such as KRAS (Kirsten rat sarcoma virus) as well as alterations in several molecular signaling pathways like Wnt signaling pathway have been reported to serve as a driving force in tumor progression and development in CRC (Migliore et al. 2011). In addition to the discussed molecular factors and pathways, intra- and extracellular effector molecules such as lysophosphatidic acid (LPA) and sphingosine-1-phosphate (S1P) have also been reported to regulate a variety of processes that will lead to onset, development, and progression of cancers such as colorectal, pancreatic, breast and lymphoma (Srivastava et al. 2021). The symptoms of early-stage patients are hard to detect, and patients reach a critical stage until diagnosis. Alterations in genes and pathways have established an urge to develop treatments specific to the patient's condition and diagnostic need. Genomic profiling of such candidate genes may be necessary for establishing a potent therapeutic approach in CRC treatment. In recent years, expression analysis of cancer-specific genes has been recognized as a potential area of clinical research for better prognostic and diagnostic analysis of cancer (Li et al. 2007). Henceforth, it is quite evident that the application of gene expression in CRC diagnosis and classification has great potential in the future as it provides an organized and unbiased approach to categorizing a variety of tumors.

In the current scenario, a wide range of computational approaches has been used for the functional evaluation of altered genes. Their role as potential diagnostic biomarkers has also been established in CRC patients. Still, the main critical problem lies in extracting and identifying such differentially expressed genes from CRC samples (Sun et al. 2021). Therefore, more effective approaches with high efficacy for CRC need to be developed. The established Computational Intelligence (CI), such as Machine learning (ML), Neural network (NN), and Artificial intelligence (AI), are constitutively being used in the analysis of high-throughput data, however, it embarks certain limitations like difficulty in categorizing cancer-related genes due to diverse data obtained from several databases, hindrance in retrieving data and compromising quality because of overfitting problems and lack in the optimum amount of expertise and knowledge in the use of computational tools by researchers and experts (Srivastava et al. 2022b). Therefore, developing such methodologies and techniques is necessary to overcome the limitations mentioned above efficiently. With the advent of NIC approaches, early diagnosis and treatment of CRC have become comparably competent and easy. Among a plethora of techniques associated with cancer diagnosis and treatment, nature-inspired techniques tend to be an efficient emerging method for cancer detection and treatment. NIC techniques involve creating advanced fundamental computational techniques and algorithms based on the concept of nature's methodology to resolve complicated challenges present in the environment. It is established on an emerging trend of deriving several conceptual ideas from nature, carefully examining nature's behavior in solving complex problems, and implementing such ideas to create ideal algorithms and computational approaches (Dewangan et al. 2014). Notably, the NIC approach involves basic principles like self-organization, dynamic systems, coherence, evolution, and immune system, which form the basis of the development of such algorithms. Several studies have reported the application of NIC-based algorithms such as Genetic algorithm (GA), Particle swarm optimization (PSO), and Ant colony optimization (ACO) algorithm for deciphering and overcoming diverse problems associated with disease diagnosis, data analysis, and gene expression profiles of several cancers including CRC with high efficiency and fidelity (Raza and Kohli 2015; Krishnaveni et al. 2019). Furthermore, it has been established that hybrid models involving clubbing of traditional approach or computational intelligence with NIC techniques can be a better and fitter model as the limitations of both the techniques can be overcome and it may help in developing a more efficient and significant model for high-throughput data analysis of cancer. Hence, this chapter aims to define the significance and role of various NIC techniques in identifying and categorizing genes and datasets associated with CRC and the future of hybrid NIC techniques in cancer detection and diagnosis.

2 Landscape of Genomic Alterations in CRC

Genetic and epigenetic alterations in potential driver genes are common causes of progression and development of CRC as these alterations lead to the transformation of normal epithelium cells into malignant forms. For the development of an efficient and effective therapeutic regime, it is quite essential to understand and characterize various molecular pathways associated with CRC progression and development (Singh et al. 2019). Three major molecular pathways established include Chromosomal Instability (CIN), Microsatellite Instability (MSI), and CpG Island Methylator-Phenotype (CIMP) (Pino and Chung 2010). Most CRC cases involve CIN pathways and chromosomal abnormalities, causing drastic changes in the functional activity of tumor suppressor genes and oncogenes. It contributes to 85% of total reported CRC cases. The genomic alterations associated with the CIN pathway include activation of oncogenes such as KRAS, c-Myc, and inactivation of tumor suppressor genes like APC, TP53, and 18qLOH. These aberrations in the CRC genes have been reported to exhibit a significant role in tumor progression and development (Esteller et al. 2001). Microsatellite instability (MSI) represents around 15% of overall CRC cases and mainly occurs due to defects in the DNA mismatch repair system (MMR), resulting in genetic hypermutability. Alterations in the MMR system mainly occur due to hypermethylation in the promoter region of MLH1 genes or mutational aberrations in MMR genes including MLH1, MSH2, and MSH6, PMS2. Accumulations of such mutations result in the development and progression of sporadic CRCs (Armaghany et al. 2012) and have also been reported to drive other cancers like sebaceous gland carcinoma (Srivastava and Sharan 2021). CpG island Methylator Phenotype (CIMP) includes hypermethylation in the CpG island region present within the promoter region resulting in inhibition of binding of a transcription factor at the promoter and alteration in histone acetylation, leading to aberrant changes in gene expression level. It has been found to be approximately 30% of total CRC cases. Genes mainly associated with CIMP include MLH1, p16, MINT 1, 2, 3 (Worthley et al. 2007).

2.1 Molecular Subtypes of CRC

CRC is a heterogeneous disease. This heterogeneity within CRC governs its prognosis and response to therapy. Therefore, it becomes evident to identify a particular molecular subtype for patient-specific proper clinical management of CRC. The past two decades have seen multiple attempts to classify CRCs (Singh et al. 2019). Guinney et al. (2015) reported a more precise classification of CRC subtypes to remove overlapping and ambiguous clinical data in the CRC classification system. This organized method is much more efficient and effective in analyzing the differences in selection criteria of the clinical data created by various traditional and computational means. CRC subtypes were classified into four molecular subtypes, and these types may exhibit differences in their genetic and molecular functioning. The following four CRC molecular subtypes are briefly described through the given Fig. 1.

3 Overview of Computational Intelligence-Based Models Used in CRC Diagnosis and Their Limitations

With the increasing incidence and burden of cancer worldwide, timely and accurate detection and prediction of cancer is a challenge. Hence, many studies have been carried out and are still ongoing, emphasizing timely diagnosis through various methods, including conventional ones (Bharadwaj et al. 2021a). A significant body of research is devoted to predicting CRC subtypes using the traditional approach, however, some deficits such as redundant features, low or inaccurate prediction, sensitivity, and specificity have also been linked (Jung et al. 2015; Chatterjee et al. 2018). The development of small molecules, antibodies, RNA, miRNA, inhibitors, and drug-based therapies targeting the genomic landscape of CRC has fueled our understanding of the development of personalized anti-cancer drugs (Cubiella et al. 2016). Moreover, high-throughput gene expression and next-generation sequencing

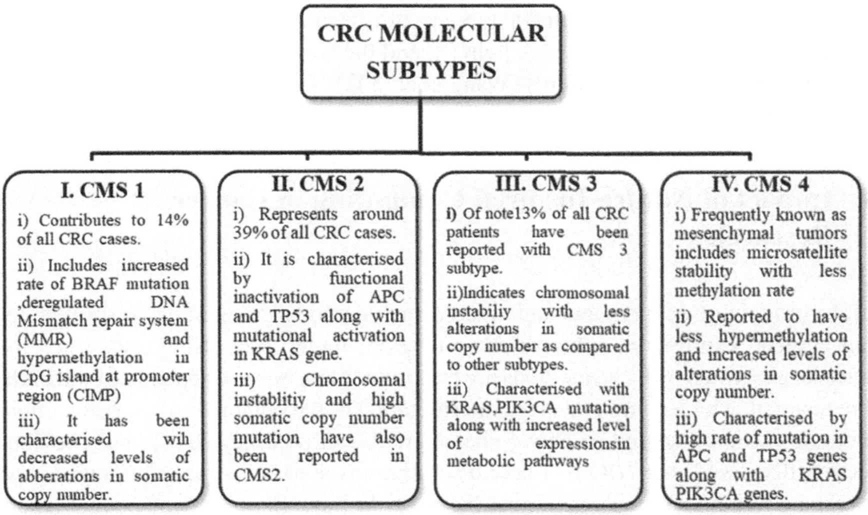

Fig. 1 Representation of CRC molecular subtypes

(NGS) technologies have attempted to introduce new information on CRC prognosis and diagnosis and offer enhanced clinical care for patients (Srivastava et al. 2022b). However, there are still enormous challenges in using the conventional technologies, mainly due to improper redundant features, data complexity, choice, and function of a predictor in CRC detection and prediction (Bolon-Canedo et al. 2014). With the challenges and limitations of accurately identifying disease pathogenesis and diagnosis, CI-based techniques are widely used for addressing the aforementioned challenges. The CI approach is commonly accepted in the present scenario and has revolutionized CRC prognosis, diagnosis, and therapy. Notably, CI, particularly ML and DL, have been shown to perform better in the clinical care of patients with CRC than the conventional methods. The increasing popularity of large-scale CI in genomic profiling of CRC detection and diagnosis has a tremendous impact on identifying the expression of gene or gene sets and is generally perceived as a potential breakthrough in understanding the genetic drivers of cancer initiation and progression (Li et al. 2004). The use of CI methods for genomic characterization of CRC, based on the differentially expressed genes between patients and control, has offered extraordinary opportunities to comprehend the natural premise of CRC to foster treatments, and intercessions, to find genomic biomarkers, and provide clinical directions for the treatment of patients. Furthermore, CI has been exploited on a larger platform by the bioinformaticians, structural and computational biologists for the design and development of vaccines, molecular modeling of unresolved proteins, to develop immunotherapeutic through genetically engineered B and T-cells epitopes, and molecular dynamic simulation of the modeled protein (Bharadwaj et al. 2021b). Consequently, it is understood that by CI methods, one can coordinate and examine the massive amount of data of cancer patients, and data mining efforts could be better implemented to enhance our understanding of cancer vulnerabilities to identify the susceptible individual, which may help expand the work to adopt the most effective therapies affected cancer patients (Wang et al. 2021; Griffith et al. 2015).

4 Impact of Nature-Inspired Computing in Cancer Research

High-throughput sequencing and microarray methods are greatly described for cancer gene expression, subtypes classification, image processing, etc. since the datasets generated by these methods are huge, complex, noisy, unbalanced, and high-dimensional, which complicates the interpretation of the outcomes, demanding more accurate methods devoid of above characteristics (Tazi et al. 2014). The previous section discussed that NIC is a new discipline inspired by nature to compute and design new computational techniques based on the naturally occurring phenomenon. Because NIC depends on natural existence in developing new processes, it is more likely to be relevant to various applications, including cancer detection and diagnosis (Wang et al. 2019a). Because of this, the rest of the discussion focuses mainly on cancer, specifically CRC, and the most innovative technique, i.e., NIC, is reviewed in detail for cancer detection and diagnosis. As far as cancer is concerned, NIC is employed to investigate cancer in great detail at the genomic level. Many researchers are exploring potential issues with the current cancer genomics methodology to guide the selection of the most appropriate diagnosis and treatments based on the genomic signature of the patients with cancer and offer the most advanced clinical benefits (Yoo et al. 2019; Ding et al. 2019). The most recent and advanced theory proposed by the group of researchers offers a new approach for better treatment and clinical efficacy. Three cancer microarray datasets were searched using Select KBest for the feature selection and a supervised machine learning approach. Moreover, three NIC techniques were also employed for cancer classification, showing better predictive action and capabilities in improved diagnosis (Barbachan et al. 2021). Since the advent of The Cancer Genome Atlas (TCGA) database and recent advances in NIC techniques, bio-inspired algorithms have been exploited in parallel to provide insight into cancer's molecular pathogenesis Fig. 2. Furthermore, the objective of precision oncology is to take this huge information and convert it into evidence of realistic value to the clinician. It is encouraging that the addition of new knowledge in healthcare, specifically oncology, through nature-inspired techniques has created an effective approach to ideally prevent cancer in its initial phase or treat it by developing a novel drug candidate which specifically targets the molecular drivers of cancer (Sarma et al. 2021).

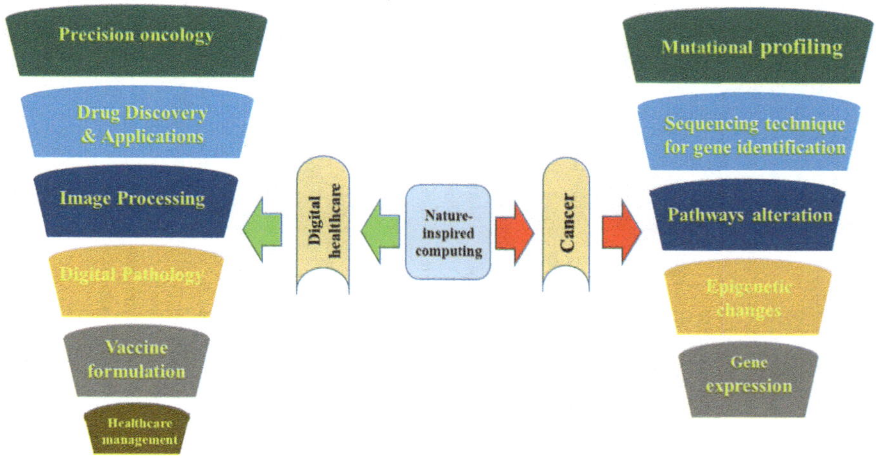

Fig. 2 Nature-inspired multidimensional approach for improved healthcare and genomics of relapsed cancer

5 Implementation of NIC Methods in CRC Detection

CRC has been reported to be a leading concern in cancer morbidity and mortality and a plethora of techniques are available for its prognosis, diagnosis, and treatment. With the advent of new techniques, algorithms, and improved clinical treatment facilities, the CRC recurrence rate has considerably decreased. Still, no such accurate method has been developed which provides ultimate accuracy and efficiency for cancer prognosis and diagnosis (Detorakis and Tambouratzis 2010). Therefore, various nature-inspired computational algorithmic methods may play a significant role in analyzing a large cohort of gene expression data for cancer diagnosis.

NIC works on the basic phenomena of retrieving ideas from the behavioral pattern of nature to resolve complex problems. These basic ideas are further implemented in developing novel computational algorithms and techniques (Agarwal and Mehta 2014). NIC-based techniques have been diversely classified based on various algorithms like bio-inspired, physical, and chemical-based algorithms. Their role in cancer diagnosis and prognosis has been well established. The classification of different NIC types has been summarized in Table 1.

5.1 Genetic Algorithm (GA)

GA was first established in the 1960s by John Holland and is described as an evolutionary algorithm in which new progenies are produced by merging various genes present in a population (Eshaghian-Wilner and Navab 2009). It is a meta-heuristic computational approach mainly involved in resolving optimization and searching

Table 1 Types of various NIC techniques and their important applications in CRC and other cancers

S. No.	Algorithm	Discovered by/Year	Function	Application in CRC	Application in other domain	References
1	Genetic algorithm (GA)	John Holland/1960s	Employed in resolving constrained and unconstrained problems related to optimization	Used as a data mining method for obtaining efficient information from Colorectal images for its diagnosis and prognosis	Used in optimizing chemotherapy in the treatment of various cancers and also categorizes different cancers for identification of biomarker genes	Agarwal and Mehta (2014), Sathya et al. (2022), Chen and Zhang (2021)
2	Particle swarm optimization (PSO)	Eberhart and Kennedy/1995	Stochastic population-based method for searching for an optimum solution in space	Used for determining the relevance of each biomarker and also significant interaction among them for survival period analysis in CRC	Involved in selecting the most relevant features, i.e., benign and malignant features or sub-features for prediction of breast cancer	Dixit et al. (2015), Shibo et al. . (2020), Vijayalakshmi et al. (2020)
3	Differential evolution algorithm (DE)	Storn/1995	It depends on mutational changes, crossing over, and population selection and evaluation	It helps in optimizing hyperparameters of CHR (complex harmonic regularization) and also predicts the survival time of CRC patients by selecting a reduced number of genes as compared to other methods like the CHR-GS (Grid search) and Elastic net	Involved in adjusting and fine-tuning the random forest technique parameters for diagnosing various cancers like lung cancer, skin cancer, blood cancer, breast cancer, diabetes, brain tumor, and Ebola	Storn and Price (1997), Wang et al. (2019b) Ahmad et al. (2021)
4	Artificial bee colony (ABC) algorithm	DervisKaraboga/2015	It is an adaptable and self-growing method following optimal direction	Used along with a Correlation-based feature selection method for determining highly informative and efficient biomarker genes in CRC datasets	Used to select significant and predictive genes from the gene expression profile obtained from microarray data of various cancers like breast, lung, and many more	Dixit et al. (2015), Alshamlan (2018), Nadira and Rustam (2018)

(continued)

Table 1 (continued)

S. No.	Algorithm	Discovered by/Year	Function	Application in CRC	Application in other domain	References
5	Ant colony optimization (ACO)	Marco Dorigo/1992	Using the weighted graph method for searching for an optimum solution	Used in combination with SVM techniques to detect diagnostically significant features of tissue Raman spectra and further develop robust analytical models with high sensitivity, specificity, and overall accuracy for CRC diagnosis	Used to eliminate irrelevant features in various cancer datasets and select critical genes using microarray datasets of different cancer patients	Agarwal and Mehta (2014), Li et al. (2014), Chiang et al. (2008)
6	Cuckoo algorithm	Xin-she Yang	Finds the whole decision space to locate optimal solution by observing the fitness value of all probable candidates	Helps in achieving an efficient, competitive implementation balancing between both diversity and convergence for multi-objective classification of various CRC gene expression data for prognosis and diagnosis	Used along with the memory-based mechanism for feature selection and keeping track of the most relevant information of different cancer datasets determined by best solutions	Rajabioun (2011), Wang et al. (2019c), Alzaqebah et al. (2021)
7	Bat algorithm	Xin-she Yang/2010	Mimics the behavioral pattern of bat and continues *until the convergence* criteria are satisfied	Used in tuning parameters of SVM classifiers for *improving the classification* proficiency in CRC diagnosis	Used for optimizing various features and eliminating irrelevant ones from the original dataset in breast cancer samples for accurate and more accessible classification	Agarwal and Mehta (2014), Xie et al. (2019), Jeyasingh and Veluchamy (2017)

(continued)

Table 1 (continued)

S. No.	Algorithm	Discovered by/Year	Function	Application in CRC	Application in other domain	References
8	Other NIC methods: i. Genetic programming ii. Firefly algorithm	i. Koza/1989 ii. Xin-She Yang/2007	i. Employs a fixed representation for programs and develops a recombination process based on this representation that could merge programs for developing offspring ii. Mimics the flashing nature of fireflies for the acquisition of food in which each firefly is attracted toward the other which is represented in their brightness pattern	i. Used to develop an unbiased selection of features and mathematically combine different expression levels for easy interpretation of various pathways deregulations and disease progression in CRC ii. Used as a weight optimization technique for identifying and classifying benign and malignant polyps in colorectal cancer	i. Used in creating a virtual paradigm for data mining by virtual reality space construction in medical data sets of breast cancer patients ii. Used as a classifier for classification of all four different types of stages in breast cancer patients and obtained increased accuracy and specificity for all the four stages	Worzel et al. (2009), Valdes et al. (2007) RS and Cao (2020), Rajaguru and Prabhakar (2018)

mechanisms problems. In contrast to other evolutionary algorithmic techniques, GA tends to have a structure with the help of which it reconstructs a new organization type from a successful pre-existing form with no loss of any crucial data related to the problem (Bruderer and Singh 1996). It is mainly based on essential components like mutational factors, genetic mechanisms, natural selection, generations, size of the population, and crossing-over level. Determining the fitness performance is vital for evaluating and selecting a population, which mainly depends on selection techniques revealed in an individual's gene (Garg and Mittal 2014). Hence, the GA-based method can differentiate benign and malignant CRC tumor groups (Taino et al. 2019) and produce accurate and efficient information for CRC diagnosis and detection. A basic process flow diagram for GA is represented in Fig. 3.

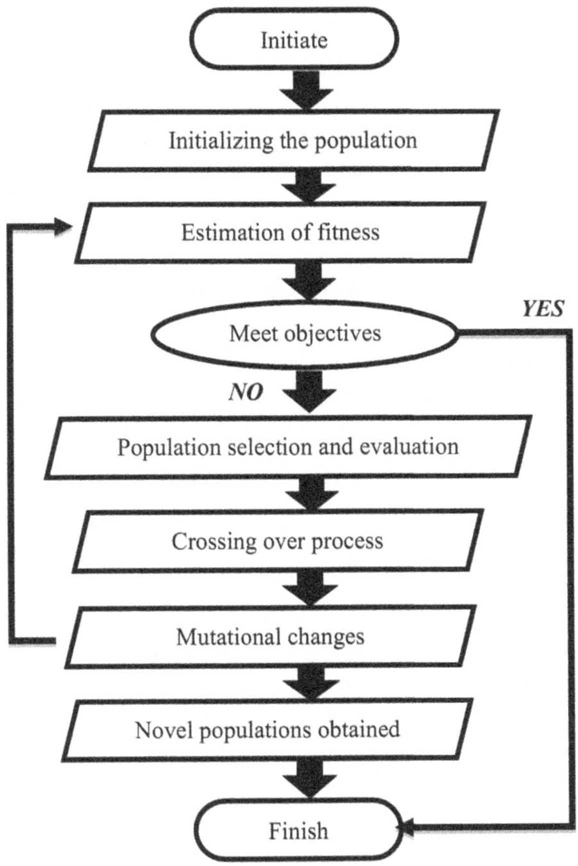

Fig. 3 Illustration of basic steps of a genetic algorithm using the flow diagram

5.2 Ant Colony Optimization (ACO)

Cancer diagnosis and marker gene selection for colorectal cancer has been a prime area of research for experts and researchers. Overfitting and dimensionality problems with the clinical data have been a major concern (Xing et al. 2001). Thus, to prevent such difficulties, it has become necessary to develop algorithmic techniques which provide accurate and helpful information. Ants come in the category of insects that are blind and live in the colonial state and tend to portray striking behavioral patterns. ACO algorithm is a vital optimization technique based on a probabilistic algorithm inspired by ants hunting and coordination activity. It is an efficient method for finding solutions to diverse problems, and it acquires simple concepts from real ants, which tend to search for the shortest route for transferring their food with the help of pheromones. They influence the behavioral pattern of other individuals with the same genotype (Agarwal and Mehta 2014) and tend to build optimal solutions to the problem, such as effective diagnosis for CRC. Yu et al. (2009), in their study, highlighted the potential of the ACO algorithm in selecting an optimal marker gene in tumor samples. However, we have shown the basic information of using the ACO algorithm with the help of Fig. 4.

5.3 Particle Swarm Optimization (PSO)

PSO is a stochastic meta-heuristic probabilistic-based algorithm created by Eberhart and Kennedy in 1995, mainly inspired by gathering patterns and biological behavior patterns of swarms of fishes or bird flocks as they migrate in search of better habitat. This algorithmic technique resolves n-dimensional difficulties and displays the result in point form or a surface type (Dixit et al. 2015). It has been applied in various areas of computational approach and research. *P* finds a suitable area for the object's functioning by altering the particles' positions. Every individual particle can provide a solution to the problems and modify its flying performance concerning the previous events and experiences of other swarm particles. Akutekwe et al. (2014), in their study, demonstrated the application of PSO for the diagnosis and detection of colorectal cancer. They intended to establish the potential role of PSO in detecting biomarkers for cancer diagnosis and treatment.

5.4 Artificial Bee Colony (ABC)

Karaboga first developed the ABC algorithm in 2015 and is entirely based on the seeking pattern of bees, making it quite an adaptable and self-developing computational technique. It is inspired by the behavior of honeybees that follows a foraging pattern to transfer a high amount of food nectar to respective sources. It aims at

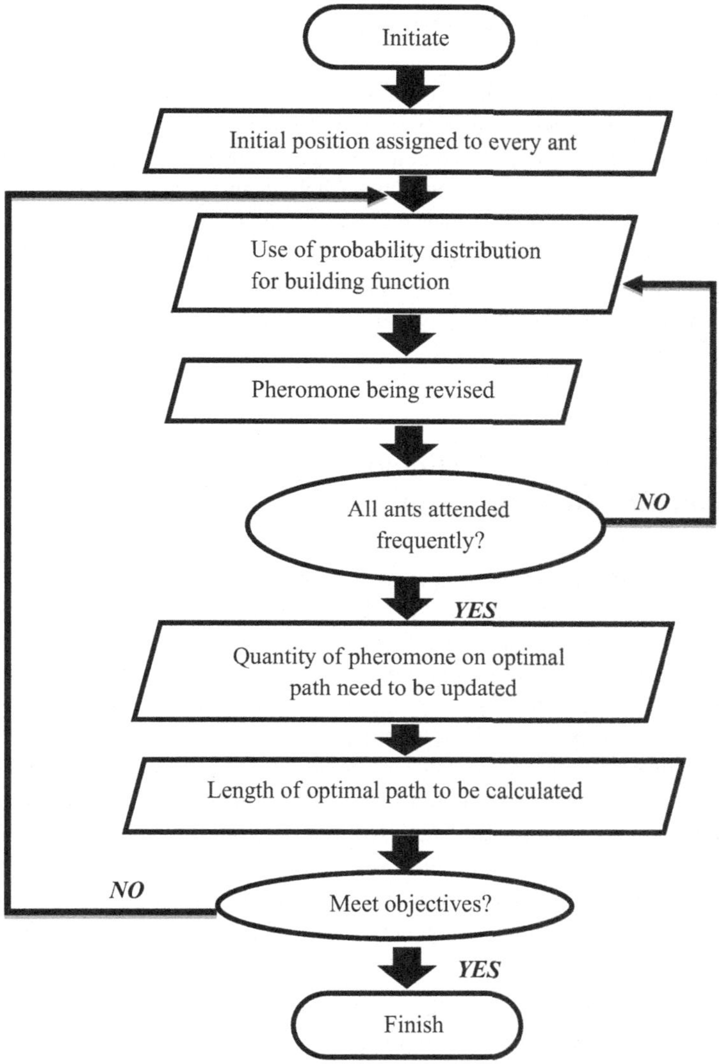

Fig. 4 Illustration of basic steps in ACO using the flow diagram

obtaining optimal solutions for problems associated with numerical optimization and revolves around the concept of resolving the issues of finding any object with the help of an optimal path (Dixit et al. 2015). It is helpful in easy execution and provides efficient gene expression analysis results. Gao et al. (2017) reported significant ABC algorithm efficiency when clubbing with the Correlation-based feature selection method. Their experimental analysis proposed that the Co-ABC algorithm provides an accurate and effective way of identifying biomarker genes using expression data of colon cancer.

5.5 *Cuckoo Search Algorithm (CSA)*

CSA imitates the parasitic nature of some cuckoo birds that use others' nests for laying their eggs by finding probable optimal solutions. The bird's search for an optimal solution entirely depends on the fitness estimate of all the possible solutions. Various alterations and modifications have helped in achieving a multi-objective cuckoo algorithm that is robust and provides an efficient approach for gene expression analysis of cancer. Wang et al. (2019a) demonstrated the role of the cuckoo search algorithm along with a multi-objective approach for CRC diagnosis and treatment. Their analysis established better diagnostic performance of this approach at the large database level.

6 Application of NIC Techniques in the Identification of Potential Biomarkers in CRC

The field of NIC techniques is emerging as an active area in high-throughput computational analysis of disease diagnostic and therapeutic, assisting in identifying specific and reliable biomarkers. This state of the art has attracted the interest of several researchers and academicians as an important discipline in custom healthcare and precision medicine in oncology. Recent development in the NIC techniques has enabled us to more precisely handle and analyze the genomic and transcriptomic datasets and provide them an accurate and meaningful expression, which could lead to the identification of the specific biomarker that has clinical implication for diagnosing congenital changes, gene therapy, targeted therapies and more importantly identification and prognosis of non-communicable diseases such as diabetes, hypertension, metabolic disorders, and cancer (Yeh 2007; Murali et al. 2018).

Here, we have described the general work-flow of the steps to identify the biomarker involved:

i. **Address the biological questions or dataset collection from (biological specimens or databases)**: The macromolecules are extracted from biological sources such as cells, tissues, organs, and fluids that are endogenous or exogenous in nature that is critical to biological processes or regulation. On the other hand, The Cancer Gene Atlas (TCGA), Gene Expression Omnibus (GEO), and cBioPortal for cancer genomics have provided more information about clinical datasets for patients with the diseases. Thus, all the samples are allowed to meet specificity, accuracy, and precision in a targeted manner by following downstream steps (Alakwaa et al. 2018).
ii. **Quantification using analytical or high-throughput techniques**: Quantifying and analyzing are essential to provide several strategies for comprehensively analyzing the collected data/samples for clinical significance. Recent advancement in analytical, omics-based, and high-throughput techniques (qPCR, microarray, and sequencing) has made it possible to provide a large amount

of quantitative data with accuracy and specificity in short interval. Nowadays, these techniques are coupled with computational algorithms that transform the data into meaningful clinical information (Xia et al. 2013).

iii. **Preprocessing or normalizing datasets to remove irrelevant biases**: This step ensures that data are free from incomplete features and biases. Moreover, preprocessing involves normalization, filtering, transformation, and scaling to remove discrepancies and improve datasets distribution and feature selection (Dieterle et al. 2006).

iv. **Construction of candidate gene panel for biomarker discovery**: Following this step, one can proceed with differential gene expression analysis by comparing the expression difference between test and standard samples. Co-expression network may be constructed to best understand the biological correlation among identified genes followed by hub-gene analysis associated with the disease's status (Hoefer et al. 2015).

v. **Model interpretation and validation through classification and feature selection using promising computational algorithms, including NIC**: Interpretation and validation are marked up by employing the advanced ML and NIC-based algorithms that have high potential and accuracy over traditional diagnostic methods for biomarker discovery.

Significantly, NIC techniques are gradually being implemented in applications such as the accurate and precise gene-based diagnosis and feature selection of cancer, using the ACO algorithm to truncate the high-dimensionality (Ke et al. 2010). This algorithm has attracted significant interest in feature selection in recent years, and further support is given by Spolaor et al. (2011). They assert that feature selection uses a multi-objective genetic algorithm (MOGA). Tazi et al. (2014) reported nature-inspired algorithms, particularly swarm intelligence algorithms (SIA), which have an extensive array of applications and are sub-divided into other algorithms such as ant colonies, bee colonies, bird flocks, and fish schools which have profoundly astonishing ability to sort out the activities, optimal solution to a variety of problems and show quick, efficient, and accurate performance compared to previously used algorithms. An effective method, namely the cuckoo search algorithm, has been proposed for classifying cancer gene expression data with reduced data noise, diagnostic utility, and imprecision (Wang et al. 2019a). The literature showed that specific insect-based algorithms, such as ACO, ABC, and firefly algorithms, are mainly optimized for detecting different stages and subtypes of cancers such as CRC. Moreover, an overview of the previous and current attempts to develop CRC screening techniques is best obtained using the GA and the algorithms mentioned above. A comparison of different bio-inspired algorithms found in Kaur and Sharma'sstudy (2017), where the author claims that the predictive accuracy of cancer (Prostate, Breast, and Lung) screening using ACO, SIA, ABC, and GA is between 70 and 99.2%, 80 and 98%, 76.4 and 99.98% and 73.5 and 99.7%, respectively.

However, current evidence supports the use of a hybrid approach algorithm with high predictive accuracy compared to the individual NIC-based computing algorithm for detecting different types of cancer, including CRC (Gautam et al. 2019). Over the past few years, the pathological evaluation performed on CRC is challenging due to its heterogeneous nature in the type of genetic changes such as mutation, epimutation, copy number alteration, fusion transcript, and changes occurring during post-transcriptional and post-translational mechanisms (Singh et al. 2017). Therefore, in the era of precision medicine, we need to adopt an integrated computational approach for pathological assessment of biomarkers to identify the genomic driver of CRC diagnosis and its relapse. In recent years, experts have started applying this emerging novel technique to the real world to discover biomarkers in CRC prognosis, diagnosis, and therapy. The recent work addressed this issue by Prabhakar et al. (2020) whose study was based on the step-by-step classification of colon cancer using the most appropriate genes and features. Classification accuracy of 99.16% was achieved by the best optimization technique, Invasive Weed Optimization (IWO), when used with the Multivariate Minimum Redundancy–Maximum Relevance (MRMR) technique, helped to reduce the dimensionality issue, and classified with Quadratic discriminant Analysis (QDA) (Prabhakar et al. 2020). The evolutionary algorithm is a meta-heuristic optimization algorithm that falls under evolutionary computation to study the various genetic factors, including single-nucleotide polymorphisms (SNPs) as several identified genetic factors are associated with CRC and are effectively used as biomarker discovery relevant for therapeutic purposes (Geng 2021).

The importance of bio-inspired algorithms has been widely emphasized in the literature regarding cancer research, mainly for diagnosing disease. The study provides extremely useful basic information on the nature-inspired techniques for genomic profiling of CRC, although a thorough study on the subject is also presented in other references. As mentioned above, the hybrid algorithm approach has high predictive accuracy compared to the individual NIC-based computing algorithm. Therefore, in the next section, examples of hybrid NIC methods in the context of CRC are discussed in detail.

7 Case Studies of Various Hybrid Techniques: A Revolutionary Change in CRC Detection and Treatment

NIC techniques have been used widely for solving nonlinear complex problems consisting mainly of substantial variables, problems that are difficult to resolve by traditional computational means, or issues that need a diversified approach for obtaining solutions. Instead, when used simultaneously with another computational intelligence tool or traditional technique, NIC-based methods provide a more efficient and accurate result for cancer diagnosis and research (Li et al. 2014). Thus, with the help of various case studies, such hybrid techniques have proved to be more relevant, redundant, and performed expression analysis of cancer-related genes with relatively increased prognostic power compared to other traditional techniques.

Case Study#1: Particle Swarm Optimization-Based Bio-Network Discovery Method for the Diagnosis of Colorectal Cancer (Akutekwe and Seker 2014).

In this study, researchers aimed to develop a two-staged computational-based approach for identifying biomarker genes associated with colorectal cancer. They have used samples of 64 patients and 48 control volunteers for their study. For the first stage, three different models were used, including Support Vector Machine Recursive Feature Elimination (SVM-RFE) and PSO. Further, in the second stage, the Daynesian Bayesian Network was implemented to develop a probable biomarker association of the samples. Using tenfold cross-validation, 50 random values of C and γ in the range of 10^{-3} to 10^{3} and 10^{-4} and 10^{4}, respectively, were investigated to determine the efficacy of the PSO for optimization of SVM hyperparameters. In order to detect a significant relationship between various features with high accuracy, G1DBN algorithm was used for DBN modeling in which initially score matrix containing the score of each DBN edge was determined, and further edge selection alpha two was used to conclude each score for developing dependencies among variables. Their result inferred that this hybrid-based approach produced more reliable and efficient results compared to the traditional approach. Also, a lesser number of biomarkers were depicted by them, which ensured better outcomes for future analysis.

Case Study#2: A Modified Ant Colony Optimization Algorithm for Tumor Marker Gene Selection (Yu et al. 2009).

In this study, researchers have used a hybrid-based model of applying a modified ACO algorithm and an SVM algorithm. ACO was used to select tumor-related marker genes, and SVM was used to assess the evaluation of obtained gene subsets. They used the ACO-SVM algorithm on 100 selected genes and searched for more efficient marker genes in which they proposed the top 10 marker genes obtained by three algorithmic-based methods further, a recognition rate of 87% of colon tumor marker genes was obtained by performing leave-one-out cross-validation (LOOCV). The determined subsets were evaluated based on their fitness function, and further probability value of pheromones left in each pathway by ants was calculated. The more pheromone left, the relevant pathway with higher probability was selected. They also developed various validation curves of CRC datasets for different algorithms to evaluate their fitness. With the help of their experiment analysis and evaluation, they established that their proposed algorithm is more effective and efficient than other computational tools for diagnosing colorectal cancer genes.

Case Study#3: Identification and Characterization of Colorectal Cancer Using Raman Spectroscopy and Feature Selection Techniques (Li et al. 2014).

The ACO algorithm is used in various optimization-based methods and several other traditional and CI-based techniques. Shaoxin et al. (Li et al. 2014), in their study, focused on the use of the SVM technique with the ACO algorithm for sorting out features related to histopathological means. Raman spectra were employed to segregate normal tissues and CRC tumor tissues, and further optimization was performed using SVM-based algorithms. ACO and SVM techniques were clubbed together to identify analytically critical characteristics and features associated with Raman spectroscopy which are significant concerning various CRC samples. For determining the diagnostic performance of this model, the receiver operating characteristic curve (ROC) was produced, which helped evaluate the function of binary classifiers, and further for obtaining maximum accuracy, the RBF kernel SVM parameters C and Gaussian radial width σ were evaluated and optimized. With the help of this statistical approach, it was established that this model is highly efficient and accurate for CRC diagnosis and provided 92.3% sensitivity and 94.2% accuracy. The proposed hybrid model was further compared with the PCA-LDA technique, and the results displayed the ACO-SVM model to be more accurate and diagnostically better than the other model as it efficiently discriminated the particular bands obtained from Raman spectra and correlated this information for diagnostic purposes.

Case Study#4: Nature-Inspired Multi-objective Cancer Subtype Diagnosis (Wang et al. 2019a).

In this study, researchers aimed to develop a multi-objective ensemble cuckoo search computational tool derived from the cuckoo search algorithm (MOESCA). The MOESCA has been proved to be an efficient and reliable tool for determining the diagnosis pattern of various cancer subtypes. They selected 35 colon adenocarcinoma datasets for establishing the potential role of the MOESCA algorithm in cancer diagnosis and treatment. The proposed algorithm was compared with other effective multi-objective algorithms and produced more accurate results on individual COAD datasets. In the MOESCA-based algorithm, a small subset of important and valuable genes was selected with the help of a binary encoding approach, and the expression of cancer gene data was recorded and translated using four different objective functions. The various labels obtained from classification algorithms were compared to determine the diagnostic results. Normalized mutual information (NMI) and normalized rand index (Rn) were used to evaluate the performance and efficacy of all compared algorithms, as they can significantly determine the similarity between growth truth labels and classification labels. At last, a cuckoo-based algorithm and an ensemble mechanism were applied to optimize the different objectives instantaneously. Thus, it was evidently proved that the MOESCA-based approach might help determine and classify the gene expression of various CRC-related genes.

Case Study #5: Cancer Classification Based on SVM Optimized by PSO and ABC (Gao et al. 2017)

In this study, researchers developed a PA-SVM approach incorporating an SVM combined with PSO and ABC-based algorithms. They addressed the disadvantages of the single computational approach, like less accuracy and low precision in resolving complex algorithmic problems. Thus, they highlighted the significance of the PA-SVM method which employed the use of FCBF (Fast Correlation-Based Feature) based selection for removing redundancy and inappropriate features, performed optimization using PSO, and the optimal solution obtained is further taken as initializing values for ABC algorithm and further modified the parameters for developing ideal results. In their experiment, they used datasets of CRC along with other various cancer types. Sixty-two colon samples containing approximately 2000 attributes were obtained from colon cancer patients and were further subjected to the PA-SVM algorithm, which provides efficient and accurate classification means using fitness function value. They calculated the selection probability of fitness value remembered by employer bees and onlooker bees. Further, the fitness values calculated before and after the search were evaluated, and the most significant one was selected. Therefore, these hybrid-based algorithms provide a more accurate and reliable diagnostic model for cancer diagnosis and treatment.

8 Smartphone Technology in Diseases Monitoring: An Intelligent Approach

Advancement in computational techniques has solved many complex computations in the context of genomic profiling of diseases. Despite significant improvements, the dependencies on the computational techniques arose several complex problems, which can be sorted by smartphone. In recent years, smartphones have a great deal with different optimization methods that have shown excellence in handling complicated efforts and are massively used to perform optimization, thereby acting as the best optimization tool (Mishra et al. 2020). Recent investigation has demonstrated that Parkinson's disease (PD) severity can be measured using smartphones and machine learning. Researchers have used the non-invasive techniques to diagnose PD by employing the smartphone application that recorded participants' responses for (finger tapping, voice, balance, gait, and reaction time) with and without PD. The measured parameters for different responses showed that smartphone assessments could enhance clinical care and offer simple, inexpensive, and widely available technologies for diagnostics purposes (Arora et al. 2015). This theory is backed up by evidence from Zhan and groups, who conclude that individuals with PD were subjected to offer tasks (stated in the previous sentence) and were assessed by a mobile application (HopkinsPD). The smartphone-derived mobile Parkinson's disease severity score was strongly correlated with intraday symptom fluctuation and response to dopaminergic medication, which could serve as novel therapeutics (Zhan et al. 2018). Similarly, in different studies, android–based application with a supervised and cloud-based machine learning approach has been demonstrated in the early detection, prevention, treatment, and survivorship of cancer (Prochaska et al. 2017; Muljo et al. 2019; Dai et al. 2019; Goceri 2021). Although smartphone technology has received global attention in recent years, domain-specific improvement is expected to result in an improved understanding of biological research, particularly for life-threatening diseases.

9 Advantages, Limitations, and Constraints of NIC Techniques in Cancer Research

Major advantages linked with the functioning of various NIC-based algorithms have been highlighted below:

- It provides better accuracy and improved flexibility for understanding clinical data.
- NIC techniques help in developing algorithmic computational tools for solving complex optimization problems and varied industrial applications.
- NIC-based algorithmic techniques have provided cost-effective, time-efficient, reliable, and efficient system management.
- NIC methods provide lower chances of possible error rates than traditional computational techniques.
- These techniques are quite versatile as they display more effective results when clubbed with other computational techniques. Many researchers worldwide are more attracted to such hybrid-based algorithmic models, and their efficacy is also very high in resolving a wide range of complex problems.
- In recent times, due to extensive research and developments, nature-inspired computational techniques have gained a lot of importance in cancer research, diagnosis, and treatment but still lack certain properties and features essential for optimum results (Yang 2020). Thus, there are certain limitations or challenges associated with NIC techniques discussed further in this section:
- The algorithms present in NIC techniques lack a proper integrated mathematical framework to evaluate clinical data. Henceforth, it doesn't have a proper understanding of the convergence system of these algorithms and also fails to estimate the rate at which convergence may occur.
- Assessment analysis of various algorithms and their studies which are mainly based on numerical experiments suggest that their comparison studies may not always be accurate.
- Mostly, NIC-based algorithms have been used for small-scale problems, but their application directly for large-scale issues has not been established yet.
- Another problem is the lack of proper training and experience of different researchers and technicians as these techniques are new and still developing in cancer research and diagnosis.

10 Conclusion and Future Perspective

With the advancement of computational techniques, particularly nature-inspired computing, over the last three decades, the diagnostic accuracy of cancer based on the genomic landscape has improved significantly. In the present review, we have looked at the use of NIC in detecting and diagnosing CRC. It can be noted that ACO and GA represented an excellent rate of classification for the diagnosis of cancer patients. Moreover, the evolutionary algorithm takes special importance for identifying genetic factors linked with the CRC phenotype. However, several researchers reveal that the hybrid approach of the NIC algorithm with soft computing techniques has a wide range of applicability and precise accuracy and is a much better option for CRC detection and diagnosis than individual algorithms. These observations are limited to the genomic and transcriptomic levels and cannot be extrapolated to the proteomic, metabolomics, and epigenetic level of CRC detection and diagnosis. This study only considers selected algorithms and can be further improved by considering other nature-inspired computational methods. Since NIC techniques have not been explored in detail in the case of CRC, this study offers a successful approach to exploiting NIC techniques in CRC screening based on the genomic landscape.

Acknowledgements The authors are thankful to the Director, Motilal Nehru National Institute of Technology, Allahabad, India, for providing research facilities. The authors are also grateful to the Ministry of Human Resource and Development, Govt. of India, New Delhi, for providing fellowship during this tenure.

References

Agarwal P, Mehta S (2014) Nature-inspired algorithms: state-of-art, problems, and prospects. Int J Comput Appl 100:14–21

Ahmad MF, Isa NAM, Lim WH, Ang KM (2021) Differential evolution: a recent review based on state-of-the-art works. Alexandria Eng J

Akutekwe A, Seker H (2014) Particle swarm optimization-based bio-network discovery method for the diagnosis of colorectal cancer. In: IEEE international conference on bioinformatics and biomedicine (BIBM), 8–13

Alakwaa FM, Chaudhary K, Garmire LX (2018) Deep learning accurately predicts estrogen receptor status in breast cancer metabolomics data. J Proteome Res 17(1):337–347

Alshamlan HM (2018) Co-ABC: Correlation artificial bee colony algorithm for biomarker gene discovery using gene expression profile. Saudi J Biol Sci 25(5):895–903

Alzaqebah M, Briki K, Alrefai N, Brini S, Jawarneh S, Alsmadi MK, Alqahtani A (2021) Memory based cuckoo search algorithm for feature selection of gene expression dataset. Inform Med Unlocked 24:100572

Armaghany T, Wilson JD, Chu Q, Mills G (2012) Genetic alterations in colorectal cancer. Gastrointest Cancer Res GCR 5(1):19–27

Arora S, Venkataraman V, Zhan A, Donohue S, Biglan KM, Dorsey ER, Little MA (2015) Detecting and monitoring the symptoms of Parkinson's disease using smartphones: a pilot study. Parkinsonism Relat Disord 21(6):650–653

Barbachan S, Henrique NP, Dorn M, Broin PO (2021) Optimization of cancer status prediction pipelines using bio-inspired computing. In: IEEE congress on evolutionary computation (CEC), 442–444

Bharadwaj KK, Rabha B, Pati S, Sarkar T, Choudhury BK, Barman A, Bhattacharjya D, Srivastava A, Baishya D, Edinur HA, Abdul KZ, Mohd NH (2021a) Green synthesis of gold nanoparticles using plant extracts as beneficial prospect for cancer theranostics. Molecules 26(21):6389

Bharadwaj KK, Srivastava, A, Panda MK, Singh YD, Maharana R, Mandal K, Manisha Singh BS, Singh D, Das M, Murmu D, Kabi SK (2021b) Computational intelligence in vaccine design against COVID-19. In: Raza K (eds) Computational intelligence methods in COVID-19: surveillance, prevention, prediction, and diagnosis. Stud Comput Intell, 923:311–329

Bolon-Canedo V, Sanchez-Marono N, Alonso-Betanzos A, Benitez JM, Herrera F (2014) A review of microarray datasets and applied feature selection methods. Inf Sci 282:111–135
Bruderer E, Singh JV (1996) Organizational evolution, learning, and selection: a genetic-algorithm-based model. Acad Manag J 39(5):1322–1349
Chatterjee S, Dey N, Shi F, Ashour AS, Fong SJ, Sen S (2018) Clinical application of modified bag-of-features coupled with hybrid neural-based classifier in dengue fever classification using gene expression data. Med Biol Eng Compu 56(4):709–720
Chen SM, Zhang JH (2021) Genetic algorithm in data mining of colorectal images. Comput Mathem Methods Med
Chiang YM, Chiang HM, Lin SY (2008) The application of ant colony optimization for gene selection in microarray-based cancer classification. In: 2008 international conference on machine learning and cybernetics, 7, pp 4001–4006
Cubiella J, Vega P, Salve M, Díaz-Ondina M, Alves MT, Quintero E, Alvarez-Sanchez V, Fernandez-Banares F, Boadas J, Campo R, Bujanda L, Clofent J, Ferrandez A, Torrealba L, Pinol V, Rodriguez-Alcalde D, Hernandez V, Fernandez-Seara J, COLONPREDICT study investigators. (2016) Development and external validation of a faecal immunochemical test-based prediction model for colorectal cancer detection in symptomatic patients. BMC Med 14(1):128
Dai X, Spasić I, Meyer B, Chapman S, Andres F (2019) Machine learning on mobile: an on-device inference app for skin cancer detection. In: 2019 fourth international conference on fog and mobile edge computing (FMEC), pp 301–305
Detorakis Z, Tambouratzis G (2010) Applying a sectioned genetic algorithm to word segmentation. Pattern Anal Appl 13(1):93–104
Dewangan S, Naik A, Agrawal A (2014) Study of nature-inspired computing. Int J Comput Sci Trends Technol (IJCST) 2(6):11–14
Dieterle F, Ross A, Schlotterbeck G, Senn H (2006) Probabilistic quotient normalization as robust method to account for dilution of complex biological mixtures. Application in 1H NMR metabonomics. Analyt Chem 78(13):4281–4290
Ding X, He M, Chan AW, Song QX, Sze SC, Chen H, Man MK, Man K, Chan SL, Lai PB, Wang X (2019) Genomic and epigenomic features of primary and recurrent hepatocellular carcinomas. Gastroenterology 157(6):1630–1645
Dixit M, Upadhyay N, Silakari S (2015) An exhaustive survey on nature inspired optimization algorithms. Int J Softw Eng Appl 9(4):91–104
Eshaghian-Wilner MM, Navab S (2009) Nanoscale image processing. Bio-Inspired Nanoscale Integr Comput 515
Esteller M, Corn PG, Baylin SB, Herman JG (2001) A gene hypermethylation profile of human cancer. Cancer Res 61(8):3225–3229
Gao L, Ye M, Wu C (2017) Cancer classification based on support vector machine optimized by particle swarm optimization and artificial bee colony. Molecules 22(12):2086
Garg R, Mittal S (2014) Optimization by genetic algorithm. Int J Adv Res Comput Sci Softw Eng 4(4):587–589
Gautam R, Kaur P, Sharma M (2019) A comprehensive review on nature inspired computing algorithms for the diagnosis of chronic disorders in human beings. Progr Artif Intell 8(4):401–424
Geng S (2021) Genome-wide association study of colorectal cancer using evolutionary computing. Doctoral dissertation, Memorial University of Newfoundland
Goceri E (2021) Diagnosis of skin diseases in the era of deep learning and mobile technology. Comput Biol Med 134:104458
Griffith M, Miller CA, Griffith OL, Krysiak K, Skidmore ZL, Ramu A, Walker JR, Dang HX, Trani L, Larson DE, Demeter RT, Wendl MC, McMichael JF, Austin RE, Magrini V, McGrath SD, Ly A, Kulkarni S, Cordes MG, Fronick CC, Wilson RK (2015) Optimizing cancer genome sequencing and analysis. Cell Syst 1(3):210–223
Guinney J, Dienstmann R, Wang X, de Reyniès A, Schlicker A, Soneson C, Marisa L, Roepman P, Nyamundanda G, Angelino P, Bot BM, Morris JS, Simon IM, Gerster S, Fessler E, De Sousa E,

Melo F, Missiaglia E, Ramay H, Barras D, Homicsko K, Tejpar S (2015) The consensus molecular subtypes of colorectal cancer. Nat Med 21(11):1350–1356

Hoefer IE, Steffens S, Ala-Korpela M, Bäck M, Badimon L, Bochaton-Piallat ML, Weber C (2015) Novel methodologies for biomarker discovery in atherosclerosis. Eur Heart J 36(39):2635–2642

Jeyasingh S, Veluchamy M (2017) Modified bat algorithm for feature selection with the Wisconsin diagnosis breast cancer (WDBC) dataset. Asian Pac J Cancer Prev APJCP 18(5):1257–1264

Jung KJ, Won D, Jeon C, Kim S, Kim TI, Jee SH, Beaty TH (2015) A colorectal cancer prediction model using traditional and genetic risk scores in Koreans. BMC Genet 16(49):1–7

Kaur P, Sharma M (2017) A survey on using nature inspired computing for fatal disease diagnosis. Int J Inf Syst Model Des (IJISMD) 8(2):70–91

Ke L, Feng Z, Xu Z, Shang K, Wang Y (2010) A multiobjective ACO algorithm for rough feature selection. In: Second Pacific-Asia conference on circuits, communications and system, 207–210

Krishnaveni A, Shankar R, Duraisamy S (2019) A survey on nature inspired computing (NIC): algorithms and challenges. Global J Comput Sci Technol D Neural Artif Intell19(3):0975–4172

Li T, Zhang C, Ogihara M (2004) A comparative study of feature selection and multiclass classification methods for tissue classification based on gene expression. Bioinformatics (Oxford, England) 20(15):2429–2437

Li J, Su H, Chen H, Futscher BW (2007) Optimal search-based gene subset selection for gene array cancer classification. IEEE Trans Inf Technol Biomed Publ IEEE Eng Med Biol Soc 11(4):398–405

Li S, Chen G, Zhang Y, Guo Z, Liu Z, Xu J, Li X, Lin L (2014) Identification and characterization of colorectal cancer using Raman spectroscopy and feature selection techniques. Opt Express 22(21):25895–25908

Migliore L, Migheli F, Spisni R, Coppedè F (2011) Genetics, cytogenetics, and epigenetics of colorectal cancer. J Biomed Biotechnol 2011:792362

Mishra S, Mishra BK, Tripathy HK (2020) Significance of biologically inspired optimization techniques in real-time applications. Robot Syst Concepts Methodol Tools Appl, 224–248

Muljo HH, Perbangsa AS, Yulius B P (2019) Improving early cancer detection knowledge through mobile learning application. iJOE 15(2):61

Mundade R, Imperiale TF, Prabhu L, Loehrer PJ, Lu T (2014) Genetic pathways, prevention, and treatment of sporadic colorectal cancer. Oncoscience 1(6):400–406

Murali R, Delair DF, Bean SM, Abu-Rustum NR, Soslow RA (2018) Evolving roles of histologic evaluation and molecular/genomic profiling in the management of endometrial cancer. J Natl Compr Canc Netw 16(2):201–209

Nadira T, Rustam Z (2018) Classification of cancer data using support vector machines with features selection method based on global artificial bee colony. In AIP Conf Proc 2023(1):020205

Pino MS, Chung DC (2010) The chromosomal instability pathway in colon cancer. Gastroenterology 138(6):2059–2072

Prabhakar SK, Rajaguru H, Kim SH (2020) An amalgamated approach to bilevel feature selection techniques utilizing soft computing methods for classifying colon cancer. Biomed Res Int 2020:8427574

Prochaska JJ, Coughlin SS, Lyons EJ (2017) Social media and mobile technology for cancer prevention and treatment. Am Soc Clin Oncol Educ Book 37:128–137

Rajabioun R (2011) Cuckoo Optimization algorithm. Appl Soft Comput 11(8):5508–5518

Rajaguru H, Prabhakar SK (2018) A study on firefly algorithm for breast cancer classification. International conference on ISMAC in computational vision and bio-engineering. Springer, Cham, pp 421–428

Raza K, Kohli M (2015) Ant colony optimization for inferring key gene interactions. In: 2015 2nd international conference on computing for sustainable global development (INDIACom), 1242–1246

RS, P., & Cao, S. Q. (2020) Analysis of CT images for detection of colorectal cancers using hybrid artificial neural networks and firefly algorithm. Procedia Comput Sci 171:1517–1526

Sarma MK, Ningthoujam R, Panda MK, Babu PJ, Srivastava A, Das M, Singh YD (2021) Translational healthcare system through bioinformatics. In: Translational bioinformatics applications in healthcare, 3–21

Sathya M, Jeyaselvi M, Joshi S, Pandey E, Pareek PK, Jamal SS, Atiglah HK (2022) Cancer categorization using genetic algorithm to identify biomarker genes. J Healthc Eng

Shibo L, Xiaojuan F, Huanyu Z, Zhengyou W Yanan Z (2020) Marker selection for predicting continuous survival period of colorectal cancer. Int J Syst Assur Eng Manag Springer Soc Reliab Eng Qual Operat Manag (SREQOM),11(4):785–791

Singh MP, Rai S, Suyal S, Singh SK, Singh NK, Agarwal A, Srivastava S (2017) Genetic and epigenetic markers in colorectal cancer screening: recent advances. Expert Rev Mol Diagn 17(7):665–685

Singh MP, Rai S, Pandey A, Singh NK, Srivastava S (2019) Molecular subtypes of colorectal cancer: an emerging therapeutic opportunity for personalized medicine. Genes Dis 8(2):133–145

Spolaor N, Lorena AC, Lee HD (2011) Multi-objective genetic algorithm evaluation in feature selection. In: Proceedings of international conference evolution MultiCriterionOptim, 462–476

Srivastava A, Sharan S (2021) Sebaceous gland carcinoma of ocular region in India: A brief literature review for disease management. Lat Am J Ophthalmol 4(4)

Srivastava A, Jit BP, Dash R, Panda MK (2021) Bioactive lipid: a novel diagnostic approach for retinoblastoma in clinical management. Int J MolImmunoOncol 6(3):136–139

Srivastava A, Rai S, Bisht D, Sachan M, Jit BP, Srivastava S (2022a) Targeting the altered tyrosine kinases in colorectal cancer: from inhibitors to drugs. In: Protein Kinase Inhibitors. Academic Press, pp 361–391

Srivastava A, Rai S, Singh MP, Srivastava S (2022b) Computational intelligence-based gene expression analysis in colorectal cancer: a review. In: Raza K (eds) Computational intelligence in oncology. Studies in computational intelligence, vol 1016, pp 387–410

Storn R, Price K (1997) Differential evolution—a simple and efficient heuristic for global optimization over continuous spaces. J Global Optim 11:341–359

Sun Q, He M, Zhang M, Zeng S, Chen L, Zhao H, Yang H, Liu M, Ren S, Xu H (2021) Traditional Chinese medicine and colorectal cancer: implications for drug discovery. Front Pharmacol 12:685002

Sung H, Ferlay J, Siegel RL, Laversanne M, Soerjomataram I, Jemal A, Bray F (2021) Global cancer statistics 2020: GLOBOCAN estimates of incidence and mortality worldwide for 36 cancers in 185 Countries. CA Cancer J Clinicians 71(3):209–249

Taino DF, Ribeiro MG, Roberto G, Zafalon GFD, do Nascimento, M.Z., Tosta, T.A., Martins, A.S., & Neves, L.A. (2019) A model based on genetic algorithm for colorectal cancer diagnosis. Lect Notes Comput Sci (Includ Subser Lect Notes Artif Intell Lect Notes Bioinform) 11896:504–513

Tazi SN, Gupta M, Jain A (2014) A survey on application of nature inspired algorithms. Int J Comput Sci Eng 4(4):33–40

Valdes JJ, Orchard R, Barton AJ (2007) Exploring medical data using visual spaces with genetic programming and implicit functional mappings. In: Proceedings of the 9th annual conference companion on Genetic and evolutionary computation, 2953–2960

Vijayalakshmi S, John A, Sunder R, Senthilkumar M, Sweta B, Rajesh K, Guang F, Usman T (2020) Multi-modal prediction of breast cancer using particle swarm optimization with non-dominating sorting. Int J Distrib Sensor Netw 16(11)

Wang Y, Liu B, Ma Z, Wong KC, Li X (2019a) Nature-inspired multi objective cancer subtype diagnosis. IEEE J Transl Eng Health Med 7:1–12

Wang Y, Liu B, Ma Z, Wong KC, Li X (2019c) Nature-inspired multiobjective cancer subtype diagnosis. IEEE J Transl Eng Health Med 7:4300112

Wang Z, Qin C, Wan B, Song WW (2021) A comparative study of common nature-inspired algorithms for continuous function optimization. Entropy 23(7):874

Wang S, Shen H-W, Chai H, Liang Y (2019b) Complex harmonic regularization with differential evolution in a memetic framework for biomarker selection. PLoS ONE 14(2)

Worthley DL, Whitehall VL, Spring KJ, Leggett BA (2007) Colorectal carcinogenesis: road maps to cancer. World J Gastroenterol 13(28):3784–3791

Worzel WP, Yu J, Almal AA, Chinnaiyan AM (2009) Applications of genetic programming in cancer research. Int J Biochem Cell Biol 41(2):405–413

Xia J, Broadhurst DI, Wilson M, Wishart DS (2013) Translational biomarker discovery in clinical metabolomics: an introductory tutorial. Metabolomics 9(2):280–299

Xie T, Yao J, Zhou Z (2019) DA-based parameter optimization of combined kernel support vector machine for cancer diagnosis. Processes 7(5):263

Xing EP, Jordan MI, Karp RM (2001) Feature selection for high dimensional genomic microarray data. Morgan Kaufmann Publishers Inc., 601–608

Yang XS (2020) Nature-inspired optimization algorithms: challenges and open problems. J Comput Sci 46:1877–7503

Yeh JY (2007) Applying data mining techniques for cancer classification on gene expression data. Cybern Syst 39(6):583–602

Yoo SK, Song YS, Lee EK, Hwang J, Kim HH, Jung G, Kim YA, Kim SJ, Cho SW, Won JK, Chung EJ (2019) Integrative analysis of genomic and transcriptomic characteristics associated with progression of aggressive thyroid cancer. Nat Commun 10(1):1–2

Yu H, Gu G, Liu H, Shen J, Zhao J (2009) A modified ant colony optimization algorithm for tumor marker gene selection. Genomics Proteomics Bioinform 7(4):200–208

Zhan A, Mohan S, Tarolli C, Schneider RB, Adams JL, Sharma S, Elson MJ, Spear KL, Glidden AM, Little MA, Terzis A (2018) Using smartphones and machine learning to quantify Parkinson disease severity: the mobile Parkinson disease score. JAMA Neurol 75(7):876–880

Potential Role of the Nature-Inspired Algorithms for Classification of High-Dimensional and Complex Gene Expression Data

Sahar Qazi, Ayesha Khanam, and Khalid Raza

1 Introduction

In the last decade, researchers find nature as a noble core of inspiration to solve complicated problems and to flourish intelligent systems. If we look at the animals, in response to evolution, they developed new skills and even new organs for existence so that they can comfortably fight for their food mates and home. These new skills and organs can be treated as a new algorithm, which can be set up by calibrating the process of evolution as a parameter (Zhang et al. 2010). Computational techniques by observing and taking natural things and processes as inspiration, the new algorithm that is being created, are called the *nature-inspired algorithm*. The reason behind the development of this nature-inspired algorithm is to improve engineering problems. In the developing world, it has become very difficult to solve different issues of industries and engineering as the data is increasing exponentially, as well as their dimension variability, storage space complexity, and time complexity. To deal with these issues, a nature-inspired algorithm is created, which can handle the dimension of large data set variables and their hard nature problem and can refine numerical benchmark functions and multi-objective functions (Agarwal and Mehta 2014; Molga and Smutnicki 2005).

In this chapter, we will discuss bio-inspired algorithms. Bio-inspired algorithms can be further categorized into three types that are algorithms inspired by the behaviour structure and evolution (Yang et al. 2016) of an organism. Figure 1 highlights the three major bifurcations of bio-inspired algorithms, namely (i) *behaviour-inspired algorithms*, (ii) *structure-inspired algorithms*, and (iii) *evolution-inspired algorithms*. Behaviour-inspired algorithms take inspiration from the natural instincts

S. Qazi · A. Khanam · K. Raza (✉)
Department of Computer Science, Jamia Millia Islamia, New Delhi, India
e-mail: kraza@jmi.ac.in

K. Raza (ed.), *Nature-Inspired Intelligent Computing Techniques in Bioinformatics*,
Studies in Computational Intelligence 1066,
https://doi.org/10.1007/978-981-19-6379-7_5

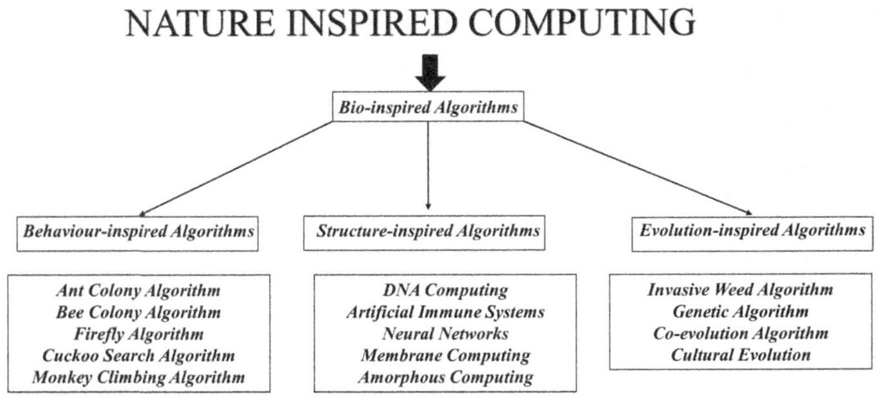

Fig. 1 Classification of nature-inspired computing (NIC) with a special interest in bio-inspired algorithms

and activities adopted by animals and insects. Some of the commonly used behaviour-inspired algorithms are—the cuckoo search algorithm, ant colony optimization, firefly algorithm, artificial bee colony, monkey climbing algorithm, etc. On the other hand, structure-inspired algorithms function based on the body arrangement and composition of human nucleic acids, cells, tissues, organs, immunity, etc. Commonly, structure-inspired algorithms are—artificial neural networks, artificial immune systems, amorphous computing, DNA computing, etc. Evolutionary-inspired algorithms are the ones that function based on Darwinian concepts of evolution and the survival of the fittest approach. These include—genetic algorithm, invasive weed algorithm, co-evolution algorithm, cultural evolution algorithm, etc.

Various behaviours of natural organisms such as flocks of birds, schools of fishes, colonies of bees, and swarms of bees have drawn interest in researchers who can apply these algorithms in solving various problems of science and engineering. Swarm intelligence, viz., based on the behaviour of birds and how they self-organize, simulates different societies and colonies of animals as in societies of animals one individual navigates another one and they work in groups to perform their task, similarly, this natural behaviour is simulated in solving real-life problems There are many features of swarm intelligence such as uniformity, speed, self-organization, configurability, and redundancy (Darwish 2018; Bonabeau et al. 1999; Tan and Gilbert 2003).

These algorithms have been in use for the last 30 years. There are various nature-inspired algorithms such as firefly algorithm, ant colony algorithm, genetic algorithm, cuckoo search algorithm, bat algorithm, immune algorithm, artificial bee colony algorithm, harmony search algorithm, and gravitational search algorithm (Wang et al. 2021). Ant-based clustering algorithm and ant-based association rule mining algorithm, both based on swarm intelligence, are useful in *gene expression analysis*. To study gene relation with a disease and determination of different gene expressions in different environments, the clustering technique is very useful. In gene expression analysis, the number of genes present in thousands and each gene consists of an absolute value. So, these genes are grouped according to some thresholds like upregulated, downregulated, and unchanged (He and Hui 2009). Gene expression data has arisen as a pivotal research area in the diagnosis of different diseases especially cancers (Raza 2014, 2016; Sahu et al. 2022). High throughput screening measures the expression level of a bunch of genes at the same time. As the gene expression data is noisy, high-dimensional and highly variable, this creates complexities in diagnosis. Thus, it requires an efficient way of classification through which genes can be classified into small subsets (Wang et al. 2019; Li et al. 2007). These genes consist of the features. So, to classify, it becomes mandatory to comprehend important features for the prediction of disease as many features possess confusion and lead to inaccurate predictions. Thus, feature selection is a challenging pre-process mechanism. It is very necessary to choose an appropriate technique for feature selection and to select high-performing genes. An appropriate method would lead to result in robust and lower errors (Alzaqebah et al. 2021).

2 Nature-Inspired Algorithms in Gene Expression Classification

2.1 Cuckoo Search Algorithm

The cuckoo search algorithm is a swarm intelligence-based nature-inspired algorithm which imitates the parasitic brooding behaviour of bird cuckoo. Cuckoos are brood parasites as they do not brood their own egg, they lay an egg and leave their egg in another bird's nest for brooding and caring for their young one. Cuckoos do not build their own nest. They lay an egg in another bird's nest and remove the one egg laid by the host. They select the host and nest very carefully so that they can mimic and match the colour and pattern of their own egg with the host egg. This is an amazing art of deceit. The timing of laying an egg should also be the same at which the host just laid the egg (Davies 1970; Shehab et al. 2017). Then the host bird drops out one egg by the nest. But in blind propelling, host bird can drop out its own egg and brood the egg of the cuckoo. After the birth of a cuckoo chick, the host bird provides food and all care to that cuckoo chick. The cunning behaviour is inheritrix in the cuckoo chick as the cuckoo chick also mimics the call of the host chick to get more food (Kaveh et al. 2012).

Besides this, if the host bird identifies the strange egg, then the host bird can simply throw out the cuckoo egg or leave its own nest and build up a new nest. In the whole procedure, one can say that the cuckoo needs more skill to mimic the egg of the host bird to find more accuracy and the host bird needs improvement in recognizing skill of the strange egg. Inspired by this natural cunning reproduction behaviour of cuckoos, Yang and Deb in 2009 proposed the cuckoo search algorithm (Yang and Deb 2009). This algorithm is based on a parasitic brooding behaviour strategy involving levy flight that is like a real-life random walk in which a bird took small steps with a heavy tail. The cuckoo search algorithm consists following three rules:

(a) The cuckoo chooses randomly a nest and lays a single egg at a time.
(b) Both the randomly chosen nest and the laid egg with the best quality of mimicry would have the ability to carry over to the new generation.
(c) Number of availabilities of the host nest is certain, and the identification of the cuckoo's egg by host bird is with a probability of $P \propto E$ (where $E = 0, 1$). If the host bird recognizes the egg is of a stranger bird, then the bird can either simply throw out the egg or leave the nest and build a nest. Then, aside n host nest will use the $P \propto$ the probability to replace the new nest, if more appropriate.

On the basis of the mentioned rules, the cuckoo search algorithm pseudocode is:

Objective function $f(x)$, $x = (x_1 .... x_d)T$
Produce starting population of n host nest X_i ($i = 1, 2, ..., n$)
While t is less than Max iterations Do
When t is less than Max Generation or it is stop criteria
Take randomly a cuckoo by Levy flight
Evaluation of quality fitness Fi
Among the n nest choose one at randomly, i.e, j
If F_j is less than F_i then
A new solution will replace j
End if
Pα a fraction of useless nests are rejected and a new nest is constructed;
Take the best nest or solution on the basis of quality;
Provide rank to the solutions and search best
End while
After the procedure results would visualize

2.2 Multi-objective Cuckoo Search Algorithm

In gene expression data, there are a huge amount of data consisting of various features. In classifying the data, there would be various objectives that are conflicting with others, so in such conditions, multi-objective approaches are required (Deb et al. 2002). Multi-objective optimization reduces the dimension of data. The multiple objective optimizations in mathematical form

$$\min\{f_1(x), ...f_m(x)\} \quad (x \in X) \tag{1}$$

Here, a number of objectives are represented by M, and the decision space for all the decision vectors of the d variable is denoted x. Here, if a probable solution x_2 can be dominated by another probable solution x_1 and $f_i(x_2) \geq f_i(x_1)$ for each i and minimum one i present there, $f_i(x_2) > f_i(x_1)$ ($i \in \{1, \ldots, M\}$). The group consisting of non-dominated solutions is termed a Pareto set. A set of non-dominated points which shows the various communicating curve of conflicting objectives is called as Pareto frontier. These non-dominated points are the image of objective vectors in a condition of a non-dominated solution in an objective space (Wang et al. 2019).

As the multiple objectives are present in multiple objective classifications so the selection of objectives is a pivotal portion of classification. Let if x_i is a set of genes or features, the target label is represented by y, and X is the subset of x_i, then there are four objective functions:

$$f_1(x) = \max\{\Sigma \ \ \mathrm{SU}(x_i, \mathrm{Y})\}, \quad x_i \in X \tag{2}$$

$$f_2(x) = \min\{\Sigma \ \ \mathrm{SU}(x_i, X_j)\}, X_i, \quad X_j \in X \tag{3}$$

$$f_3(x) = \min\{|X|\} \tag{4}$$

$$f_4(x) = \max \frac{t_p + t_n}{t_p + t_n + f_p + f_n} \tag{5}$$

Here, SU (a, b) is symmetric uncertainty among two data sets a and b. The t_p represents true positive, t_n stands for true negative, f_p is for false positive and f_n for false negative.

To decompose multi-objective classification into a single scalar classification, Tchebycheff method is used. In this approach, we add a weight vector

$$\lambda = \{\lambda_1, \ldots, \lambda_M\} \text{ and } \sum M \ \lambda_K = 1, \ \lambda_K \succcurlyeq 0, \quad K = 1 \tag{6}$$

Here, number of objectives is represented by M, then to calculate single objective function, algorithm will be

$$\text{gte}\left(x|\lambda, z^*\right) = \max\{\lambda_k | f_k(x) - z \times k|\}, \quad 1 \preccurlyeq k \preccurlyeq M \tag{7}$$

Here, $z^* = \{ z^*_1, \ldots, z^*_M\}$ is reference point, each objective is represented by $f_k(x)$, and best value for each objective is z^*x.

Then, binary encoding is used for computing the fitness of the objective function.

$$h = \left|\frac{2}{\pi}\text{archton}\left(\frac{\pi}{2}s_i^k(t)\right)\right| \tag{8}$$

where binary code,

$$s_i^k(t) = \begin{cases} 0 & \text{if}(\varphi < h) \\ 1 & \text{if}(\varphi \geq h) \end{cases} \tag{9}$$

Here, h is the fitness of objective function, φ is extracted from Gaussian distribution, where $\varphi \sim N$ (0.5, 0.12), $s_i^k(t)$ represents the decimal dimension for the individual vector of the population.

After the selection of a feature or gene, population-based two stochastic search methods which are inspired by the differential evolutionary algorithms are applied to explore and exploit features or genes. Ensemble mechanism is approached to choose one stochastic method for application. Ensemble mechanism updates these two methods with earlier experience.

2.3 Artificial Bee Colony (ABC)

It abides by the swarm intelligence optimization approaches. This algorithm was first proposed by Karaboga (2005) to study the potential of genes that cause cancers in humans. It is a selection feature extensively deployed for retrieving the best solutions to optimization problems. The main inspiration for this algorithm is bees and their mode of finding food. In the ABC algorithm, bees are categorized as—onlooker's bees, employed bees, and scout bees. Employed bees start to find the best food sources and bring the food (nectar) to their nest. Onlooker bees, on the other hand, monitor the food (nectar) that has been brought by the employed bees. While scout bees are responsible for finding random alternative food sources when food sources from employed bees have run out. Eventually, scout bees become employed bees (Utami and Rustam 2019). This way, the optimal solutions retrieved by the employed bees have a greater probability of selection when compared to the scout bees. The search space in the algorithm is large, however, due to this, the convergence of results is slow. This algorithm has been used to identify genes that can be used as biomarkers for screening for various cancers for identifying the best possible solutions in numerical optimization. This algorithm removes non-serving or least potent solutions and provides the best and dominant ones only (Moosa et al. 2016).

The convergence speed of the ABC algorithm can be improved by hybridizing it with existing classifiers to bring out the best results at a faster speed. There have been various extensions of the ABC algorithm that is currently used for gene expression data classification such as—correlation artificial bee colony (Co-ABC) (Alshamlan 2018), minimum redundancy maximum relevance artificial bee colony (mRMR-ABC) (Alshamlan et al. 2015), artificial bee colony used with support vector machines (ABC-SVM) (Alshamlan et al. 2016), i.e. a hybrid model deployed for faster convergence of results and discretization of food sources for an artificial bee colony (DfABC). A new classification rule was generated for improving and guiding bees to potential areas of search space known as—BeeMiner. It works by deploying an information-theoretic heuristic function (IHF) (Talebi and Abadi 2014).

Pseudocode for ABC Miner and SVM-based approach would be as follows:

- *Set the parameter: Colony size, Max cycles, limit*
- *Start food source*

- *Through SVM classifier classification accuracy is computed and on the basis of fitness food source is evaluated*
- *Cycle = 1*
- *While cycle is less than max cycles DO*
- *With the help of employed bees' new solutions are generated*
- *Evaluation of new solution by computing the fitness i.e. the classification accuracy produced by SVM classifier*
- *Add greedy selection procedure*
- *Through fitness value compute the probability values*
- *On the basis of probability of food source with the help of onlooker bees, generate new solution*
- *Evaluation of new solution by computing the fitness i.e. the classification accuracy produced by SVM classifier*
- *Add greedy selection procedure.*

2.4 AdaBOOST Algorithm

AdaBoost algorithm is a statistical metaheuristic algorithm introduced in 1996 by Freund and Schapire (1996). AdaBoost algorithm is short of adaptive boosting a type of ensemble learning (machine learning). AdaBoost is an adaptive boosting algorithm as it combines with succeeding weak learners to produce a robust final out by modifying the data that were misclassified earlier. At first, AdaBoost algorithm creates an N number of decision trees with the given data set. The misclassified data is used as input for the next model or tree during the first decision tree construction. This process is repeated until a base learner is created. AdaBoost algorithm created the tree as a stump having only one node with two leaves. These stumps are weak learners. The AdaBoost sequence of stumps is important. Every stump is influenced by its previous stump. The output of data set could be in the form of binary or categorical this leads to classification. The data set consists of features and records. Stumps are created on the basis of features, how much the number of features and the same number of stump features will be created. If the data set contains 4 types of features, then 4 stumps base learner model will be created. Leaves of the stumps contain the record of misclassified or right classified data. Entropy will be computed according to the record. The model consisting of the least entropy will be selected as the base learner. All the records will be credited with a sample weight. Here, formula for a sample weight $W = 1/N$, where N is the number of features in the data set. So, in this case, as the data set contains 7 types of features, then 7 stumps base learner model will be created and the sample weight 1/7 will be added to all the records assigned in the first learner model. Initially, an equal weight of X_i is assigned to the stump or model. In proceeding to the next stage or model, we have to update the sample weight. As the sample weight is not updated and the same weight is assigned with the next stage, then the output will be from the previous stump or stage. For update, the sample weight for the correctly classified record is decreased, and the weight for the misclassified record is increased. The weight is updated according to the performance of stumps (https://www.mygreatlearning.com/blog/adaboost-algorithm).

$$\text{New sample weight} = \text{sample weight} \times \mathrm{e}^{\text{performance}} \tag{10}$$

where

$$\text{performance} = \frac{1}{2}\log\frac{[1 - \text{TE}]}{\text{TE}} \tag{11}$$

Here, TE = total error found in the stump that sum all errors in the classified data or record for the sample weight and the log is the natural log.

Then the algorithm will proceed to the next stump or model and will repeat the process till it passes through all the model's stumps and will find fewer errors in comparison with the normalized weight of the first stage. The final output is created based on the weighted vote, that is, if there are 3 stumps and 2 stumps have generated 0 as output and one stump has given 1 as output, then AdaBoost will go with the majority vote and will generate 0 as output (https://www.mygreatlearning.com/blog/adaboost-algorithm).

Pseudocode for AdaBoost algorithm can be written as follows [22]:

1. *Let input training is x,y, number of iteration is j an integer*
2. *The X_i weight will be assigned to model*
3. *For each iteration i = 1.....j*

   {

   Produce a classifier Ci(x) with reduced weighted error over the X

   Weight of X_i will be updated

   }
4. *The weighted vote of individual $C_i(x)$ leads to the construction of $C^*(x)$ final classifier on the basis of the accuracy of weighted vote on weighted training set.*
5. *Output will be $C_*(x)$*

2.5 Support Vector Machines

The support vector machine (SVM) is deployed for regression and classification problems. Here, the SVM classifier has a hyperplane wherein the data is classified as either class 0 or class 1. Since SVMs belong to supervised machine learning, they already have some input labels based on which they classify the data (Vanitha et al. 2015). In gene expression data too, they take the gene function knowledge to associate genes that tend to have similar functioning in disease, thus, classifying highly expressed genes from poorly expressed ones. SVMs are deployed for microarray gene expression classification as they are not rigid like unsupervised learning algorithms. They are flexible in selecting the functioning of similar genes, can handle scattered data points that are observed with large data sets, and have a good capacity to rectify outliers in a data set. They have shown better performance in accuracy, specificity, and sensitivity when compared to other traditional algorithms such as neural networks and self-organizing maps (SOMs) (Bhat 2017). Researchers have developed several extensions for SVMs to further improve their performances. Once such hybrid extension is the *support vector machine recursive feature elimination procedure* (SVM-RFE) and its variants. Both the algorithms that have been combined are greedy, referring to their significantly high performances and effectiveness for binary classifications in biological data sets. Recently, a new R-package called "*sigFeature*" has been developed that provides an automatic selection of features for binary classification and is based on SVM and *t* statistics. The research group tested the R-package on six microarray data sets that were downloaded from Gene Expression Omnibus (GEO) consisting of varied biological attributes. They concluded that the features selected by sigFeature showcased greater classification accuracy for gene expression classification of microarray data (Das et al. 2020).

2.6 Boruta Algorithm

A wrapper algorithm that works around random forest (RF) is called as *Boruta algorithm.* It provides some shadow features as it increases randomness in a data set. This algorithm is known to manage associations between variables along with the alternating nature of random forest measures. Boruta algorithm tends to relate all features associated with the outcome predictor variable. Therefore, the Boruta algorithm is used to identify good solutions with small to no error in small subsets (Ragunthar and Selvakumar 2019). The algorithm has 3 main steps:

Step 1 Randomness is added to the data set by forming shadow features so as to initialize the algorithm, which is then trained on a random forest (RF) classifier on the data set. It applies a parameter called *mean decrease accuracy* (MDA) to check for the significance of each feature.

Step 2 After every cycle, the algorithm screens whether any feature has a greater significance (greater Z-score value) than its shadow features. It sideways also eliminates features that have less to no significance.

Step 3 The algorithm is terminated once all the features are selected or declined or reach a threshold point of random forest cycle runs.

2.7 K-Nearest Neighbour Algorithm

KNN is another approach that is deployed for gene expression data classification. The main characteristics of the algorithm consist of *time computation, power,* and *easy data interpretation* (Mei et al. 2009). Although it is known for its practicality while handling data, its main limitation lies in its speed. The humongous the data, the slow is KNN's processing speed. Therefore, with gene expression classification or any biological data classifications, KNN algorithms are usually avoided. Although, they work great when the data size is small (Parry et al. 2010).

The major steps involved are

(a) Loading the data and starting with the value of k. In order to retrieve the predicted class, run cycles from 1 to a total number of data points.
(b) The stretch between each row of training data and the test data is computed by Euclidean distance. The calculated distances are then arranged in an ascension. From the sorted array, the most recurrent rows are screened, thus, the best possible predicted classes are obtained.

2.8 Random Forest Algorithm

Random forests (RFs) are simply ensemble approaches that were first proposed by Leo Breiman in the early 2000s (Breiman 2001) wherein the base learners are trees that are combined by averaging. This algorithm is a very handy and useful method its parameters that give out good predictions and solutions to complex problems. Random forests can be utilized as a classifier and a regressor model as it tends to predict rich results for both problems (Díaz-Uriarte and Alvarez de Andrés 2006). In this algorithm, trees are produced on bootstrapped samples from training data that are not correlated. By then combining these trees, averages are taken that allow for efficient results to be screened for a problem (Ragunthar and Selvakumar 2019). Random forest (RF) should be considered the gold standard toolbox for microarray gene expression analysis because of its magnanimous performance and ease of utility (Díaz-Uriarte and Alvarez de Andrés 2006).

3 Pros and Cons of Nature-Inspired Algorithms for Gene Classification

In Sect. 2, we explained all the commonly deployed algorithms for gene expression data sets. However, each of the algorithms discussed has some pros and cons too! Based on various research studies and experimental analysis, it is known that random forests (RF) are the best when we talk about classifying large data sets along with support vector machines (SVMs) and then KNN models. The RF classification algorithm provides better solutions but its performance can be improved by ensemble methods such as bagging, boosting, and stacking. Gene expression data analysis is usually huge, and it is best if such humongous data is dealt with the stacking method, to ensure better and more accurate classification. This way, the diseases can be diagnosed easily and the diagnosis would also become easier in its early stages (Ragunthar and Selvakumar 2019). Furthermore, it is important to note that the type of feature selection method, the sample type and size, number of genes, etc. everything plays an important role in deciding which nature-inspired computation (NIC) model should be used in which case. On the basis of accuracy, specificity, and sensitivity, RF and SVMs have been discerned to be the best first choices when it comes to a classification problem. In fact, in bioinformatics too, biological data classification is heavily dependent on either SVMs or RF models only. For small-size data sets, the KNN model or ABC models are generally used (Pirooznia et al. 2008).

Although NIC is widely used today in numerous domains of science besides engineering and research, it is noteworthy to mention here that these algorithms have some limitations too. The common limitations of these NIC models are listed as follows (Agarwal and Mehta 2014):

(i) *Curse of Dimensionality*: Search space is one of the common problems when it comes to NIC algorithms. NIC algorithms such as swarm intelligence and evolutionary algorithms have been tested on high dimensions (more than 1000), while newer bio-inspired algorithms such as artificial bee colony (ABC) algorithm or cuckoo search optimization algorithms (CSOA) have been checked for only a few dimensions, less than 40. Therefore, their efficiency cannot be completely relied upon when dealing with classification of biomedical/biological data sets.

(ii) *No "One-Algorithm-Solves-All"*: Different NIC models exist today because there are myriad kinds of complexities and problems in real life. Therefore, there is no such algorithm that can provide efficient and robust solutions to all sorts of problems.

(iii) *Expertise:* Nature-inspired computing (NIC) models are works of good and time taking programming. Not everyone holds up good programming skills and expertise, and it is not feasible for every researcher to devote time to evaluating and screening these NIC models.

4 Future Prospects

As the gene expression data is noisy, high-dimensional and highly variable, this creates complexities in diagnosis. Various nature-inspired algorithms can lead to an efficient way of classification through which genes can be classified into small subsets. NIC algorithms have been heavily deployed for complex optimization problems. Although the performance is tremendously effective when compared to traditional algorithms, however, the problem of time complexity and speed still exists. One way to improve the existing loopholes in NIC algorithms is to combine them with quantum computing and chaos theory approaches (Darwish 2018).

Cuckoo search algorithm along with multi-objective classification with ensemble mechanism can make better classification of gene expression data. An AdaBoost algorithm is also an ensemble-based approach. Ensemble mechanism generates multiple models or search methods and produces a result with high accuracy. In future, ensemble-based different nature-inspired methods can be explored more in gene expression data analysis. Currently, there is no single platform to evaluate and assess all the nature-inspired algorithms in one go over some benchmark functions. Thus, this is an open opportunity for researchers to develop a single toolbox assessment or a platform that can help bioinformaticians/biological researchers/experts to select the best NIC model for solving their problem based on sample size, nature of sample, and complexity of the problem. Furthermore, newer algorithms such as ABC or CSOA must be tested on greater dimensions to understand their true classification potential under different situations and environments.

5 Conclusion

Nature-inspired computing (NIC) takes inspiration from the behaviour, structure, and evolution of humans, animals, birds, and insects. The NIC algorithms are greatly used in different domains of scientific research today—from engineering to medical research to even art! Most tedious and important research analysis in bioinformatics deals with gene expression data classification analyses. The NIC algorithms such as cuckoo search optimization algorithm (CSOA), artificial bee colony (ABC), KNN, support vector machines (SVM), Boruta algorithm, and random forest (RF) are heavily used today. Depending on the type of problem, data set size and nature, these models are selected for finding optimal solutions. Gene expression classification is an important step towards a better diagnosis of a disease, eventually leading to appropriate and specific treatment. It is so because with the data set classified, the number of highly expressed genes can be identified easily, that are considered prognostic markers for disease. Although there are a few limitations of these NIC models that are being used for gene expression classification, however, the future seems prosperous for the NIC fraternity.

Acknowledgements SQ is supported by the DST-INSPIRE fellowship (DST/INSPIRE/03/2018/000623) provided by the Department of Science & Technology (DST), Govt. of India.

References

Agarwal P, Mehta S (2014) Nature-inspired algorithms: state-of-art, problems and prospects. Int J Comput Appl 100(14):14–21

Alshamlan HM (2018) Co-ABC: Correlation artificial bee colony algorithm for biomarker gene discovery using gene expression profile. Saudi J Biol Sci 25(5):895–903. https://doi.org/10.1016/j.sjbs.2017.12.012

Alshamlan HM, Badr GH, Alohali YA (2016) ABC-SVM: artificial bee colony and SVM method for microarray gene selection and multi class cancer classification. Int J Mach Learn Comput 6(3):184–190. https://doi.org/10.18178/ijmlc.2016.6.3.596

Alshamlan H, Badr G, Alohali Y (2015) mRMR-ABC: a hybrid gene selection algorithm for cancer classification using microarray gene expression profiling. Biomed Res Int 2015:1–15. https://doi.org/10.1155/2015/604910

Alzaqebah M, Briki K, Alrefai N, Brini S, Jawarneh S, Alsmadi MK, Mohammad RMA, ALmarashdeh I, Alghamdi FA, Aldhafferi N, Alqahtani A (2021) Memory based cuckoo search algorithm for feature selection of gene expression dataset. Inf Med Unlocked 24:100572

Bhat FH (2017) Evaluating SVM algorithms for bioinformatics gene expression analysis. Retrieved 23 Mar 2022, from http://www.ijcse.net/docs/IJCSE17-06-02-023.pdf

Bonabeau E, Theraulaz G, Dorigo M, Theraulaz G, Marco DDRDF (1999) Swarm intelligence: from natural to artificial systems (No. 1). Oxford University Press

Breiman L (2001) Random forests. Mach Learn 45(1):5–32. https://doi.org/10.1023/a:1010933404324

Darwish A (2018) Bio-inspired computing: algorithms review, deep analysis, and the scope of applications. Future Comput Inform J 3(2):231–246

Das P, Roychowdhury A, Das S, Roychoudhury S, Tripathy S (2020) sigFeature: novel significant feature selection method for classification of gene expression data using support vector machine and t statistic. Front Genet 11. https://doi.org/10.3389/fgene.2020.00247

Davies GH (1970) The life of birds, parenthood. www.pbs.org/lifeofbirds/home/ index.html

Deb K, Pratap A, Agarwal S, Meyarivan TAMT (2002) A fast and elitist multiobjective genetic algorithm: NSGA-II. IEEE Trans Evol Comput 6(2):182–197

Díaz-Uriarte R, Alvarez de Andrés S (2006) Gene selection and classification of microarray data using random forest. BMC Bioinform 7(1). https://doi.org/10.1186/1471-2105-7-3

Freund Y, Schapire RE (1996) Experiments with a new boosting algorithm. In: International conference on machine learning, Bari, 3–6 July 1996, 148–156. http://dl.acm.org/citation.cfm?id=3091696.3091715

He Y, Hui SC (2009) Exploring ant-based algorithms for gene expression data analysis. Artif Intell Med 47(2):105–119

Karaboga D (2005) An idea based on honey bee swarm for numerical optimization. Technical Erciyes University, Engineering Faculty, Computer Engineering Department, Tech. Rep. 1–10. http://citeseerx.ist.psu.edu/viewdoc/download;jsessionid=F362BB83638092677F80C098EDD725D5?doi=10.1.1.714.4934&rep=rep1&type=pdf

Kaveh A, Bakhshpoori T, Ashoory M (2012) An efficient optimization procedure based on cuckoo search algorithm for practical design of steel structures 2(1):1–14

Li J, Su H, Chen H, Futscher BW (2007) Optimal search-based gene subset selection for gene array cancer classification. IEEE Trans Inf Technol Biomed 11(4):398–405

Mei Z, Shen Q, Ye B (2009) Hybridized KNN and SVM for gene expression data classification. Life Sci J 6(1):61–66. http://www.lifesciencesite.com/lsj/life0601/12_life0601_61_66_Hybridized.pdf

Molga M, Smutnicki C (2005) Test functions for optimization needs. Test Funct Optim Needs 101:48

Moosa JM, Shakur R, Kaykobad M, Rahman MS (2016) Gene selection for cancer classification with the help of bees. BMC Med Genom 9(S2). https://doi.org/10.1186/s12920-016-0204-7

Parry RM, Jones W, Stokes TH, Phan JH, Moffitt RA, Fang H, Shi L, Oberthuer A, Fischer M, Tong W, Wang MD (2010) k-Nearest neighbor models for microarray gene expression analysis and clinical outcome prediction. Pharmacogenomics J 10(4):292–309. https://doi.org/10.1038/tpj.2010.56

Pirooznia M, Yang JY, Yang MQ, Deng Y (2008) A comparative study of different machine learning methods on microarray gene expression data. BMC Genom 9(S1). https://doi.org/10.1186/1471-2164-9-s1-s13

Ragunthar T, Selvakumar S (2019) Classification of gene expression data with optimized feature selection. Int J Recent Technol Eng (IJRTE) 8(2):1–7

Raza K (2014) Clustering analysis of cancerous microarray data. J Chem Pharm Res 6(9):488–493. https://www.jocpr.com/articles/clustering-analysis-of-cancerous-microarray-data.pdf

Raza K (2016) Analysis of microarray data using artificial intelligence based techniques. Handbook of research on computational intelligence applications in bioinformatics. IGI Global, USA, 216–239. https://doi.org/10.4018/978-1-5225-0427-6.ch011

Sahu A, Qazi S, Raza K, Singh A, Verma S (2022) Machine learning-based approach for early diagnosis of breast cancer using biomarkers and gene expression profiles. In: Computational intelligence in oncology, studies in computational intelligence (SCI). Springer, vol 1016, 285–306. https://doi.org/10.1007/978-981-16-9221-5_17

Shehab M, Khader AT, Al-Betar MA (2017) A survey on applications and variants of the cuckoo search algorithm. Appl Soft Comput 61:1041–1059

Talebi M, Abadi M (2014) BeeMiner: a novel artificial bee colony algorithm for classification rule discovery. In: 2014 Iranian conference on intelligent systems (ICIS), 2014, pp 1–5. https://doi.org/10.1109/IranianCIS.2014.6802576

Tan AC, Gilbert D (2003) Ensemble machine learning on gene expression data for cancer classification

Utami DA, Rustam Z (2019) Gene selection in cancer classification using hybrid method based on particle swarm optimization (PSO), artificial bee colony (ABC) feature selection and support vector machine. In: AIP conference proceedings. https://doi.org/10.1063/1.5132474

Vanitha CDA, Devaraj D, Venkatesulu M (2015) Gene expression data classification using support vector machine and mutual information-based gene selection. Procedia Comput Sci 47:13–21. https://doi.org/10.1016/j.procs.2015.03.178

Wang Y, Liu B, Ma Z, Wong KC, Li X (2019) Nature-inspired multiobjective cancer subtype diagnosis. IEEE J Transl Eng Health Med 7:1–12

Wang Z, Qin C, Wan B, Song WW (2021) A comparative study of common nature-inspired algorithms for continuous function optimization. Entropy 23(7):874

Yang SX, Luo C, Li H, Ni J, Zhang J (2016) Theory and applications of bioinspired neural intelligence for robotics and control. Comput Intell Neurosci

Yang XS, Deb S (2009) Cuckoo search via Lévy flights. In: 2009 world congress on nature and biologically inspired computing (NaBIC), 2009, pp 210–214. https://doi.org/10.1109/NABIC.2009.5393690

Zang H, Zhang S, Hapeshi K (2010) A review of nature-inspired algorithms. J Bionic Eng 7(4):S232–S237

Optimized Nature-Inspired Computing Algorithms for Lung Disorder Detection

V. Lakshman Narayana, R. S. M. Lakshmi Patibandla, V. Pavani, and P. Radhika

1 Introduction

Asthma, chronic obstructive (COPD), infections including tuberculosis and influenza, lung cancer, pneumonia, and other breathing issues are all included under the umbrella term "lung disease." The symptoms of lung disease can vary depending on the type of disease that is affecting the patient. A cough that won't go away, blood or mucus being coughed up, and soreness when blowing in or out are some of the most common symptoms. To make images of the organs for diagnostic and analytical reasons, medical imaging is utilized. Diseases of both the airways and other lung structures are referred to as lung disorders or respiratory diseases (Lakshmanaprabu et al. 2019). Pneumonia, TB, and coronavirus disorder 2019 are all examples of respiratory illness (COVID-19). Tuberculosis kills 3 million people each year, lung cancer kills 1.6 million people each year, as does pneumonia, according to the Forum of National Respiratory Societies. Millions of people around the world were infected, and healthcare systems were overwhelmed by the COVID-19 epidemic (Shakeel et al. 2019; Zhou et al. 2021). One of the biggest causes of morbidity and mortality on this planet is lung illness. To increase long-term survival statistics, it is critical to get an early diagnosis (Zhang and Hong 2019; Sharifrazi et al. 2021). Skin tests, blood tests, sputum sample tests, chest X-ray examinations, and imaging techniques (CT) scans have traditionally been used to detect lung disease (Song et al. 2019; Guo and Razmjooy 2019). In medical imaging for disease identification, deep learning has

V. Lakshman Narayana · V. Pavani · P. Radhika
Vignan's Nirula Institute of Technology and Science for Women, Peda Palakaluru, Guntur, Andhra Pradesh, India
e-mail: radhika@vignannirula.org

R. S. M. Lakshmi Patibandla (✉)
Department of IT, VFSTR, Guntur, Andhra Pradesh, India
e-mail: patibandla.lakshmi@gmail.com

K. Raza (ed.), *Nature-Inspired Intelligent Computing Techniques in Bioinformatics*, Studies in Computational Intelligence 1066,
https://doi.org/10.1007/978-981-19-6379-7_6

recently shown remarkable potential when used. As a branch of machine learning, "deep learning" deals with algorithms that are inspired by the structure and function of the human brain. The discovery, quantification, and categorization of patterns in image data are supported by recent breakthroughs in machine learning, notably deep learning (Narin et al. 2021). Due to its capacity of deep learning could learn characteristics based on data, rather than domain-specific expertise, these breakthroughs were achievable.

Deep learning is increasingly gaining traction as a cutting-edge technology that can enhance a wide range of medical functions. As a result, these developments make it easier for doctors to identify and classify various medical disorders (Liu et al. 2020). Big data analysis in the healthcare business has been hailed as an essential tool for spotting trends and patterns that can be used to improve patient care. Cancer is the leading cause of mortality in India, according to the National Cancer Institute. Diagnosis of lung cancer is crucial. Due to the short time span between diagnosis and death, lung cancer came in second place among all malignancies. Only 15% of patients identified after five years will make it (Faramarzi et al. 2020). Cancer incidence in India is increasing at a rate of 11% each year. Over the course of a year, this caused harm to 2.5 million individuals and the death of 4 lakh people. Tobacco-related malignancies kill 20% of males in India between the ages of 30 and 69, and tumor is among them. Lung cancer early detection is unquestionably beneficial to society (Yoda et al. 2017). Data mining is useful in the medical field for a variety of purposes, including predicting disease and prescribing medications based on a patient's individual circumstances, and assessing the efficacy of specific treatments. By utilizing data mining, healthcare organizations can better understand their patient's needs while also reducing treatment costs. Cancer disrupts the human body's natural process of cell death, causing cells to thrive and proliferate when they shouldn't. The early detection tool for lung cancer aids society, and thus, researchers are interested in establishing a mechanism for predicting the disease.

Clustering and segmentation of lung cancer datasets using k-means, foggy k-means, ANN, and fuzzy neural networks have been described in the literature, with the goal of detecting lung cancer in its earliest stages. Foggy K-means, a new method for dealing with clustering in lung cancer datasets, was presented in Lakshmanaprabu et al. (2019). Foggy k-means clusters data for two clusters based on the classification numbers, and each attribute characteristic is selected for clustering in order to determine the characteristics that are most important for clustering. Foggy k-means outperformed standard k-means in terms of cluster validity scale Dunn index connection and Silhouette (Fulladosa et al. 2018). Outliers in a large dataset are a limitation of the work. As a result, domain experts might use the results to help them develop their strategy. Images of X-rays, CT scans, MRI scans, etc., can be classified as cancerous or non-cancerous using the BPNN model presented in Wang et al. (2019). In addition, a genetic algorithm is utilized to extract features based on fitness. Solutions based on the hidden layers are significantly more viable when dealing with large volumes of data and trying to tackle an unsupervised learning assignment with a recurrent neural network. The reason behind this is that there are a large number of undefined parameters in a neural network with a high degree of complexity. It can help patients have a better prognosis by detecting lung cancer at an early stage. With this information, a proper judgment may be made based on the current status of a patient. The technique is time-consuming and requires a high level of sophistication, which are both drawbacks of the paper (Shakeel et al. 2019).

Lung common cancer cancer-related death in humans. In order to come up with new strategies for preventing lung cancer, it is critical that both hereditary and environmental factors be uncovered. 400 cancer and non-cancer patient data were obtained from various diagnostic facilities and preprocessed using a K-means algorithm for clustering in order to select relevant and non-related data first. AprioriTID but instead decision tree algorithms are used to discover the next significant frequent patterns. Both techniques are effective in finding patterns in large datasets that have been grouped. It was possible to use this model to estimate a person's risk of developing lung cancer. The decision tree approach used to mine frequent patterns and clustered data has fewer parameters for a dataset than other machine learning methods. Making huge, multi-level decision trees is difficult and time-demanding, even if they are simple to use when compared to other choice models. It's going to take a long time. Because the amount of data in the medical sector is so big, using an algorithm to create a decision tree results in a bloated, unintelligible decision tree.

The preprocessing step utilized to aid create precise segmentation of sputum color pictures is hence a filtering algorithm. In all cases, Hopfield neural network extracted features outperform fuzzy C-means clustered in terms of accuracy and reliability. Nuclei and cytoplasm areas were identified and segmented by the Hopfield neural network. Fuzzy C-means, on the other hand, was unable to locate the nuclei, and what little it did locate was incomplete. Due to its greater convergence-segmentation error than the Hopfield neural network, the fuzzy C-means is also less sensitive to changes in intensity than the Hopfield neural network. Fuzzy logic and neural networks were utilized by Zhang and Hong (2019) to develop a lung cancer detection system. Neural networks were also used to categorize normal and abnormal photographs. Other parameters (symptoms) can be used as input to the fuzzy logic system in the event of an abnormal outcome in order to determine the patient's case based on the membership function of the inputs. It makes it possible to represent dynamic systems in a more understandable manner.

Real-time difficulties can be solved by expanding rough estimates into a fuzzy environment. In other words, adjusting system parameters will generally only modestly boost system performance in many optimization issues. Depending on the technology utilized and the number of detectors purchased, this strategy may not be cost-effective for the average lung cancer patient. Because it is a labor-intensive procedure, it cannot be used for the early identification of lung cancer. In order to help clinicians manage incidental or screen found ambiguous pulmonary nodules, Sharifrazi et al. (2021) introduced machine learning-based lung cancer prediction models. It is possible that such methods could help reduce the number of benign tumors that are followed or worked up unnecessarily, as well as improve decision making. There are many different ways to predict lung cancer, and this article provides an exhaustive list of ones and highlights their strengths and weaknesses. They describe the most widely used methods for classifying lung nodules and predicting lung cancer based on CT imaging data. For large datasets in health care, the methodology of clustering will have a direct impact on the results of clustering, which is one of the key analytical approaches in data mining.

Detailed descriptions of clustering algorithms, such as k-means, enhanced k-means, hierarchical and partitioned clustering, as well as their respective strengths and shortcomings, were provided in the review studies (Song et al. 2019, 2021; Guo and Razmjooy 2019; Narin et al. 2021; Shakeel et al. 2020; Ranjbarzadeh et al. 2021) for both the medical and analogous huge data domains. Adaptive k-means, fuzzy c-means, and hybrid and integrated techniques for data mining and machine learning are illustrated in the papers (Tahir et al. 2021; Merriënboer et al. 2015). Although these articles have been carefully examined, there is still a need to improve the classification or clustering performance for lung cancer datasets in order to aid medical practitioners make accurate predictions. Because of this proposed technique in the following section, we'll discuss how to combine a clustering method with a classification method in order to find the most important factors in the development of lung cancer.

1.1 Introduction on Nature-Inspired Optimization Algorithms

Swarm robotics, biological systems, physiochemical systems, and other natural phenomena are all examples of what are known as "nature-inspired optimization algorithms" (NIOAs). Since they are defined in diverse ways, most NIOAs share a similar structure. There are 11 NIOAs that will be examined in this section, each with their own set of guiding principles laid out and debated within the framework of the section's common methodology. The unified model makes it easy to compare and contrast various algorithms' similarities and differences.

Figure 1 depicts a typical NIOA process, which can be broken down into four stages. Population and other factors are set up in step S1. To guarantee that as many solutions as possible are covered, the initial number is usually produced using random methods; population size is set depending on expert opinion and individual requirements, but typically should be as large. Since iterative approaches are popular in NIOAs, initializing the algorithm's maximum iteration duration and the precision threshold is an important step in the process.

The target functional (i.e., the BBOB variables) creates the fitness function, which typically has an optimal value, to reflect the effectiveness of each particular solution. A local optimal solution exists for each individual, while a global ideal exists for the entire population. In step S2, the population's fitness values are calculated, and if the globally best solution meets the termination requirements, NIOAs will display the findings (in step S4). The alternative is to use step S3, in which important activities (specified by multiple elements or operators) are carried out in order to share knowledge across the entire population so order to develop great individuals. It then moves

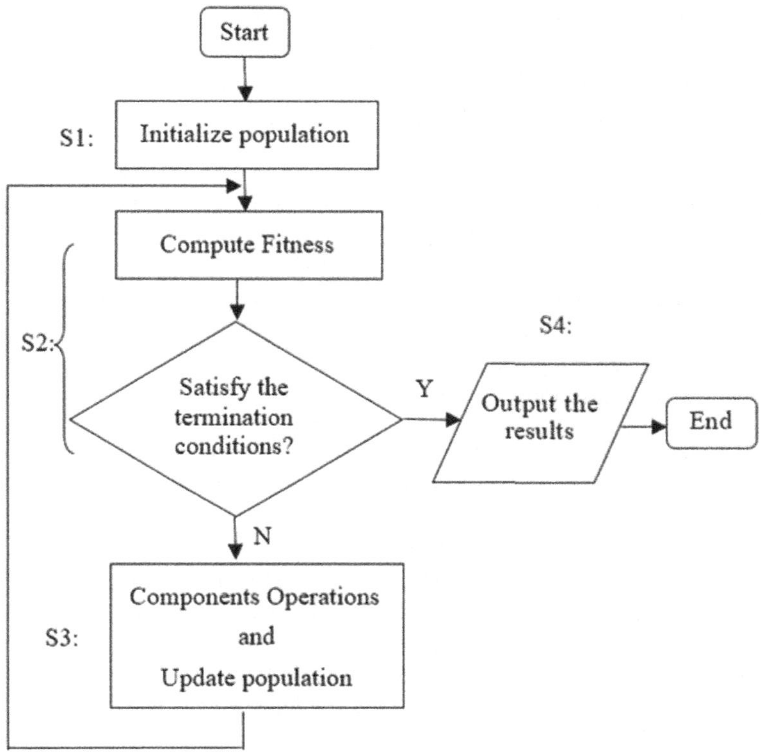

Fig. 1 Common process of NIOAs

on to the following iteration, which begins with a population update and moves to step S2. It follows this procedure that the 11 NIOAs are described using common symbols, where D represents the overall dimension number of optimization methods, M is the particular number of each NIOA, and N represents the total iteration times of all NIOAs.

1.2 Common Characteristics

When it comes to population simulation, there are many alternative approaches, but the Reynolds model (Brubaker et al. 2018) explains the essential rules for flock aggregation movements of the simulated flock formed by a distributed behavioral model (Brubaker et al. 2018) as demonstrated in Sect. 2.

Randomicity Uncertainty about an occurrence with a predetermined probability enhances an individual's ability to conduct a global search (Chou et al. 2016). Individuals are randomly seeded in all 11 NIOAs in order to cover as much of the possible space as possible, but some additional mechanisms have been implemented in them to enhance their local and global search abilities, such as mo in GA, IA, and DE, rand1 and rand 2 in PSO, random parameters rand(0,1) and, rand(0,1) and, Lévy flight in CS, randj and vella in GSA, r1, r2, A1, A2, A3 and C1, C1, C2, C3 in GWO, $u(1,1)$ in HS,

Information Interactivity A better chance of finding the global optimal is increased when members of NIOAs communicate with each other directly or indirectly. Each particle in PSO updates its position using the global optimization pg(t) in the same way that the employed forager or onlooker in ABC updates their angular velocity v_i, $d(t + 1)$ or uses another different spot x_j, $d(t)$; species in BA use the optimal global pg(t) to update there own positions; as to FA, firefly information is mixed between $x_i(t)$ and $x_j(t)$ in order to move toward $x_j(t)$.

Boost the effectiveness of NIOAs is, in fact, an extremely complicated issue, with the approaches of parameter selection, topology structure, and teaching strategy all having a significant impact. Preliminary conclusions are drawn in this study in order to give a reference framework again for the selection and enhancement of NIOAs. Many metaheuristic algorithms and their modifications have been published in the last 30 years and provide efficient and productive answers to optimization issues in the areas of AI (more than 120 according to our statistics). NIOAs have come a long way, and they've been put to good use in a variety of industries, but they still have a long way to go, as evidenced by the four key challenges listed below.

On the problem of expansion of fundamental continuous NIOAs into diverse optimization problems such as COP and MOOP, the investigation is not sufficient. COPs and MOOPs can be addressed in a variety of ways based on this analysis of more than 120 MHAs' learning processes and topological structures. Research into mathematical abstractions in the preceding (2) and problem extensions is currently limited to only GA, PSO, and DE; hence, it is necessary to expand the scope of research.

(4) NIOA's research visualization platforms are yet another issue. There appear to be only a few conversations on this topic, apart from a simple one. As a result, only a small number of benchmark tests are suited to specific optimization challenges, such as parameter tweaking. An additional optimization challenge arises since it is difficult to discern the characteristics of distinct NIOAs (the algorithms all appear quite similar) due to limited and poor theoretical examination of the NIOAs. Furthermore, this is the driving force for our efforts to evaluate and study the theoretical and experimental properties of 11 typical NIOAs.

The other problem is that there aren't enough NIOA-related disciplines being combined. Because of this, parallel processing technologies like GPU-based hardware, robot swarms, and cloud platforms receive less attention when used in conjunction with NIOAs. NIOAs have a lot of room for multidisciplinary study, as evidenced by a few studies.

Fourth, there has been a lack of effort to apply NIOAs to a wide range of problem domains (Tuan et al. 2016). However, it is difficult to generate an NIOA that can address all of the probable use cases. Some NIOAs, on the other hand, may be more useful in dealing with specific types of difficulties. An in-depth & targeted examination of the reasons for the enhanced methods is lacking in the existing enhanced NIOA methods. It's not obvious why this NIOA was picked over other NIOAs that have been used for the same application challenge.

As a result, we believe that researchers in the NIOAs should focus on the following three issues going forward. The NIOAs study will focus on these three issues in the future. Two. Creating new NIOAs to solve complex optimization challenges. There are a number of difficult real-world optimization issues, such as those involving many models and multiple objectives, limited or uncertain situations, and large-scale optimization challenges. The development of more precise and effective NIOAs to address the aforementioned issues is an important new approach.

2 Literature Survey on Lung Disorder Classifications

Many people suffer from lung ailments, which are among the most common illnesses in the world. In the USA, tens of millions of people were affected by lung illness. Most lung disorders are caused by smoking, infections, and heredity. Asthma, cough (COPD), infections including tuberculosis and influenza, lung cancer, pneumonia, and other breathing issues are all included in the term "lung disease." According to the specific condition being impacted, the signs and symptoms of lung disorders might be rather different: mucous, discomfort, or pain on inhalation or exhalation. It is necessary to have an environment that allows for accessing data, data analysis, processing, and algorithm creation in order to perform medical picture analysis and process. When it comes to creating photographs of the human body for diagnostic and scientific purposes, medical imaging is the technique and procedure used. For the classification of lung illnesses using CT images, an automated approach was proposed in this work. This study uses a hybrid genetic particle–swarm optimization algorithm approach to choose the extracted characteristics of lung CT images, and classifiers are utilized to classify the information into relevant datasets. Section 2 of the report is devoted to a literature review. Problem definition and literature gaps are discussed in Sect. 3. Section 4 explains the purpose and, finally, the strategy that is being offered.

Optimizer (GA) was used in conjunction with KNN in order to increase classification accuracy. For the classification of test samples, the GA was used instead of picking k-neighbors after evaluating all of the training sets and then calculating distance. Rough set theory hybridized with bee colony optimized (BCO) was used in their earlier work to obtain the reduced features and functionality before categorization. Classifiers such as KNN, CART, and SVM were used to compare their results. It was in October 2013 (Wang et al. 2019) that Mahdi Setayesh and his colleagues presented a unique restricted optimization model for finding smooth texture in noisy images. To find the best solution, two optimization technique techniques are used. The punishing and preservation approaches of constraint handling are used in these two algorithms (Serra et al. 2017). Edge detection strategies are tested on two image pairs with varying types and amounts of noise, and a customized version of the Charming algorithm is used as the Gaussian filter and the resilient rank order (RRO) algorithm is used as the statistical edge detector. It has been used to compare various algorithms based on Pratt's figure of merit. New edge detectors are more durable under noisy situations and outperform Canny or RRO algorithms for images with impulsive or Gaussian noise, according to experimental results. Because it uses the penalizing approach rather than the preservation method, the new algorithm works more quickly. To detect and classify lung disorders, Shakeel et al. (2019) devised a feature extraction and feature selection method that relied on moment invariants, digital storytelling by genetic algorithm, and Naive Bayes and logistic regression classifiers. Feature extraction was used to extract the image's most valuable elements, while noise removal was used to clean up the image's appearance. Then, the image's most important elements were selected using a feature selection technique, and the classifiers were used to classify the results. The logistic regression classifier outperforms the naive Bayes classifier in terms of accuracy.

Genetic algorithm (GA) was employed by Gunavathi and Premalatha in 2014. The t-Statistics, signal-to-noise ratio (SNR), and FTest values were used to identify informative genes. The top-m informative genes yielded the first set of GA potential solutions. The k-nearest neighbor (kNN) approach classification accuracy was utilized as a fitness function for GA. The classifiers utilized in this study were kNN and SVM. It was observed that the proposed work was suitable for optimized feature selection based on the experimental results. In four datasets, the GA-kNN approach achieves 100% accuracy while the GA-SVM method achieves in five of the ten datasets with the help of both the selected genes. Accurate microarray-based tumor classification can be achieved using the GA plus kNN and SVM algorithms. Global optimization is not guaranteed by the genetic algorithm. While particle swarm optimization guarantees global optimization, it nevertheless suffers from the selection of initial particles. Poor research is the result of particle selection errors. • The most of research does not use a hybrid genetic algorithm/particle swarm optimization method. Using medical CT imaging in clinical diagnosis has been increasingly common during the past few years. Detection and location of pathogenic changes are improved by this method of analysis. In CT images, the gray levels of different tissues are shown differently depending on their location in the body. Computer-aided diagnostics systems rely on accurate segmentation of lung illnesses using CT scans. A new technique for classifying lung CT scans using image-based feature extraction is presented here. We are going to utilize a new method for feature selection that combines genetic and particle swarm optimization (PSO).

A deep learning-based technique for the identification of chest illness using X-ray scans. Using an X-ray picture dataset, they created and tested an automated network model for the detection of chest diseases. As compared to other soft computing techniques, the CNN model has a substantial advantage in terms of testing accuracy and training time. In Varela-Santos and Melin, machine learning and soft computing approaches were used to detect pneumonia in chest X-ray pictures (2020). After X-ray pictures had been segmented, texture features were retrieved, and then a neural network was used to classify the images. In Lin et al., CNN was used to detect pneumonia in chest X-ray pictures (2019). The X-ray image was fed into a ConvNet that was built using data from the Kaggle repository. Wang et al. suggested another lung cancer detection method based on computer vision and a cooperative CNN model (2019). Local and global characteristics were retrieved from lung scans using a segmentation technique to identify ROIs in the pictures (Xie and Yuille 2017). The categorization was intended to be performed by a cooperative CNN model. VGG16, a form of CNN, was employed in their deep learning-based approach to the categorization of pneumonia in X-ray chest images. During the learning process, they applied a blend of learning algorithms and fine-tuning techniques to maximize results. Pneumonia can be detected using X-ray imaging and CNN. The CNN was trained to recognize normal and pneumonic X-ray images. The ability to accurately and efficiently detect pneumonia from chest X-ray pictures (2020).

COVID-19 illness diagnosis utilizing chest X-ray pictures has recently been the subject of a number of deep learning-based investigations. A deep learning model called DeTraC was used to classify COVID-19 illness from X-ray pictures. Even in the face of data inconsistencies, the DeTraC model holds up admirably. In a similar utilizing deep learning to detect COVID-19 illness in chest X-ray pictures (2021). Images of COVID-19 from normal patients were taken using X-rays. Data augmentation and preprocessing techniques were used. Automatic feature extraction was used to create a CNN model for classification.

A study of the many CNN models used for the classification of lung diseases into COVID-19, bacterial meningitis. By means of CT scans of the chest, they developed a CNN model for classifying pneumonia with COVID-19 lung infections. Technique of detection and categorization of COVID-19 infection into bacterial meningitis, viral pneumonia, and also the ordinary class was proposed. They used a deep transfer learning strategy to apply the proposed methodology to a variety of chest X-ray datasets of varying sizes. For the identification of COVID-19 illness, the double deep action recognition systems were developed. For improved detection, they used pre-trained models. COVID-19, bacterial infection, and viral pneumonia were all found.

Only the properties of infected lung areas are relevant for diagnoses in existing CNN-based efforts, which have taken the entire lung image. Because of the paucity of ROI estimates in X-ray pictures, features that are not relevant to classification are extracted (Wagner et al. 2017). Because there are no ROI-specific features, it also reduces the severity of disease analysis. Using CNNs for early lung disease diagnosis is difficult because of the long training times required. On small X-ray samples, COVID-19 and tuberculosis detection methods utilizing deep convolution neural networks have been tested on 10–15% samples for testing and 80–85% samples for training and validation, respectively. Such models necessitate a higher ratio of training to testing if they are to be considered efficient and reliable.

The feature extraction methodologies of CNN depended on pre-training models that used irrelevant ImageNet datasets for lung disease predictions. A paucity of patient records in CNN or before models prompted incorrect feature extraction due to the recent emergence of the COVID-19 pneumonia virus.

3 Contribution of the Chapter

Preprocessing the raw image using median censoring, histogram equalization, and intensity adjustment are used to improve the quality of the X-ray images. Then, we locate the lung ROI by extracting ROI-specific features and applying the computational techniques for automatic classification (Asghari and Navimipour 2018). Finally, we apply the soft computing methods to automatically classify the X-ray images. The enhanced image depicts the affected areas, allowing for better forecasting. Additionally, the improved X-ray images were put to use in ROI estimation via an adaptive segmentation technique. Before feature extraction, a segmentation technique removes non-lung-specific parts from the dataset (Krawczyk et al. 2012). The image has been segmented and a robust set of characteristics such as the histogram of oriented grain (HOG), texture, invariant moment, and ROI intensity properties have been retrieved. After this, we use the feature scaling (normalization) strategy to improve detection accuracy. Medical professionals can utilize the normalized features to assess the severity of the disease.

LSTM network, fully connected network, SoftMax, and classification layer are created successively in the RNN deep learning model. The F-RRN-LSTM model was developed with the goal of reducing calculation time and increasing precision (Okulewicz and Mandziuk 2017). Two datasets were used to evaluate the suggested computer vision method: 70–30 percent for training and testing. The results were acquired and presented for the developed framework utilizing the suitable training/testing ratio of 70–30 percent. It has been shown that the current proposal F-RRN-LSTM performs better than the current state of the art.

Recognition and prevention of COVID-19 illnesses are the primary goals of the planned research project. Soft computing and machine learning approaches may be used to analyze and predict lung diseases effectively, and then a deep learning-based strategy, F-RNN-LSTM, is proposed for reliable prediction (Serra et al. 2017). CAD tools need to be able to assess the severity of a disease before they can be used to help find it. Raw breast X-ray images are enhanced, and infected areas are extracted; i.e., ROI and important features are extracted using the suggested model which is superior than the high-dimensional features extracted of CNN and also its variants (Mirjalili 2015). Another important part of the proposed approach is feature scaling or normalization, which not only enhances the accuracy of classification but also aids in illness severity estimates.

4 Proposed Model

Imaging of an organ with a CT scan is possible but not seen on a standard X-ray, which allows for an earlier diagnosis (Song et al. 2019). As more and more cases of lung cancer are discovered too late, treatment options become more limited and patients' chances of survival plummet (Guo and Razmjooy 2019). A new and robust approach to the early detection of lung cancer is critical (Narin et al. 2021). CT scan pictures are utilized for a cancer diagnosis; radiologists use them to distinguish between malignant and non-malignant nodules (Shakeel et al. 2020). These procedures necessitate the use of highly skilled radiologists, who may be difficult to locate in rural areas. Aside from the possibility of human error, optimization-based solutions are needed to assist radiologists in diagnosing and reducing the prevalence of false results in manual testing (Ranjbarzadeh et al. 2021).

CT scans can be processed using image processing techniques to identify nodules, their shape, size, and other features. Medical image analysis has been widely and rapidly deployed in order to build expert support systems for the detection of various diseases, including lung cancer identification (Romanowski et al. 2015). Nodules that determine a patient's fate exist, but their shape and size vary from slice to slice, making it difficult to identify them. They are often linked to other pulmonary structures, such as arteries or bronchial tubes (Song et al. 2021). The hue of the images on CT scans can also be altered as a result of this. Their definition is made more complex by these variables.

Segmentation divides the pixels of an image into portions that are linked to the image's subject matter. Almost all computer vision systems begin with this stage. Segmentation algorithms typically use pixel intensities as a starting point. All algorithms have a threshold parameter that must be set before they can work (Liu et al. 2020). Segmentation increases when the threshold is set correctly. Given the intensity values, the threshold value is determined. Otsu thresholding techniques are employed in this study to attain the best possible threshold value. The process is discussed clearly.

Step-1: NCI Lung Cancer Database Consortium provides 120 lung CT images from which to draw. Once lung CT pictures have been gathered, they are normalized using both tumored and non-tumored CT images from the cancer imaging database.

Step-2: Clinical research using neuroimaging often includes image preprocessing as a critical step in ensuring the quality of the resulting images. A Wiener filter is used on the image to reduce the mean square error of the denoising process. The statistical approach to minimizing blurring and smoothing in images is known as Wiener filtering.

Step-3: Thirdly, image segmentation is the process of breaking up an image into several sub-images. For a threshold T, the global thresholding approach divides the image into sections based on the intensity of the gray-level pixels in each section.

Step-4: The region of interest is smoothed by applying a certain structuring element to every available site in order to estimate morphological operations.

Step-5: The most critical step in acquiring the segmented nodule's pattern information is feature extraction. Each segmented nodule is split into 60 different geometric, statistical, texture, and structure aspects. Gray-level co-occurrence matrix (GLCM) is used as a second-order statistics approach.

5 Results

CT imaging is regarded as the best method for assessing and visualizing abnormalities because of its reduced distortion. A picture's quality and distortion can be improved and reduced with image preprocessing techniques such as normalization and enhancement (Zhang and Hong 2019). Image-enhancing techniques use a variety of filters in both the spatial and frequency domains. The proposed model is implemented in python and executed in Google Colab. The proposed model is compared with the existing CNN-based lung cancer detection (CNN-LCD) model, and the results represent that the proposed model's performance levels are high. Figure 2 represents the image analysis accuracy levels.

The process of converting raw converted into information features that may be processed while still maintaining the integrity of the information contained in the original dataset is referred to as feature extraction. When compared to applying machine learning straight to the raw data, this method produces superior outcomes. The feature extraction accuracy levels of the existing and proposed models are shown in Fig. 3.

Clustering is the process of dividing a population or sets of pieces of data into a variety of groups in such a way that the data points in each group are more similar to one another than they are to the data points in any other groups. To put it another way, the goal is to divide people into groups based on the characteristics they share and then place those groups into clusters. The clustering accuracy levels are clearly illustrated in Fig. 4.

The capacity of the lungs to hold air, move air, and absorb oxygen is one of the things that doctors may measure in order to diagnose lung problems. These examinations, which are known as pulmonary function tests, are of the utmost use in determining both the general kind of lung illness and the degree to which it manifests itself. Other diagnostic procedures, such as further chest imaging, bronchoscopy, and thoracoscopy, give medical professionals the ability to zero down on the root cause of a lung issue. Figure 5 represents the lung disorder detection accuracy levels of the traditional and proposed model.

Fig. 2 Image analyzing accuracy levels

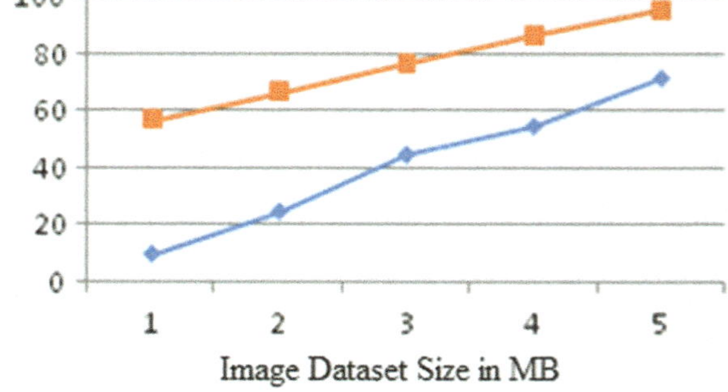

Fig. 3 Feature extraction accuracy levels

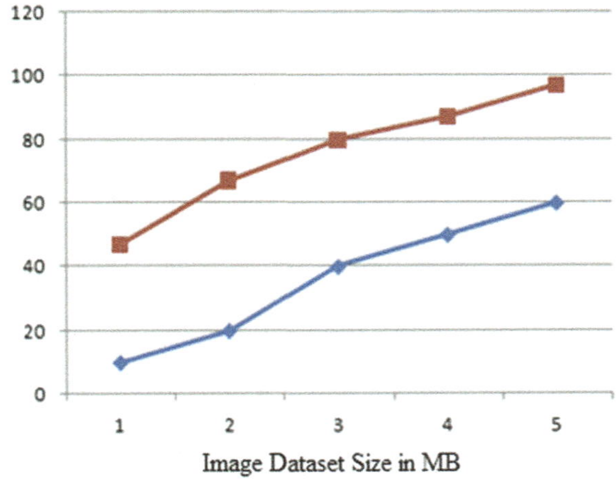

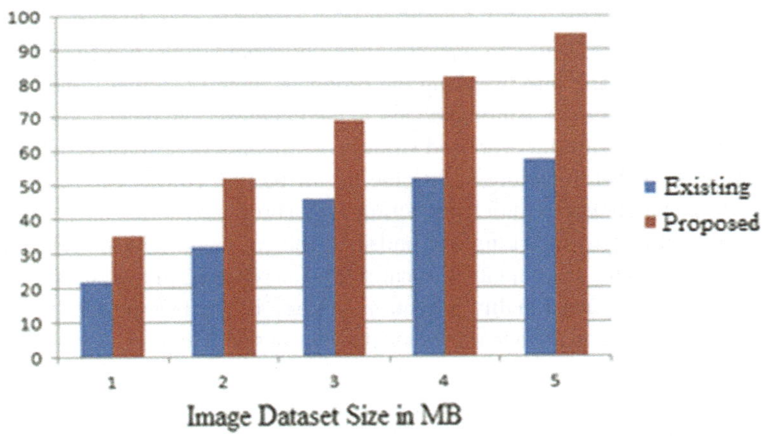

Fig. 4 Clustering accuracy levels

6 Conclusion

Cancer rates have grown dramatically due to declining living standards, such as sedentary lives, bad diets, and an increase in smoking, pushing researchers and scientists to take action to treat this hazardous disease. This condition can be treated more easily and with a lower death toll if caught early, according to scientific research. We'll examine a novel way of detecting lung cancer early in this post. CT-based lung pictures can be diagnosed using an automatic method based on deep neural networks, as demonstrated here. The marine predator's algorithm, a metaheuristic technique used in the suggested method, was used to find the best possible layout and network accuracy. High-level characteristics extracted from deep networks were used in the suggested strategy, resulting in accurate classification and diagnosis. In addition, the

Fig. 5 Lung disorder detection accuracy levels

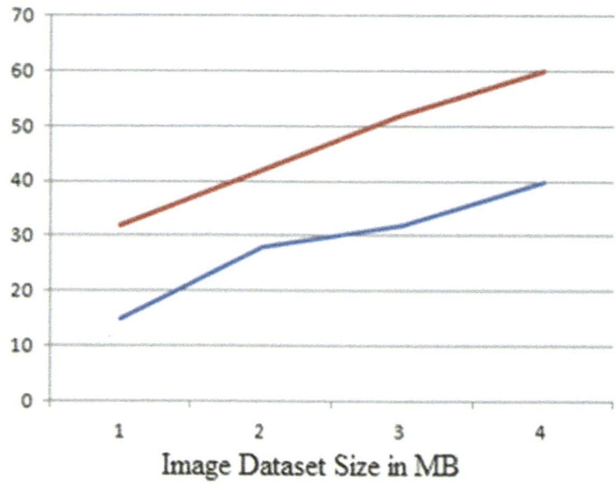

size of the feature vector was lowered while the precision of the early steps was raised, resulting in a reduction in space savings while increasing speed and accuracy. In the future, we'll look into tweaking the proposed method so that it may be used in a real-time environment. Automated aid in medical tests can detect both diseased and healthy tissues with the proposed strategy. It was possible to discover areas of tissue variation in X-ray pictures using the bio-inspired method proposed for lung segmentation, which was used in conjunction with a proposed composition. When an X-ray image with aggregated lung area and spots indicating possible diseased tissues is received, the doctor must analyze it carefully for possible lung disorders. In the future, the feature dimensionality reduction strategy is applied so that the training time will be reduced and the accuracy can also be improved by considering the most relevant features.

References

Asghari S, Navimipour NJ (2018) Nature inspired meta-heuristic algorithms for solving the service composition problem in the cloud environments. Int J Commun Syst e3708

Brubaker K, Garewal A, Steinhardt RC, Esser-Kahn AP (2018) Bioinspired counter-current multiplier for enrichment of solutes. Nat Commun 9(1):736

Chou K-Y, Lin C-S, Chien C-H, Chiang J-S, Hsia C-H (2016) Using statistical parametric contour and threshold segmentation technology applied in X-ray bone images. In: 2016 International symposium on intelligent signal processing and communication systems (ISPACS). IEEE, pp 1–5

Faramarzi A, Heidarinejad M, Mirjalili S, Gandomi AH (2020) Marine predators algorithm: a nature-inspired metaheuristic. Expert Syst Appl 152 (Article ID 113377) (View at: Publisher Site, Google Scholar)

Fulladosa E, Austrich A, Munoz I, Guerrero L, Benedito J, Lorenzo J, Gou P (2018) Texture characterization of dry-cured ham using multi energy X-ray analysis. Food Control 89:46–53

Guo G, Razmjooy N (2019) A new interval differential equation for edge detection and determining breast cancer regions in mammography images. Syst Sci Control Eng 7(1):346–356, 2019 (View at: Publisher Site, Google Scholar)

Krawczyk B, Filipczuk P, Wozniak M (2012) Adaptive splitting and selection algorithm for classification of breast cytology images. Lect Notes Comput Sci 7653:475–484. https://doi.org/10.1007/978-3-642-34630-949

Lakshmanaprabu S, Mohanty SN, Shankar K, Arunkumar N, Ramirez G (2019) Optimal deep learning model for classification of lung cancer on CT images. Future Gener Comput Syst 92:374–382 (View at: Publisher Site, Google Scholar)

Liu Z, Liao R, Ma H et al (2020) Classification of marine microalgae using low-resolution Mueller matrix images and convolutional neural network. Appl Opt 59(31):9698–9709 (View at: Publisher Site, Google Scholar)

Van Merriënboer B, Bahdanau D, Dumoulin V et al Blocks and fuel: frameworks for deep learning. https://arXiv.org/abs/1506.00619 (View at: Google Scholar)

Mirjalili S (2015) The ant lion optimizer. Adv Eng Softw 83:80–98. https://doi.org/10.1016/j.advengsoft.2015.01.010

Narin A, Kaya C, Pamuk Z (2021) Automatic detection of coronavirus disease (COVID-19) using X-ray images and deep convolutional neural networks. Pattern Anal Appl 24:1–14 (View at: Publisher Site, Google Scholar)

Okulewicz M, Mandziuk J (2017) The impact of particular components of the PSO-based algorithm solving the dynamic vehicle routing problem. Appl Soft Comput 58:586–604

Ranjbarzadeh R, Kasgari AB, Ghoushchi SJ, Anari S, Naseri M, Bendechache M (2021) Brain tumor segmentation based on deep learning and an attention mechanism using MRI multi-modalities brain images. Sci Rep 11(1):1–17 (View at: Publisher Site, Google Scholar)

Romanowski J, Korytkowski M, Scherer R (2015) Efficient bone detector and geometric descriptor for X-ray imaging. In: 2015 IEEE symposium series on computational intelligence. IEEE, pp 1337–1342

Serra S, Malpezzi L, Bedeschi A, Fuganti C, Fonte P (2017) Final demonstration of the co-identity of lipiarmycin A3 and tiacumicin B (fidaxomicin) through single crystal X-ray analysis. Antibiotics 6(1):7

Shakeel PM, Burhanuddin MA Desa MI (2019) Lung cancer detection from CT image using improved profuse clustering and deep learning instantaneously trained neural networks. Measurement 145:702–712 (View at: Publisher Site, Google Scholar)

Shakeel PM, Burhanuddin M Desa MI (2020) Automatic lung cancer detection from CT image using improved deep neural network and ensemble classifier. Neural Comput Appl 1–14 (View at: Publisher Site, Google Scholar)

Sharifrazi D, Alizadehsani R, Roshanzamir M et al (2021) Fusion of convolution neural network, support vector machine and Sobel filter for accurate detection of COVID-19 patients using X-ray images. Biomed Signal Process Control 68(Article ID 102622) (View at: Publisher Site, Google Scholar)

Song S, Jia H, Ma J (2019) A chaotic electromagnetic field optimization algorithm based on fuzzy entropy for multilevel thresholding color image segmentation. Entropy 21(4):398 (View at: Publisher Site, Google Scholar)

Song Y, Zheng S, Li et al L (2021) Deep learning enables accurate diagnosis of novel coronavirus (COVID-19) with CT images. IEEE/ACM Trans Comput Biol Bioinf (View at: Publisher Site, Google Scholar)

Tahir MB, Khan MA, Javed et al K (2021) Recognition of apple leaf diseases using deep learning and variances-controlled features reduction. Microprocess Microsyst Article ID 104027 (View at: Publisher Site, Google Scholar)

Tuan TM et al (2016) A cooperative semi-supervised fuzzy clustering framework for dental X-ray image segmentation. Expert Syst Appl 46:380–393

Wagner N, Sahin CS, Pena J, Streilein WW (2017) A nature-inspired decision system for secure cyber network architecture. In: 2017 IEEE symposium series on computational intelligence (SSCI). IEEE, pp 1–8

Wang X, Chen H, Gan et al C (2019) Weakly supervised deep learning for whole slide lung cancer image analysis. IEEE Trans Cybern 50(9):3950–3962 (View at: Publisher Site, Google Scholar)

Xie L, Yuille A (2017) Genetic CNN. In: Proceedings of the 2017 IEEE international conference on computer vision, pp 1379–1388, Venice, Italy, Oct 2017 (View at: Publisher Site, Google Scholar)

Yoda T, Shibuya K, Miura K, Myoubudani H (2017) Characterization of the adsorption ability of silk-derived activated carbon fibers using X-ray analysis and camera imaging methods. Measurement 101:103–110

Zhang Z, Hong W-C (2019) Electric load forecasting by complete ensemble empirical mode decomposition adaptive noise and support vector regression with quantum-based dragonfly algorithm. Nonlinear Dyn 98(2):1107–1136 (View at: Publisher Site, Google Scholar)

Zhou Y, Lu Y, Pei Z (2021) Accurate diagnosis of early lung cancer based on the convolutional neural network model of the embedded medical system. Microprocess Microsyst 81(Article ID 103754) (View at: Publisher Site, Google Scholar)

Overview and Classification of Swarm Intelligence-Based Nature-Inspired Computing Algorithms and Their Applications in Cancer Detection and Diagnosis

Fatima Nazish Khan, Mohammad Asim, and Mohammad Irfan Qureshi

1 Introduction

In recent years, with the enhancement of the popularity of machine learning technologies, there is a great interest by researchers in nature-inspired computing (NIC) techniques (Lindfield and Penny 2017; Fan et al. 2020; Carbas et al. 2021). Alongside, this has also been a way to enhance the computing power available to individuals as well as for small/medium-sized businesses (Mason et al. 2018). An increasing number of core computer hardware and graphics processing units have gained more popularity in this area. With the emergence of nature-inspired computing techniques, researchers have understood and modeled solutions for complex as well as real problems. NIC is a field of artificial intelligence (AI) and research-based exploration area where there is transferring of knowledge to engineered systems from natural phenomenon (Gautam et al. 2019). NIC has a wide sphere of applicability in computational intelligence, query optimization, evolutionary computation, robotics, disease diagnosis, agriculture applications, resource management, sentiment analysis, data mining, stock prediction, task scheduling, vehicle routing, and transportation. NIC techniques have been effectively used in solving several realistic problems such as diagnosis of disease (Poo et al. 2016), query optimization (Sharma et al. 2021), feature selection (Arora et al. 2019), sentiment analysis (Pandey et al. 2017), task

F. N. Khan · M. I. Qureshi (✉)
Department of Biotechnology, Jamia Millia Islamia, New Delhi, India
e-mail: miqureshi@jmi.ac.in; mirfanq@gmail.com

F. N. Khan
e-mail: fatimanazishkhan@yahoo.com

M. Asim
Department of Clinical and Experimental Medicine, University of Surrey, Guildford GU27WG, UK
e-mail: m.asim@surrey.ac.uk

K. Raza (ed.), *Nature-Inspired Intelligent Computing Techniques in Bioinformatics*, Studies in Computational Intelligence 1066,
https://doi.org/10.1007/978-981-19-6379-7_7

scheduling (Sharma et al. 2019), stock prediction (Kazem et al. 2013), clustering (Nanda and Panda 2014), etc. Although there are many techniques to be used in disease diagnosis, NIC algorithms are very efficient and have gained more attention to problems of modern research. In the last few years, various NIC approaches (inspired by natural phenomenon, for example, humans, birds, insects, water, animals, etc.) were utilized for the diagnosis of human diseases effectively. Chronic diseases for example cancers are a group of long-lasting debilitating, and often lethal diseases which usually affect both the mental and physical status of an individual and led to human disabilities (Raghupathi and Raghupathi 2018). Tumors primarily form when cells in organs continuously undergo repeated and uncontrolled divisions and lead to various types of cancers in different organs (Fig. 1). Cancerous cells generally move through blood, reach various body parts and destroy tissues, having symptoms like fever, excessive sweating, weight loss, fatigue, anemia, excessive hoarseness as well as cough, and regular headache. Cancer is mainly classified into five groups such as sarcomas (cancer of muscles, bones, and connective tissues), carcinomas (such as lung and colon cancer), leukemia (bone morrow initiation), adenomas (present in pituitary and thyroid glands), lymphomas (found in tissues of the immune system). Most of the cancers seen in males are prostate, lung, colon, stomach, liver, rectum, etc., while, women are mostly affected by breast, cervix, rectum, colon, lung, and stomach cancer. There are several risk factors such as lifestyle (smoking, poor diet, etc.), increased genetic susceptibility, and environmental factors underlying the diagnosis of cancers, and often distinct symptoms can help in prompting further investigations to diagnose cancers on a global scale (Fig. 2) (Gautam et al. 2019). Therefore, to diagnose different types of cancers various NIC algorithms have been utilized (Gautam et al. 2019; Sheikh et al. 2022).

In the last few decades, there has been a large growth of biological data such as genomes, protein sequences, and gene expression data in various scientific areas. NIC has opened a new way for its immense simplicity and parallelism in computation, and therefore many algorithms were invented to handle many real-world complex problems. For example, neural computing mimics the biological neuron to ensure computational distributed and parallel processing. There are lots of modern vaccinations as

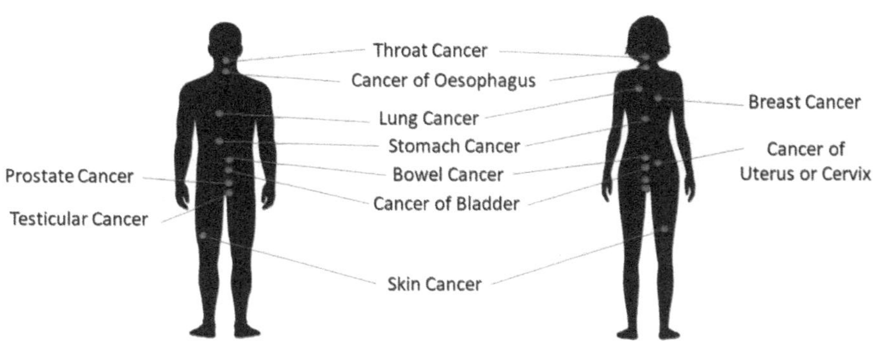

Fig. 1 Some examples of different types of cancers in males and females

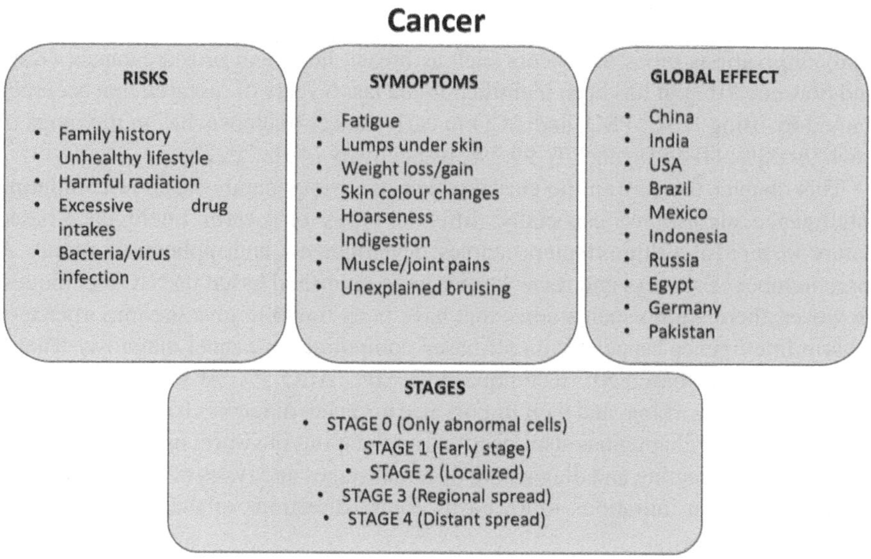

Fig. 2 Cancer: risks, symptoms, global effect, and stages

well as other ways to treat fatal diseases, however, many chronic diseases don't seem to vanish yet. And hence, these critical diseases can be easily handled when they are properly diagnosed. Recently, nature-inspired-based algorithms called Swarm Intelligence (SI) have been mostly used by researchers to solve bioinformatics-related problems around the world (Zhang et al. 2013). SI is a modern and recent optimization technique taking inspiration from birds, insects, fish, and animals. It is a field of computer science that finds NP-hard problem solutions with the help of mutual interaction of agents. SI has recently emerged as one of the natures inspired algorithms, having the ability to give fast, accurate, and low-cost solutions to complex problems. SI algorithms are majorly motivated by the collection of social insects' behavior such as termites, bees, wasps, etc. These insects (few individual capacities) can cooperatively do many complex tasks essential for their livelihood. Swarm intelligence involves ant colonies, bacterial development, bird flocking, fish schooling, and animal herding (Dadaneh et al. 2016). Particle Swarm Optimization (PSO), is mostly utilized in comparison to other existing selection methods. PSOs have potential benefits for the selection of features like it has the highest ability to search for an ideal cure; it is beneficial in selecting functionality due to the ability to have memory and understanding the solutions of problems; low cost and easy implementation of PSO; it connects ideas from different areas into the best solution; it represents binary and discrete data; it is more efficient and faster than other techniques of feature selection; it uses a simple algorithm for practical, effective and promising results. Different survey research papers have been published for the diagnosis of disease (Al-Absi et al. 2011; Junoha et al. 2017; Khan et al. 2021a). Similarly, a review of studies on the diagnosis of fatal disorders with the help of NIC algorithms like ACO,

PSO, ABC, and GA observed that these algorithms could be effectively used for analyzing various types of cancers such as breast, lung, and prostate cancer (Kaur and Sharma 2017). It has been identified in the last 6 years of research that accuracy gained by using ABC, PSO, and ACO in early cancer diagnosis lies in the range of 76.4–99.98%, 80–98%, and 70–99.2%, respectively (Aziz 2022).

This chapter focuses on the classification of various nature-inspired computing intelligence algorithms, especially different types of swarm intelligence-based nature-inspired algorithms their principles, developments, and applications scopes. A large number of surveys and reviews have been published based on NIC techniques. However, there are no such studies that have been found to give the importance of Swarm Intelligence especially insect-based computing for cancer diagnosis. Therefore, many insect-based NIC techniques like ABC, ALO, FA, ACO as well as GSO, also principles, working, and their importance in cancer diagnosis has been discussed here. Further, the chapter has also described applications of swarm intelligence-based algorithms in detecting and diagnosing different stages and types of cancers. Finally, it has focused on imitations followed by future directions of these techniques in cancer diagnosis.

2 Classification of Nature-Inspired Computing Algorithms

Based on various biological inspirations, various NIC algorithms were developed. Swarm-based algorithms, as well as evolutional-based algorithms, were the two most accepted NIC algorithms that were inspired by collective animals' behavior and natural evolution, respectively. Based on an understanding of these two categories, Binitha and Sathya divided NIC algorithms into three categories such as smarm-based algorithms (Zainal et al. 2015), evolutionary-based algorithms (Nicklow et al. 2010), and ecology-based algorithms (Binitha and Sathya 2012), which cover the majority of NIC algorithms. But there are also some limitations to using these 3 NIC algorithms, and henceforth category known as multi-objective algorithms has also been used (Sayers et al. 2019). The classifications used are presented in Fig. 3.

2.1 Evolutionary-Based Nature-Inspired Computing Algorithms

2.1.1 Artificial Neural Network (ANN)

Artificial neural network (ANN) is a machine learning algorithm, which can model the structure of biological brains, especially neurons, and use different learning mechanisms like the back-propagation approach (gradient descent-based method) (Parker

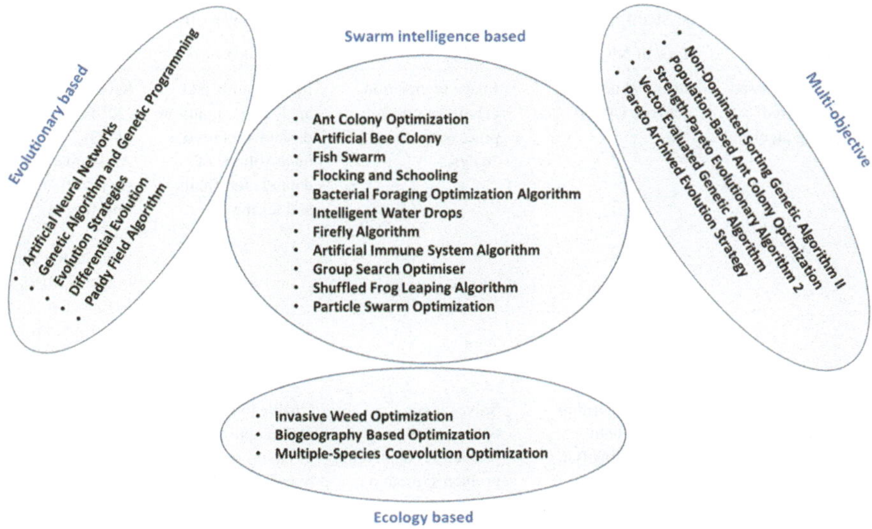

Fig. 3 Classification of commonly used nature-inspired computing algorithms

1985), as well as feed-forward networks (Hornik et al. 1989). It is a kind of algorithm that processes information and has features of generalisability and fault tolerance ability, as well as several types of algorithms based on neural networks that were optimized for a large number of applications. Earlier, ANN mainly focused on simple networks having one hidden layer, but in more complex and modern approaches several layers and deeper networks have been developed and utilized. ANNs were applied to a large number of problem domain applications (Table 1). Since the publication of research on a mathematical model by McCulloch and Pitts in the 1940s, ANNs and their derivative precursors have been used to mimic the function of biological neurons, leading to the development of ideas of perceptron model and other learning rules (Khan et al. 2021b; Khan et al. 2022). However, the learning rule and perceptron model are unable to train layered networks, i.e., the network cannot classify non-linear problems, and hence in the 1980s, there is the development of the back-propagation method as a learning approach for multi-layer networks that can easily handle non-linear problems (Parker 1985). Also, with developing computational power, deep neural networks were used with the consumption of less time and provided better results (LeCun et al. 2015).

2.1.2 Genetic Algorithm and Genetic Programming

Genetic algorithms (GAs) are based on natural selection, in which organisms having useful traits will exist along with pre-existing organisms. With mutation, the important elements are initially produced from pre-existing elements before passing to the

Table 1 Applications, strengths, and limitations of some useful NIC algorithms

NIC algorithms	Type of NIC algorithms	Applications/strengths	Limitations	References
Evolutionary-based nature-inspired computing algorithms	Artificial neural network (ANN)	Image recognition, classification, and processing; time-series analysis	Identification and analyzing outputs are difficult and need a huge volume of datasets for training and testing	Sayers et al. (2014), Gu et al. (2018), Arulkumaran et al. (2019), Yusoff et al. (2019), Whittington and Bogacz (2019), LeCun et al. (2015), Arora et al. (2020), Gu et al. (2018)
	Genetic algorithm (GA); Genetic programming (GP)	Solve classic optimization problems; engineering problems; protein folding	Unable to find the global optima instead of having many hyper-parameters; time taken to complete the process with insufficient computing power; sometimes could produce infeasible or inefficient solutions	Sayers et al. (2014), Paul et al. (2015), Islam et al. (2019)
	Evolution strategies (ES)	Hot water flashing nozzle optimization; engineering problems; optimization problems for training ANN that predict utilization of CPU host in the cloud; construction problems optimization; optimization and problems modeling	Does not take into account some particular problems and investigate optimization performance	Klockgether and Schwefel (1970), Bäck et al. (1991), Mason et al. (2018), Lin et al. (2019), Liu and Zhang (2020)
Ecology-based nature-inspired algorithms	Invasive weed optimization (IWO)	Solve scheduling problems, capacitated vehicle routing problems, optimize wireless network problems, and power systems economic dispatch	Need initialization of hyper-parameters before starting an iterative process, and if it is not accomplished the algorithm will never get converged or sometimes converge to a local minimum	Barisal and Prusty (2015), Velmurugan et al. (2016)
	Biogeography-based optimization (BBO)	Planning of outdoor wireless system; fuel cells and solar parameter estimation; scheduling problems; chaotic systems parameter identification; cloud computing	Limitations with already made software packages in the application of BBO algorithm; hyper-parameters need refinement for complex problems	Simon (2008), Wang and Xu (2011), Kim et al. (2014), Niu et al. (2014), Goudos et al. (2015), Lin (2015)

(continued)

Table 1 (continued)

NIC algorithms	Type of NIC algorithms	Applications/strengths	Limitations	References
Multi-objective nature-inspired algorithms	Non-dominated sorting genetic algorithm II (NGSA-II)	Widely used in industries, especially in the engineering field; optimization of risk of flood as well as water distribution network	Risk of losing the benefit of this algorithm and identifying solutions would be non-clear for analysis	Jourdan et al. (2004), Sayers et al. (2019)
	Population-based ACO (P-ACO)	Used in civil-engineering sector; traveling salesman problem; quadratic assignment problem	Suffers from algorithm complexity	Dorigo and Stützle (2019)

next generation, i.e., from parent to child. Therefore, the "genome" treated as a solution to the problem helps in modeling genetic algorithms that can further be mutated and recombined. GA can be considered as an optimization algorithm and are suitable for non-convex, discrete, non-linear, and multi-modal problems. This method has been coming into existence in 1975 as an efficient optimization approach. Holland (Holland 1992) gave the concept of adaptive algorithms that use selection, mutation, and crossover, while De Jong (De Jong 1975) suggested that GA performed well on noisy as well as discontinuous datasets. The genetic algorithm process is divided into four sub-processes viz. generation, selection, crossover, and mutation. Genetic programming (GP) based on the GA approach, applies the same method as GA to program generation, where programs are considered as a tree of operands and operators that can be joined and mutated for optimization (Koza and Poli 2005). Genetic algorithms have many applications in research and real-world problems (Table 1).

2.1.3 Evolution Strategies

Evolution Strategies (ES), derived from "evolutionary algorithms", work based on GA and GP, and hence utilize and applied recombination, mutation, and selection on individuals' populations. In comparison to genetic algorithms, ES evolves both parameters and characteristics, and also each individual in ES is considered a single point in exploring in search space (Beyer and Schwefel 2002). This method came into existence at the Berlin's Technical University in the 1960s initially by Schwefel and Rechenberg in their research (Schwefel 1977). ES has a number of applications in various fields of science and engineering (Table 1).

2.1.4 Other Evolutionary-Based Nature-Inspired Algorithms

Instead of ANN, GA, GP, and ES, many other evolutionary-based nature-inspired algorithms such as Differential Evolution (DE) that work on the same principle of genetic algorithms and evolutionary strategies, a population-based meta-heuristic search algorithm which generally optimizes a problem by improving solution candidate iteratively on the basis of the evolutionary method. There are several DE variants such as the adaptive DE with optional external archive (JADE), standard DE, self-adaptive DE (JDE and SADE), and composite DE (CODE) that help in solving optimization problems. Another is Paddy Field Algorithm (PFA), which takes inspiration from rice spreading into paddy fields, which is a random and fast algorithm that can search problems globally (Kong et al. 2012).

2.2 Ecology-Based Nature-Inspired Algorithms

2.2.1 Invasive Weed Optimization

Invasive Weed Optimization (IWO) works on the principle of the behavior of spreading weeds (problem solutions) across an area or terrain, and hence the offspring die when the seeds of weeds fall on a particular area (Velmurugan et al. 2016). Simply, in the IWO algorithm, each weed is considered as the problem solution, the weeds population is termed as the set of all weeds, and the algorithm proceeds iteratively with several steps. Initially, the weeds population has been initialized having random numbers which are distributed normally with a mean of zero within the search space. Next, the seeds will be produced more when the fitness of weeds is higher, and this is governed by an equation having fitness of current weed, fitness in current weed population (maximum as well as a minimum), and a number of weeds produced (maximum as well as a minimum). Seeds that are generated will normally be distributed near to parent weed on the remaining search space, and further ranked and those having weaker performance will be eliminated from the population. IWO has many applications in research and real-world problems (Table 1).

2.2.2 Biogeography-Based Optimization

Biogeography-Based Optimization (BBO) works on the principle of living things' distribution and their spreading within a geographical area. In this, there is the migration of living things between different geographical areas leads to the generation of new species for filling habitats and niches where unexploited resources are present, and also some of the species will be extinct due to competition if some other species came into that habitat. A large number of species live in optimal habitats (Simon 2008). Biogeography-based optimization starts with population solutions initialization, which is possible and allowed in problem space. Also, at this stage, many hyper-parameters are defined (Simon 2008). In this algorithm the problem solution that has been optimized is represented by "habitat", and hence there is the initialization of a large number of these habitats. The fitness, i.e., index of habitat suitability is calculated for each habitat, and also from these species, emigration and immigration rates are calculated, which are further utilized for probabilistically modification of each habitat. This cycle is repeated to calculate the fitness of each habitat (Simon 2008). BBO has many applications in research and real-world problems (Table 1).

2.2.3 Further Ecology-Based Nature-Inspired Algorithms

Instead of IWO as well as BBO, one majorly used ecology-based nature-inspired algorithm is also there, i.e., multi-species coevolution optimization (PS^2O) that works on multi-species coevolution represented by problem space within an environment.

2.3 Multi-Objective Nature-Inspired Algorithms

2.3.1 Non-Dominated Sorting Genetic Algorithm II

Non-dominated sorting genetic algorithm II (NGSA-II), type of multi-objective optimization algorithm, and related with possible solutions, population generated approximated Pareto front (Coello and Zacatenco 2005). These solutions are evaluated against many objectives, i.e., ranging from 1 to "*n*" that are marked for minimization or maximization. Through the iterative method, there is the generation of approximate Pareto front by NSGA-II, where initially there is random creation of a population of specific size n for pre-seeding having good candidates. Further, there is a calculation of two values, for each solution p, in which one value is called domination count (i.e., solutions number that dominate p) and the second value is solutions set which dominates p. Every solution present in the first non-dominated rank zero count domination, and for each solution each dominated set member was visited, and hence dominating count decreased by one. When the set dominating count is decreased to zero, then the set is added to the second non-dominated rank, and the process is repeated until rank is assigned to each solution (Table 1). Further for each solution fitness value is assigned which is equal to its rank. Many algorithms such as MOGA-ANN (Behzadian et al. 2009), LEMMO-ANN (Sayers et al. 2019), and LEMMO (Jourdan et al. 2004) worked on the principle of the NSGA-II algorithm.

2.3.2 Population-Based ACO

The ACO, already known as a single-objective nature-inspired algorithm, is now modified for supporting a Pareto-optimal and multiple objectives approach (Dorigo and Stützle 2019). Application of ACO for multi-objective optimization problems called P-ACO, which is built based on the process of ACO (Jing et al. 2018). It uses pheromones matrices, a "lifespan" concept, random weights, and a pheromone decoding scheme. The ACO process is divided into two stages, i.e., evaluation and construction, which are managed by data structures namely solution archive and pheromone matrix. In starting, there is nothing in the solution archive, which is initialized to minimum levels. To the solution archive, the best solution is added after each iteration, and the pheromone matrix will get updated for the new solution (Dorigo and Stützle 2019). After a large number of iterations, various possible solutions will be generated, and each novel solution generally replaces the previously existing solution (Table 1) (Jing et al. 2018).

2.3.3 Further Multi-Objective Nature-Inspired Algorithms

Instead of NGSA-II as well as P-ACO various other multi-objective nature-inspired algorithms are be utilized such as Strength-Pareto Evolutionary Algorithm 2, SPEA2 (use NSGA-II and evolutionary process and the fitness is calculated by summing of objectives (fitness) and the number of dominated solutions (strength)); Vector-Evaluated Genetic Algorithm, VEGA (use fitness function that returns a vector, splits the population and permits various populations to optimize) and Pareto Archived Evolution Strategy, PAES (based on evolutionary strategies, and originally based on the $(1 + 1)$ strategy, all further versions developed work on basis of $(\mu + 1)$ and $(\mu + \lambda) - \text{ES})$.

2.4 *Swarm Intelligence-Based Nature-Inspired Algorithms*

2.4.1 Ant Colony Optimization

Ant Colony Optimization (ACO) algorithm works on the principle that workers will be sent out by ant colony for randomly hunting in search of food, who generally left behind pheromone trails. If worker ants found food, they further backward to nest and their pheromone trail has been reinforced. When two paths were there to reach food, all ants randomly took the path at the starting point, but further, as time goes, a shorter path has reinforced. Similarly, other ants are also attracted to take paths of stronger pheromone trails, and by this approach, large ants colony found food, and get huge workers for efficient paths (Dorigo and Stützle 2019). ACO approaches were initially developed in the early 1990s into progressed and meta-heuristic algorithms. ACO algorithm usually works by several artificial ants that will create a path at each iteration via space problem, consisting of nodes (i.e., solutions) as well many edges lying in between the nodes that are pointing toward the objective. Once artificial ants visited nodes, they were then not revisited, and each step was selected through a stochastic mechanism biased by a variable "pheromone" linked with an individual edge that can be read by artificial ants. One of the common examples of ACO algorithms is the Best–Worst Ant System (BWAS), where there are considerations of both best and worst-performing solutions to reinforce edges present in better solutions, and edges that were penalized present in poor solutions. Also, this method includes a restart mechanism for any optimization process which got stuck, as well as a mutation system that, has many solutions generated with the addition of randomness. ACO can be potentially utilized along with other search algorithms, in which the hyper-parameters can be manageable with the iteration of algorithms. ACO is considered a meta-heuristic ACO applied in many important fields such as traveling salesman problems (Dorigo and Stützle 2019), engineering problems (Sayers et al. 2014), and machine learning related optimization, vehicle routing problems, and bioinformatics problems. There is a problem with using ACO, just like genetic algorithms, it is effective if enough time and computing power are there for problem-solving (Bhardwaj et al. 2022).

2.4.2 Artificial Bee Colony

Artificial Bee Colony (ABC) algorithm works on the principle of bee's behavior in a hive communicating with each other and searching for the food source. Here, artificial bees are divided into three categories such as onlookers, scouts, and employed. For optimization problems, potential solutions are "sources of food" having solution fitness as "nectar quantity". The new sources of food have been identified by scouts. Employed bees generally help in evaluating the quality of existing sources of food, whereas, onlookers have the responsibility to get the information on the quality of food source that has been evaluated and checked by employed bees. Finally, onlookers evaluate the source of food in comparison to neighbor ones for discarding solutions and replacing them with scouted solutions (Yang 2010). The working of the algorithm initially starts with random identification of the source of food (solution) for the individual employed bee. Then this source of food has been evaluated and modified by each employed bee. If the new source of food is more potent than the previous one, then they remember the new one and forget the previous food source. The employed bee then gives this information to onlooker bees, who further select the source of food based on fitness informed by every employed bee and modify the food source. Better the modification and replacement of original food sources take place, and also replacing of abandoned sources with novel food sources is accomplished by scouts (Yang 2010). ABC algorithm has been used in several applications such as in comparison of the performance of artificial test cases to other state-of-the-art algorithms (Pham and Castellani 2015), and in-network communications optimization (Ning et al. 2016) and for coloring of image segmentation (Sağ and Çunkaş 2015). Instead of many applications, there are limitations in using the ABC algorithm due to the lack of realistic test cases as well as computational time is the limiting factor in utilizing these applications (Sağ and Çunkaş 2015).

2.4.3 Fish Swarm Algorithm

Fish Swarm Algorithm (FSA) is a type of swarm intelligence algorithm related to artificial fishes that will swarm collectively toward sources of food, i.e., each fish individually finds the areas where food is abundant and move to that area. At this time in searching for food, the fish don't move to the overcrowded areas due to high competition for the food source. And also, they follow a grouping system that is not overcrowded to search for areas rich in food (Li 2002). The FSA algorithm initiates with a random distribution of artificial fish throughout the problem space, following particular behaviors like "following", "preying", "swarming", and "moving". There is an iteration of behaviors of the fish initiated with "following" then "moving" each time until an improved position is not found. If it is found the improvement in position, produced by any behavior will be taken, and if there is no improvement seen on any position after several iterations, then there will be the execution of "movement" behavior (Yu and Li 2014). For the best availability of food (i.e., objective function evaluation), fish check its vision range for other fish in "following" behavior, and hence, when it finds particular fish and if it has a large amount of food as well as some other fish near to it than the current fish, then these current fish reach toward that fish. Otherwise, "searching" behavior is used by the current fish (Yu and Li 2014).

The "preying" behavior is the identification of center position by fish of nearby fish present in its appropriate range. And if that point has been found and there is larger availability of food than the current fish as well as a point was not overcrowded, then fish will reach that particular spot. In "searching" behavior, the new state that is within the range of vision is selected and evaluated by the fish, and hence if the state is better the fish move to that state. It will randomly move if a limited number of evaluations have been reached. In "movement" behavior, there is a selection of random directions and movement in that particular direction by that fish. Artificial FSA has been used in many areas such as in the detection of structural damage and identification of damage severity; applied in redundancy and reliability problems that showed its performance in terms of accuracy as well as computational efficiency (He et al. 2015); and in optimization of industrial standard (Neshat et al. 2012). Various issues have been persisting in using the FSA algorithm due to the computational time and effort needed for the production of high-quality solutions (Neshat et al. 2012).

2.4.4 Glow-Worm Swarm Optimization

Glow-worm swarm optimization (GSO) is a type of insect-based optimization method inspired by lightning bugs (glow-worms) glow and their biological behavior like mates attracting prey. Glow-worms utilize their bioluminescence light for various tasks, and hence each glow-worm has luciferin (a fluorescent) (Kalaiselvi et al. 2017). Glow-worms update chemical intensification, i.e., the emission of luciferin (similarly to pheromones generated by ants), which helps glow-worms to brighten at various intensities. At the time of glowing all the glow-worms communicate with each other and during reproduction, they attract prey. The larger the luciferin quantity, the brighter the glow-worm looks, and furthermore, mates are attracted (Chakraborty and Kar 2017).

2.4.5 Firefly Algorithm

Firefly algorithm (FA), a type of insect-based meta-heuristic technique developed by Yang, that includes adaptive firefly algorithm, chaotic firefly algorithm, multi-objective firefly algorithm, memetic algorithm based on firefly algorithm, hybrid firefly algorithm, Lagrangian firefly algorithm and discrete firefly algorithm (Yang 2010; Yang and He 2013). This algorithm is mainly related to fireflies' mutual attraction, courtship behavior, and flashing. Fireflies in algorithm act as characteristics such as swarm intelligence self-organization and decision-making capability, and the process starts with the flashing behavior of fireflies helping to attract mating partners. Brightness, i.e., illumination as well as attraction are the most important parameters present in fireflies. The fireflies flashing intensity helps in the attraction of more fireflies, but an algorithm, there is mutual attraction of fireflies to each other due to the unisex characteristics of fireflies. Brightness and attraction are directly proportional to each other, i.e., less bright fireflies will be attracted to brighter ones (Kaushal et al. 2021). But there is random if there is the absence of a brighter firefly. FA has been used in many applications like load dispatch problems, solving NP-hard problems, stock market price forecasting, energy conversation, and image compression (Sasikala et al. 2020; Yang and He 2013).

2.4.6 Ant Lion Optimizer

Ant Lion Optimizer (ALO), a type of insect-based NIC technique designed by Mirjalili, inspired by biological as well as insects' social behavior (Mirjalili 2015). Ant lion's life cycle majorly has a larvae phase, and larvae are known as doodlebugs while adult antlions are dragonflies found in minimum availability of space. The antlion larvae generally make pits of different shapes and sizes by digging, and if the prey sits on the trap, the antlion wisely grasps the prey by adding some sand to the pit's bottom (Mani et al. 2018). Traps building, ants' random walking, re-building of traps, and grasping for food are major points of ants hunting.

2.4.7 Other Swarm Intelligence-Based Nature-Inspired Algorithms

Instead of ACO, ABC, ALO, FA, GSO, and FSA other swarm intelligence algorithms are also there such as Flocking and schooling in birds (inspired by birds' flocking behavior, can search for food, nesting areas or mates); Bacterial Foraging Optimization Algorithm, BFOA (inspired by bacterial foraging behaviors); Intelligent Water Drops algorithm, IWD (inspired movement of a pattern of movement of water in rivers, seas and lakes); Firefly Algorithm, FA (works on the principle of swarming behavior of fireflies); Artificial Immune System Algorithm, AISA (works on the response of immune systems to received threats to the system); Group Search Optimizer, GSO (inspired by searching behavior of animals); Shuffled Frog Leaping Algorithm, SFLA (inspired by transfer of culture (i.e., frogs) in a population) and Particle Swarm Optimization, PSO (based on genetic algorithm approach, swarms are moved).

3 NIC Techniques (Insect-Based) to Diagnose Human Disease

In the previous two decades, a huge number of NIC techniques come into existence for solving different optimization problems. In Fig. 4, some NIC insect-based approaches used for different disease diagnoses are mentioned. Insect-based optimization techniques, types of NIC methods, taking inspiration by the behavior of various types of insects such as wasps, ants, fireflies, bees, and ant lions. These algorithms are widely utilized in solving many real-life problems like disease diagnosis, clustering feature selection (Karaboga and Ozturk 2011); sentiment analysis (Kumar and Khorwal 2017), and query optimization (Sharma et al. 2017).

4 Role of ABC, ACO, GSO, ALO, and FA in the Diagnosis of Cancers

Different survey research papers have been published for the diagnosis of disease. Al-Absi et al. gave importance to soft computing approaches in the field of medical sciences, and in their study, the authors mainly focused on ANN, GA, and fuzzy logic approaches. They observed that linking GA with fuzzy logic and ANN can produce better results in medical sciences (Al-Absi et al. 2011). Kharya et al. (2012) have used many data mining methods like support vector machine (SVM), decision tree, and naive Bayes classifier as well as logistic regression for early detection as well as diagnosis of cancer (especially breast cancer), and authors used UCI machine learning and SEER datasets and observed, decision tree serves as the best approach for cancer prediction having an accuracy of 93.62% (Kharya 2012). For detecting WBCs in leukemia by using a microscopic image, researchers used the ACO technique, and the author found that no studies have been found which utilized the K-means clustering method to detect leukemia, and hence authors gave K-means cluster algorithm for WBCs identification and accurate prediction (Bukhari et al. 2021; Khan et al. 2021a; Junoha et al. 2017). Similarly, a review of studies on the diagnosis of fatal disorders with the help of NIC algorithms like ACO, PSO, ABC,

Nature-inspired computing techniques (insect-based) that diagnose different diseases	
Algorithm	**Disease diagnosed**
Wasp swarm optimization (WSO)	Heart diseases, cancer
Ant colony optimization (ACO)	Cancer, diabetes
Bee system (BS)	Cancer, thyroid
Honey bee foraging algorithm (HBA)	Heart disorders, cancer, diabetes
Bees' algorithm (BA)	Cancer, heart diseases, thyroid
Artificial bee colony (ABC)	Diabetes, heart diseases, cancer
Bee colony optimization (BCO)	Cancer, thyroid, heart disorders
Virtual bee algorithm (VBA)	Heart disorders
Glow-worm swarm optimization (GSO)	Hypertension, diabetes cerebrovascular, disease
Termite algorithm (TA)	Dengue, heart diseases, gallbladder diseases
Honey bees mating optimization (HBMO)	Heart disorder, cerebrovascular disease, multiple sclerosis, cancer
Fast marriage in honey bee's optimization (FMHBO)	Heart disorders, cancer
Firefly algorithm (FA)	Cancer, diabetes, heart diseases
Bee collecting pollen algorithm (BCPA)	Dengue, cancer
Roach infestation optimization (RIO)	Liver disorders, motor neuron disease, heart diseases
Bumblebees mating optimization (BBMO)	Hepatitis, heart diseases
Bees swarm optimization (BSO)	Heart diseases, cancer, Parkinson's disease
Fruit fly optimization algorithm (FFOA)	Diabetes, thyroid, Parkinson's disease
Bees' life algorithm (BLA)	Heart disorders
Dragonfly algorithm (DA)	Alzheimer, cancer, gallbladder and retinal disease
Moth-flame optimization (MFO)	Heart disorders, Parkinson's disease, Alzheimer
Ant lion optimizer (ALO)	Cancer, epilepsy, autism, depression, heart disorders, Alzheimer
Grasshopper optimization algorithm (GOA)	Epilepsy, heart disease, cancer

Fig. 4 NIC (insect-based) techniques for the diagnosis of different diseases. Modified from Gautam et al. 2019

and GA observed that these algorithms could be effectively used for analyzing various types of cancers such as breast, lung, and prostate cancer (Kaur and Sharma 2017).

Five NIC algorithms such as GSO, ACO, FA, ABC, and ALO are mostly used in the early detection and diagnosis of different types of diseases like cancer, as well as their working is shown (Figs. 5, 6, and 7). ACO is a meta-heuristic probabilistic algorithm emphasized by the foraging behavior of ants that solves many optimization problems. It generally took an idea from real ants and finds the shortest path who move to food resources from their nest. Also, ants will be communicated by pheromone (chemical substance). ABC is a self-growing and flexible algorithm that depends on bees foraging behavior and hence the problem is solved by object searching followed by an optimal path. GSO is a type of insect-based optimization method inspired by lightning bugs (glow-worms) glow and their biological behavior like mates attracting prey. FA is a type of insect-based meta-heuristic technique mainly related to fireflies' mutual attraction, courtship behavior, and flashing. ALO is also a type of insect-based NIC technique inspired by biological as well as insects' social behavior. FSA, an algorithm taking inspiration from fish's behavior, has a high fault tolerance, flexibility, and convergence rate. This works very similar to GA, but it does not have mutation and crossing over parameters. It has been identified from the last 6 years pf research that accuracy gained by using ABC, PSO, and ACO in early cancer diagnosis lies in the range of 76.4–99.98%, 80–98%, and 70–99.2%, respectively. Also, ABC, PSO as well as ACO were observed as best utilized for diagnosing breast, prostate, and lung cancer, respectively. Chiang et al., in their study, developed a hybrid model that combines ACO, SVM, and MLP to detect cancer patient's datasets mainly for lung and prostate cancer. The classification rate achieved by using this hybrid model was 97% (Chiang et al. 2008). Sivakumar and Karnan, in their research, utilized an extended ABC optimization method to predict earlier signs and symptoms of breast cancer. They used right and left mammographic images of breast (bilateral subtraction) taken from the MIAS (Mammography Image Analysis Society) database. Authors achieved a predictive accuracy rate of 96.4% with help of this method (Sivakumar and Karnan 2012). Similarly, Shah et al., developed another hybrid classifier to detect breast cancer. In this study, the authors did a comparison between ABC and Global guided ABC (GGABC) performance to diagnose breast cancer risk in females. Here, the authors also did a comparative study of ABC, guided ABC, and global ABC, and they found that the accuracy rate was 99.98% by using GGABC which was better than ABC (Shah et al. 2019).

Moosa et al., have developed another cancer classifier known as modified ABC (mABC) to identify the genes used for early cancer diagnosis by using more than ten different datasets. It was interestingly found, mABC is a better classifier for cancer diagnostic compared with ACO, ABC, and GA (Moosa et al. 2016). Nazarian et al. used the ABC method for predicting and classifying breast cancer, and it has been found that the accuracy rate was higher in using the ABC algorithm when compared to NN, SVM naive Bayes, and C4.5 (Nazarian et al. 2013). Alshamlan et al., in their study, predict colon cancer using the ABC algorithm, and the accuracy rate was 76.8% (Alshamlan et al. 2015). Kumari (2017), developed a hybrid method using ANN as well as ABC to diagnose breast cancer and achieved an accuracy of 95.5% (Kumari 2017). Parveen and Kavitha, in their study, designed FSFCM (a hybrid model) by combining FA as well as fuzzy C-means (FCM) to diagnose lung cancer. In the model, FA was utilized to search the initial cluster center, and FCM was used for improving the final results. The FSFCM performance has been compared to FCM and K-means performance for detecting a similar type of problem. The authors utilized normal, malignant, and benign images of lung cancer and accuracy were calculated based on the solidity, perimeter, and area of each image. The

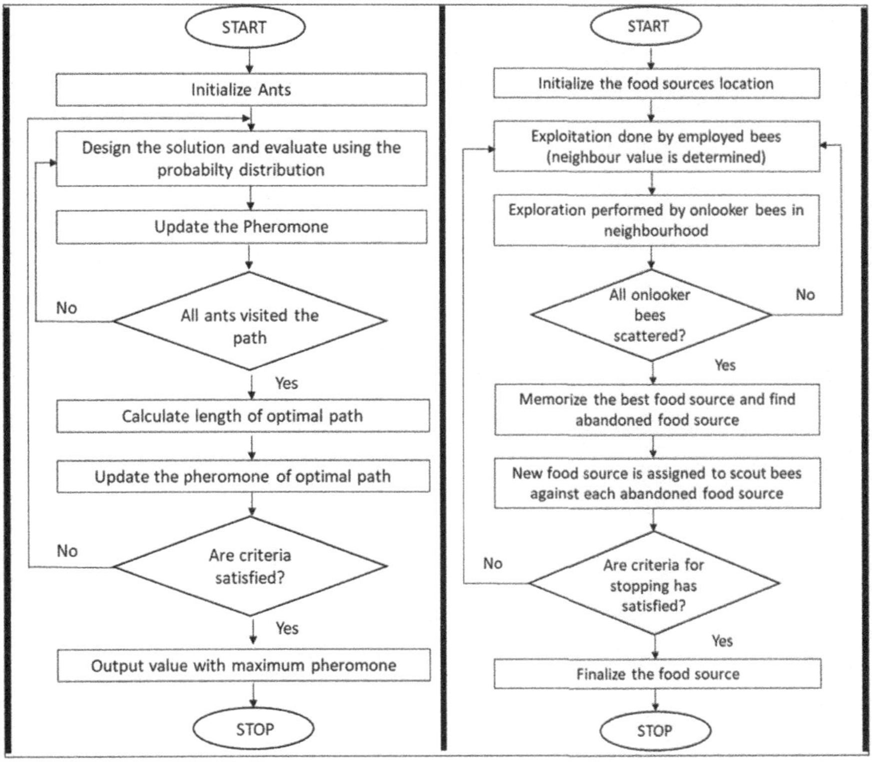

Fig. 5 General workflow of ACO (left) and ABC (right)

hybrid model achieved an accuracy of 97.56% which was better than FCM (90.27%) and K-means (79.5%) (Parveen and Kavitha 2015). Similarly, Kalavathi and Dhavapandiammal, in their research developed another model known as FA-FCM that combines two clustering methods, i.e., Fuzzy C-means and FA to detect lung cancer. Images segmentation and selection of pixels are performed by using FA-FCM, and the analysis is performed on 50 images of lung cancer by using the Vision and Image Analysis group/International Early Lung Cancer Action Program (VIA/I-ELCAP) (Kalavathi and Dhavapandiammal 2016).

Zamani and Nadimi-Shahrak, in their study, diagnose breast cancer by using and checking the performance of four various techniques of swarm intelligence like FA, PSO, invasive weed optimization (IWO) as well as imperialist competitive algorithm (ICA) by taking a dataset of Wisconsin breast cancer. Authors found that FA produced better results in comparison to PSO (0.73%), IWO (3.65%), and ICA (2.19%) (Zamani and Nadimi-Shahraki 2016). Singh and Kumar gave their algorithm namely, the error-guided ABC (E-ABC) algorithm (ABC variant) for ovarian cancer diagnosis. In their study, the authors did a comparison of E-ABC performance, i.e., preciseness and correctness, in ovarian cancer diagnosis with several well-designed NIC techniques. In comparison to other methods, E-ABC achieved the highest MSE

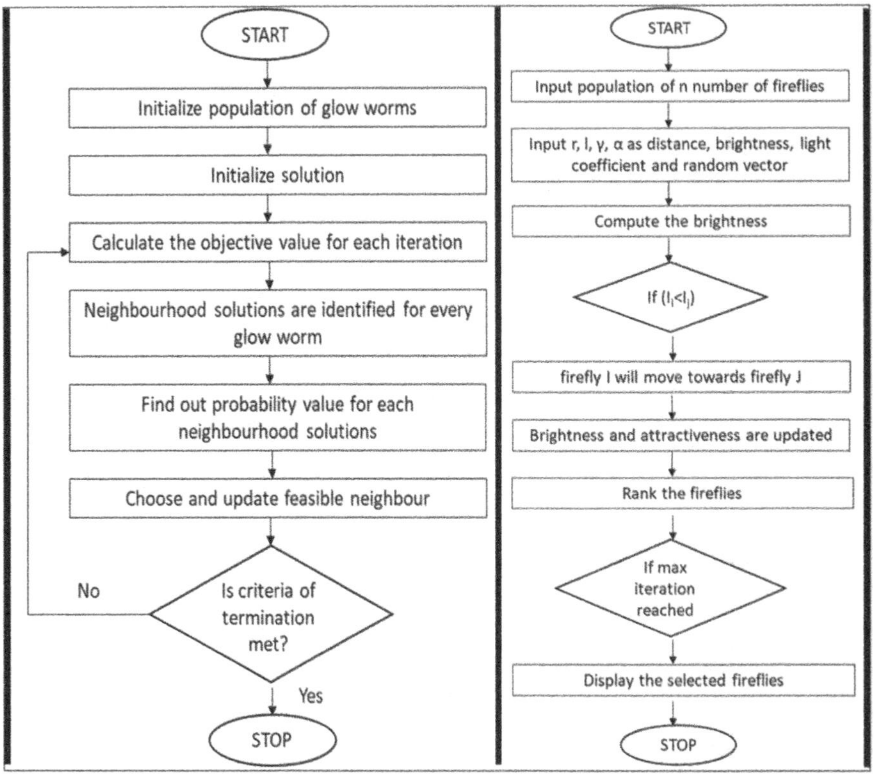

Fig. 6 General workflow of GSO (left) and FA (right)

(0.0947), and also the convergence rate of E-ABC was better than other methods. The classification rate to diagnose benign and malignant tumors was achieved as 97.4% and 99.6%, respectively (Singh and Kumar 2017). Some used combined features of FA and decision bagged tree to evaluate breast cancer risk, and the results of using this combination techniques are more promising in breast cancer classification (Hariprasath et al. 2021).

Banu and colleagues in their research used a combination of two clustering techniques, i.e., fuzzy and FA to detect cancers in human beings. This method used five different datasets and identify highly expressed and suppressed genes. Fuzzy firefly method outperforms FA, PSO as well as fuzzy PSO (Banu et al. 2017). Umamaheswari and Sumathi gave a combined solution (FA-ANNIS) that uses two-hybrid methods such as adaptive neuro-neutrosophic inference system (ANNIS) and enhanced firefly algorithm (EFA) that diagnose circulating tumor cells (CTCs) for breast cancer analysis. In this method, EFA helps in finding perceptive qualities that are used in the early detection of signs and symptoms of breast cancer. The predictive accuracy rate is 92% achieved in very less time (0.72 s) (Umamaheswari and Sumathi 2019).

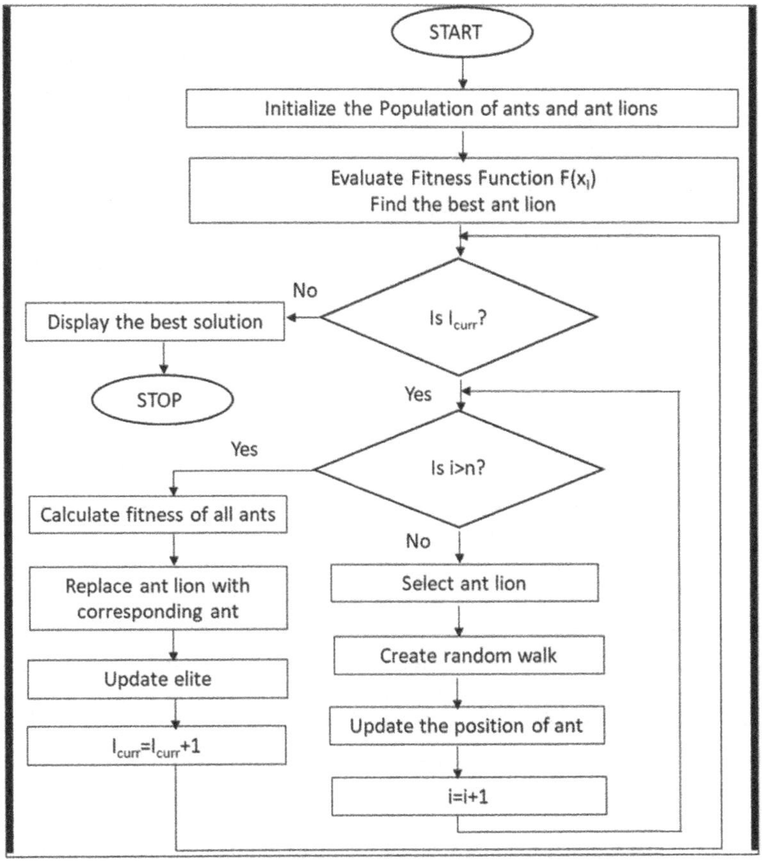

Fig. 7 General workflow of ALO

Deoskar et al., in their study, used ACO for diagnosing lung cancer, as they took UCI datasets of lung cancer to perform their experiment. Each symptom of lung cancer served as an ant and probably gave a feasible partial solution. With this optimization algorithm, the authors achieved an accuracy of 84% (Deoskar et al. 2013). Shukla and Motwani, in their study, used ACO along with the association rule for predicting cancer patients (Shukla and Motwani 2014). Patankar et al., used ACO based edge detection method to diagnose breast cancer, which determines the object as well as their connected boundaries (Patankar et al. 2014). Similarly, Christopher and Jamera, in their study also utilized the ACO algorithm for the early-stage detection of patients having lung cancer (Christopher and Jamera 2015). Rashmi (2017), in her study, used SVM and ACO to develop a hybrid method for diagnosing many diseases such as cancer, heart disease, and diabetes. This hybrid technique used MATLAB and diagnose breast cancer with an accuracy rate of 96.5% (Rashmi 2017). PSO, mostly used algorithms for diagnosing the various types of cancers, and hence Mohanty et al., for breast cancer diagnosis developed discrete PSO, and accuracy was found to be 97% (Mohanty et al. 2011). Vimaladevi and Kalaavathi, in their study, compared a hybrid of Back Propagation Neural Network (BPNN) and PSO as well as GA-BPNN for diagnosing three cancerous types, and the study showed that accuracy was more than 80% with GA, NN, and PSO hybrid method but it was below 80% in GA-BPNN in each type of cancer (Vimaladevi and Kalaavathi 2014). Shahbeig et al. developed and used a hybrid (fuzzy logic and PSO) method for breast cancer diagnosis in female patients. They found that the sensitivity, specificity, and accuracy of using this approach were 90.55%, 93.33%, and 91.88%, respectively (Shahbeig et al. 2017). PSO was also found to achieve the highest predictive accuracy in prostate cancer diagnosis.

5 Limitations

NIC algorithms have many applications that solve problems in the field of computational biology, engineering, logistics, and telecommunications. Wang and Beni (1989) proposed an SI algorithm for the development of cellular robotic systems by using many basic features applied in several applications (Krishnaveni et al. 2019; Rai 2022). The major challenges of NIC techniques are present in computational fluid dynamics, clustering of the dataset, feature selection, detection of hand gestures, machine learning, image processing, path planning, neural networks training, etc. (Darwish 2018; Chakraborty and Kar 2017). These techniques mainly focus on image analysis, and there are many problems in handling images of different diseases, and hence to achieve better results the image should be taken in different stages and from different tissues.

It was observed that a large number of hybrid methods of NIC algorithms (insect-based) were used in the diagnosis of cancer. Yet, several hybridization approaches still needed to be explored to detect and diagnose different cancerous types. Although many studies showed that ACO and PSO are the most used algorithms, there is a need to explore the performance of hybrid as well as individual approaches for cancer diagnosis (Ehteram et al. 2021). Swarm intelligence-based approaches are very helpful in pattern detection, and genes and protein function identification, yet several issues should be addressed in the future such as updating mechanisms and identifying appropriate value ranges by the researchers to explore more NIC algorithms. Also, the algorithms like ACO and PSO have the highest degree of randomness as well different results are generated with different runs of the same program, therefore, it is essential to add problem-specific domain information in tools and software of SI which helps in the reduction of computational time and randomness. There is a problem with using ACO, just like genetic algorithms, it is effective if enough time and computing power are there for problem-solving (Dorigo and Stützle 2019).

Similarly, the ABC algorithm also have limitation is using due to the lack of real-world test cases as well as computational time is the limiting factor in utilizing these applications (Oyelade and Ezugwu 2022; Ning et al. 2016). Another issue is that not all the algorithms act similarly to different types of issues, for example, based on the nature of the problem, a suitable algorithm will be chosen for better results. Although fuzzy c-means and K-means are the most popularly used clustering methods in many swarm intelligence algorithms applications and have hence been used for creating hybrid models and performance comparison of swarm-based approaches, more research has been done for better classification and accuracy rate (Figueiredo et al. 2019).

6 Future Directions

In previous decades, NIC techniques were evolved for screening and diagnosis of diseases, including cancers, and hence predict diseases at the earliest before they spread and caused serious illness. Further, the presence of a huge amount of cancer-related datasets made these techniques more popular in screening, classification, diagnosis, and prognosis of cancer. There are many applications of various NIC algorithms to date, yet more research is needed in the future.

- It has been observed that a large number of hybrid methods of insect-based NIC algorithms were used in the diagnosis of cancer. For example, it has been discussed that a hybrid model of neural network and guided ABC more precisely detect breast cancer. Yet, several hybridization approaches still needed to be explored to detect and diagnose different cancerous types.
- Similarly, for the detection of leukemia, the combination of SVM and discrete FA and their optimization will be used for classifying data (Siddique and Adeli 2015).
- Although many studies showed that ACO and PSO are the most used algorithms, there is a need to explore the performance of hybrid as well as individual approaches for cancer diagnosis.
- NIC methods will also be useful for researchers in the future who want variants of swarm intelligence-based algorithms to study more human disorders.
- Instead of using these algorithms in diagnosing cancer, NIC methods can also be used for solving advanced problems in the field of feature selection, agriculture, query optimization, social networks, astronomy, finance, bioinformatics, etc.
- NIC techniques can be used in the discovery of new biomarkers based on personalized treatment (also called precision medicine) in lung cancer, and also integrate diagnostic imaging datasets with new biomarkers.
- NIC algorithms along with AI and ML can help physicians to know the future events at advanced levels and can draw conclusions using a large number of clinical datasets.
- In later years, an extensive analysis can be done which helps in the role identification of different NIC techniques to diagnose various human disorders. However, the type of dataset effect, sampling method, pre- and post-processing as well as

NIC accuracies normalization in prediction and diagnosis of the disease can be determined.

7 Conclusion

NIC techniques are computational techniques that are generally inspired by the natural behavior of organisms that solve real-world problems. In the last three decades, many NIC techniques were widely used in various reliability research such as in diagnosing human diseases. For the last ten years, these algorithms have been majorly used in query optimization problems, diagnosis of disease, mining of stock data, and query sentiment analysis. This chapter presented the classification of various NIC intelligence algorithms focusing on swarm intelligence-based nature-inspired algorithms and their applications in detecting and diagnosing different types of cancers. Swarm intelligence-based algorithms, motivated by the collection of social insect behavior such as termites, bees, wasps, etc. as well as bird flocking, ant colonies, bacterial development, fish schooling, and animal herding to solve complex bioinformatics-related problems around the world. Further, the role of five insect-based NIC algorithms was discussed in the diagnosis of different types of cancer (breast, prostate, lung, and ovarian). These methods will also be useful for the researchers in the future who want variants of swarm intelligence-based algorithms to study more human disorders and solve advanced problems.

References

Al-Absi HR, Abdullah, A, Hassan MI, Bashir Shaban K (2011). Hybrid intelligent system for disease diagnosis based on artificial neural networks, fuzzy logic, and genetic algorithms. In: International conference on informatics engineering and information science. Springer, Berlin, Heidelberg, pp 128–139

Alshamlan H, Badr G, Alohali Y (2015) mRMR-ABC: a hybrid gene selection algorithm for cancer classification using microarray gene expression profiling. Biomed Res Int

Arora M, Dhawan S, Singh K (2020) Data-driven prognosis of cervical cancer using class balancing and machine learning techniques. EAI Endorsed Trans Energy Web 7(30):e2

Arora S, Singh H, Sharma M, Sharma S, Anand P (2019) A new hybrid algorithm based on grey wolf optimization and crow search algorithm for unconstrained function optimization and feature selection. IEEE Access 7:26343–26361

Arulkumaran K, Cully A, Togelius J (2019) Alphastar: an evolutionary computation perspective. In: Proceedings of the genetic and evolutionary computation conference companion, pp 314–315

Aziz RM (2022) Application of nature inspired soft computing techniques for gene selection: a novel frame work for classification of cancer

Bäck T, Hoffmeister F, Schwefel HP (1991) A survey of evolution strategies. In: Proceedings of the fourth international conference on genetic algorithms

Banu PN, Azar AT, Inbarani HH (2017) Fuzzy firefly clustering for tumour and cancer analysis. Int J Model Ident Control 27(2):92–103

Barisal AK, Prusty RC (2015) Large scale economic dispatch of power systems using oppositional invasive weed optimization. Appl Soft Comput 29:122–137

Behzadian K, Kapelan Z, Savic D, Ardeshir A (2009) Stochastic sampling design using a multi-objective genetic algorithm and adaptive neural networks. Environ Model Softw 24(4):530–541
Beyer HG, Schwefel HP (2002) Evolution strategies—A comprehensive introduction. Nat Comput 1(1):3–52
Bhardwaj T, Mittal R, Upadhyay H, Lagos L (2022) Applications of swarm intelligent and deep learning algorithms for image-based cancer recognition. In: Artificial intelligence in healthcare. Springer, Singapore, pp 133–150
Binitha S, Sathya SS (2012) A survey of bio inspired optimization algorithms. Int J Soft Comput Eng 2(2):137–151
Bukhari SNH, Jain A, Haq E, Mehbodniya A, Webber J (2021) Ensemble machine learning model to predict SARS-CoV-2 T-cell epitopes as potential vaccine targets. Diagnostics 11(11):1990
Carbas S, Toktas A, Ustun D (2021) Nature-inspired metaheuristic algorithms for engineering optimization applications. Springer, Singapore
Chakraborty A, Kar AK (2017) Swarm intelligence: a review of algorithms. In: Nature-inspired computing and optimization, pp 475–494
Chiang YM, Chiang HM, Lin SY (2008) The application of ant colony optimization for gene selection in microarray-based cancer classification. In: 2008 international conference on machine learning and cybernetics, vol 7. IEEE, pp 4001–4006
Christopher T, Jamera BJ (2015) A study on mining lung cancer data for increasing or decreasing disease prediction value by using ant colony optimization techniques. In: Proceedings of the UGC sponsored national conference on advanced networking and applications
Coello CAC, Zacatenco CSP (2005) Twenty years of evolutionary multi-objective optimization: a historical view of the field. CINVESTAV-IPN Evolutionary Computing Group
Dadaneh BZ, Markid HY, Zakerolhosseini A (2016) Unsupervised probabilistic feature selection using ant colony optimization. Expert Syst Appl 53:27–42
Darwish A (2018) Bio-inspired computing: algorithms review, deep analysis, and the scope of applications. Future Comput Inf J 3(2):231–246
Deoskar P, Singh DD, Singh DA (2013) An efficient support based ant colony optimization technique for lung cancer data. Int J Adv Res Comput Commun Eng 2(9)
De Jong KA (1975). An analysis of the behavior of a class of genetic adaptive systems. University of Michigan
Dorigo M, Stützle T (2019) Ant colony optimization: overview and recent advances. In: Handbook of metaheuristics, pp 311–351
Ehteram M, Ferdowsi A, Faramarzpour M, Al-Janabi AMS, Al-Ansari N, Bokde ND, Yaseen ZM (2021) Hybridization of artificial intelligence models with nature inspired optimization algorithms for lake water level prediction and uncertainty analysis. Alex Eng J 60(2):2193–2208
Fan X, Sayers W, Zhang S, Han Z, Ren L, Chizari H (2020) Review and classification of bio-inspired algorithms and their applications. J Bionic Eng 17(3):611–631
Figueiredo E, Macedo M, Siqueira HV, Santana CJ Jr, Gokhale A, Bastos-Filho CJ (2019) Swarm intelligence for clustering—A systematic review with new perspectives on data mining. Eng Appl Artif Intell 82:313–329
Gautam R, Kaur P, Sharma M (2019) A comprehensive review on nature inspired computing algorithms for the diagnosis of chronic disorders in human beings. Progr Artif Intell 8(4):401–424
Goudos SK, Plets D, Liu N, Martens L, Joseph W (2015) A multi-objective approach to indoor wireless heterogeneous networks planning based on biogeography-based optimization. Comput Netw 91:564–576
Gu J, Wang Z, Kuen J, Ma L, Shahroudy A, Shuai B, Liu T, Wang X, Wang G, Cai J, Chen T (2018) Recent advances in convolutional neural networks. Pattern Recogn 77:354–377
Hariprasath K, Tamilselvi S, Saravana Kumar NM, Kaviyavarshini N, Balamurugan S (2021). Performance analysis of nature-inspired algorithms in breast cancer diagnosis. In: Nature-inspired algorithms applications, 267–294
He Q, Hu X, Ren H, Zhang H (2015) A novel artificial fish swarm algorithm for solving large-scale reliability–redundancy application problem. ISA Trans 59:105–113

Holland JH (1992). Adaptation in natural and artificial systems: an introductory analysis with applications to biology, control, and artificial intelligence. MIT Press
Hornik K, Stinchcombe M, White H (1989) Multilayer feedforward networks are universal approximators. Neural Netw 2(5):359–366
Islam ML, Shatabda S, Rashid MA, Khan MG, Rahman MS (2019) Protein structure prediction from inaccurate and sparse NMR data using an enhanced genetic algorithm. Comput Biol Chem 79:6–15
Jing L, Zhuo-qun Z, Li-li Z, Kang-jie S (2018) Multi-objective ant colony optimization algorithm based on discrete variables. In: IOP conference series: earth and environmental science, vol 189, no. 4. IOP Publishing, p 042031
Jourdan L, Corne D, Savic D, Walters G (2004) Hybridising rule induction and multi-objective evolutionary search for optimising water distribution systems. In: Fourth international conference on hybrid intelligent systems (HIS'04). IEEE, pp 434–439
Junoha AK, Alshormana MA, Muhamada WZAW, Zakariaa MH, Desaa AM (2017) Multi algorithms for improving leukemia images edge detection. Int J Appl Eng Res 12(18):7402–7425
Kalaiselvi T, Nagaraja P, Basith ZA (2017) A review on glowworm swarm optimization. Int J Inf Technol (IJIT) 3(2):49–56
Kalavathi P, Dhavapandiammal A (2016) Segmentation of lung tumor in CT scan images using FA-FCM algorithms. Res Gate 18(5):74–79
Karaboga D, Ozturk C (2011) A novel clustering approach: artificial Bee Colony (ABC) algorithm. Appl Soft Comput 11(1):652–657
Kaur P, Sharma M (2017) A survey on using nature inspired computing for fatal disease diagnosis. Int J Inf Syst Model Des (IJISMD) 8(2):70–91
Kaushal C, Kaushal K, Singla A (2021) Firefly optimization-based segmentation technique to analyse medical images of breast cancer. Int J Comput Math 98(7):1293–1308
Kazem A, Sharifi E, Hussain FK, Saberi M, Hussain OK (2013) Support vector regression with chaos-based firefly algorithm for stock market price forecasting. Appl Soft Comput 13(2):947–958
Khan FN, Ahmad S, Raza K (2021a) Clinical applications of next-generation sequence analysis in acute myelogenous leukemia. In: Translational bioinformatics applications in healthcare. CRC Press, pp 41–66
Khan, F. N., Khanam, A. A., Ramlal, A., & Ahmad, S. (2021b). A review on predictive systems and data models for covid-19. In *Computational Intelligence Methods in COVID-19: Surveillance, Prevention, Prediction and Diagnosis* (pp. 123–164). Springer, Singapore.
Khan FN, Yousef M, Raza K (2022) Machine learning-based models in the diagnosis, prognosis and effective cancer therapeutics: current state-of-the-art. In: Computational intelligence in oncology. Springer, Singapore, pp 19–54
Kharya S (2012) Using data mining techniques for diagnosis and prognosis of cancer disease. arXiv: 1205.1923
Kim SS, Byeon JH, Yu H, Liu H (2014) Biogeography-based optimization for optimal job scheduling in cloud computing. Appl Math Comput 247:266–280
Klockgether J, Schwefel HP (1970) Two-phase nozzle and hollow core jet experiments. In: Engineering aspects of magnetohydrodynamics
Kong X, Chen YL, Xie W, Wu X (2012) A novel paddy field algorithm based on pattern search method. In: 2012 IEEE international conference on information and automation. IEEE, pp 686–690
Koza JR, Poli R (2005) Genetic programming. In: Search methodologies. Springer, Boston, MA, pp 127–164
Krishnaveni A, Shankar R, Duraisamy S (2019) A survey on nature inspired computing (NIC): algorithms and challenges. Glob J Comput Sci Technol: D Neural Artif Intell 19(3)
Kumar A, Khorwal R (2017) Firefly algorithm for feature selection in sentiment analysis. In: Computational intelligence in data mining. Springer, Singapore, pp 693–703

Kumari DJ (2017) Structural redesign of artificial neural network for predicting breast cancer with the aid of artificial bee colony. Ind J Sci Technol 10(15):1–8
LeCun Y, Bengio Y, Hinton G (2015) Deep learning. Nature 521(7553):436–444
Li XL (2002) An optimizing method based on autonomous animats: fish-swarm algorithm. Syst Eng-Theory Practice 22(11):32–38
Lin J (2015) A hybrid biogeography-based optimization for the fuzzy flexible job-shop scheduling problem. Knowl-Based Syst 78:59–74
Lin Y, Yang Q, Guan G (2019) Scantling optimization of FPSO internal turret area structure using RBF model and evolutionary strategy. Ocean Eng 191:106562
Lindfield G, Penny J (2017) Introduction to nature-inspired optimization. Academic Press
Liu K, Zhang J (2020) Nonlinear process modelling using echo state networks optimised by covariance matrix adaption evolutionary strategy. Comput Chem Eng 135:106730
Mani M, Bozorg-Haddad O, Chu X (2018) Ant lion optimizer (ALO) algorithm. In: Advanced optimization by nature-inspired algorithms. Springer, Singapore, pp 105–116
Mason K, Duggan M, Barrett E, Duggan J, Howley E (2018) Predicting host CPU utilization in the cloud using evolutionary neural networks. Futur Gener Comput Syst 86:162–173
Mirjalili S (2015) The ant lion optimizer. Adv Eng Softw 83:80–98
Mohanty AK, Sahoo S, Pradhan A, Lenka SK (2011) Breast cancer assessment and diagnosis using particle swarm optimization. Int J Comput Sci Technol 2(3):37–41
Moosa JM, Shakur R, Kaykobad M, Rahman MS (2016) Gene selection for cancer classification with the help of bees. BMC Med Genomics 9(2):135–165
Nanda SJ, Panda G (2014) A survey on nature inspired metaheuristic algorithms for partitional clustering. Swarm Evol Comput 16:1–18
Nazarian M, Dezfouli MA, Haronabadi A (2013) Classification of breast cancer samples through using the artificial bee colony algorithm. Int J Comput Appl Technol Res 2(5):522–525
Neshat M, Adeli A, Sepidnam G, Sargolzaei M, Toosi AN (2012) A review of artificial fish swarm optimization methods and applications. Int J Smart Sens Intell Syst 5(1)
Nicklow J, Reed P, Savic D, Dessalegne T, Harrell L, Chan-Hilton A, Karamouz M, Minsker B, Ostfeld A, Singh A, Zechman E, ASCE Task Committee on Evolutionary Computation in Environmental and Water Resources Engineering (2010) State of the art for genetic algorithms and beyond in water resources planning and management. J Water Resour Plan Manage 136(4):412–432
Ning J, Zhang C, Zhang B (2016) A novel artificial bee colony algorithm for the QoS based multicast route optimization problem. Optik 127(5):2771–2779
Niu Q, Zhang L, Li K (2014) A biogeography-based optimization algorithm with mutation strategies for model parameter estimation of solar and fuel cells. Energy Convers Manage 86:1173–1185
Oyelade ON, Ezugwu AE (2022) Characterization of abnormalities in breast cancer images using nature-inspired metaheuristic optimized convolutional neural networks model. Concurr Comput: Practice Experience 34(4):e6629
Pandey AC, Rajpoot DS, Saraswat M (2017) Twitter sentiment analysis using hybrid cuckoo search method. Inf Process Manage 53(4):764–779
Parker DB (1985) Learning logic technical report tr-47. Center of Computational Research in Economics and Management Science, Massachusetts Institute of Technology, Cambridge, MA
Parveen SS, Kavitha C (2015) Segmentation of CT lung nodules using FCM with firefly search algorithm. In: 2015 international conference on innovations in information, embedded and communication systems (ICIIECS). IEEE, pp 1–6
Patankar V, Nawgaje D, Kanphade R (2014) A implementation of ant colony optimization technique for cancer diagnosis. Int J Current Eng Technol 4:568–570
Paul PV, Moganarangan N, Kumar SS, Raju R, Vengattaraman T, Dhavachelvan P (2015) Performance analyses over population seeding techniques of the permutation-coded genetic algorithm: an empirical study based on traveling salesman problems. Appl Soft Comput 32:383–402
Pham DT, Castellani M (2015) A comparative study of the Bees Algorithm as a tool for function optimisation. Cogent Eng 2(1):1091540

Poo MM, Du JL, Ip NY, Xiong ZQ, Xu B, Tan T (2016) China brain project: basic neuroscience, brain diseases, and brain-inspired computing. Neuron 92(3):591–596
Raghupathi W, Raghupathi V (2018) An empirical study of chronic diseases in the United States: a visual analytics approach to public health. Int J Environ Res Public Health 15(3):431
Rai R (2022) Swarm intelligence and bio-inspired computation. In: Applied soft computing: techniques and applications, pp 1–22
Rashmi SS (2017) Hybrid model using unsupervised filtering based on ant colony optimization and multiclass SVM by considering medical data set. Int Res J Eng Technol 4(6):2565–2571
Sağ T, Çunkaş M (2015) Color image segmentation based on multiobjective artificial bee colony optimization. Appl Soft Comput 34:389–401
Sasikala S, Ezhilarasi M, Arun Kumar S (2020) Detection of breast cancer using fusion of MLO and CC view features through a hybrid technique based on binary firefly algorithm and optimum-path forest classifier. In: Applied nature-inspired computing: algorithms and case studies. Springer, Singapore, pp 23–40
Sayers W, Savić DRAGAN, Kapelan Z, Kellagher R (2014) Artificial intelligence techniques for flood risk management in urban environments. Procedia Eng 70:1505–1512
Sayers W, Savic D, Kapelan Z (2019) Performance of LEMMO with artificial neural networks for water systems optimisation. Urban Water J 16(1):21–32
Schwefel HP (1977) Evolutionsstrategien für die numerischeoptimierung. In: NumerischeOptimierung von Computer-Modellenmittels der Evolutionsstrategie, Birkhäuser, Basel, pp 123–176
Shah H, Chiroma H, Herawan T, Ghazali R, Tairan N (2019) An efficient bio-inspired bees colony for breast cancer prediction. In: Proceedings of the international conference on data engineering 2015 (DaEng-2015). Springer, Singapore, pp 597–608
Shahbeig S, Helfroush MS, Rahideh A (2017) A fuzzy multi-objective hybrid TLBO–PSO approach to select the associated genes with breast cancer. Signal Process 131:58–65
Sharma M, Singh G, Singh R (2017) Stark assessment of lifestyle based human disorders using data mining based learning techniques. IRBM 38(6):305–324
Sharma M, Singh G, Singh R (2019) A review of different cost-based distributed query optimizers. Progr Artif Intell 8(1):45–62
Sharma M, Singh G, Singh R (2021) Clinical decision support system query optimizer using hybrid firefly and controlled genetic algorithm. J King Saud University-Comput Inf Sci 33(7):798–809
Sheikh K, Ramlal A, Khan FN (2022) Computational resources for oncology research: A comprehensive analysis. In: Computational intelligence in oncology, pp 65–92
Shukla R, Motwani D (2014) Cancer detection using frequency pattern ant colony optimization
Siddique N, Adeli H (2015) Nature inspired computing: an overview and some future directions. Cogn Comput 7(6):706–714
Simon D (2008) Biogeography-based optimization. IEEE Trans Evol Comput 12(6):702–713
Singh A, Kumar D (2017) Novel ABC based training algorithm for ovarian cancer detection using neural network. In: 2017 International conference on trends in electronics and informatics (ICEI). IEEE, pp 594–597
Sivakumar R, Karnan M (2012) Diagnose breast cancer through mammograms using EABCO algorithm. Int J Eng Technol 4(5):302–307
Umamaheswari TS, Sumathi P (2019) Enhanced firefly algorithm (EFA) based gene selection and adaptive neuro neutrosophic inference system (ANNIS) prediction model for detection of circulating tumor cells (CTCs) in breast cancer analysis. Clust Comput 22(6):14035–14047
Velmurugan T, Khara S, Nandakumar S, Saravanan B (2016) Seamless vertical handoff using invasive weed optimization (IWO) algorithm for heterogeneous wireless networks. Ain Shams Eng J 7(1):101–111
Vimaladevi M, Kalaavathi B (2014) Cancer classification using hybrid fast particle swarm optimization with back-propagation neural network. Int J Comput Commun Technol 3(11)
Wang J, Beni G (1989) Cellular robotic system with stationary robots and its application to manufacturing lattices. In Proceedings. IEEE International Symposium on Intelligent Control, pp 132–137

Wang L, Xu Y (2011) An effective hybrid biogeography-based optimization algorithm for parameter estimation of chaotic systems. Expert Syst Appl 38(12):15103–15109

Whittington JC, Bogacz R (2019) Theories of error back-propagation in the brain. Trends Cogn Sci 23(3):235–250

Yang XS (2010) Nature-inspired metaheuristic algorithms. Luniver Press

Yang XS, He X (2013) Firefly algorithm: recent advances and applications. Int J Swarm Intell 1(1):36–50

Yu L, Li C (2014) A global artificial fish swarm algorithm for structural damage detection. Adv Struct Eng 17(3):331–346

Yusoff NIM, Alhamali DI, Ibrahim ANH, Rosyidi SAP, Hassan NA (2019) Engineering characteristics of nanosilica/polymer-modified bitumen and predicting their rheological properties using multilayer perceptron neural network model. Constr Build Mater 204:781–799

Zainal N, Zain AM, Sharif S (2015) Overview of artificial fish swarm algorithm and its applications in industrial problems. In: Applied mechanics and materials, vol 815. Trans Tech Publications Ltd., pp 253–257

Zamani H, Nadimi-Shahraki MH (2016) Swarm intelligence approach for breast cancer diagnosis. Int J Comput Appl 151(1):40–44

Zhang Y, Agarwal P, Bhatnagar V, Balochian, Yan J (2013) Swarm intelligence and its applications. Sci World J

Nature-Inspired Computing: Scope and Applications of Artificial Immune Systems Toward Analysis and Diagnosis of Complex Problems

K. R. Dasegowda, Akshar Radhakrishnan, Majji Rambabu, Sameera Peri, Karthick Vasudevan, H. Prabhavathi, and Mohammed Abdul Kareem

1 Introduction

Nature has always been a thing of awe since the dawn of humanity. Humans have always sought inspiration from their surroundings, despite being the most intelligent species on the planet. Since time immemorial humans have been observing nature, trying not just to comprehend, but also to replicate it. There are still driving forces

Authors K. R. Dasegowda, Akshar Radhakrishnan, and Majji Rambabu contributed equally to this work.

K. R. Dasegowda · A. Radhakrishnan · M. Rambabu · S. Peri · K. Vasudevan
Department of Biotechnology, School of Applied Sciences, REVA University, Bangalore, Karnataka, India
e-mail: dasegowda@gmail.com

A. Radhakrishnan
e-mail: akshar15rk@gmail.com

M. Rambabu
e-mail: majjirambabu8@gmail.com

S. Peri
e-mail: sameeraperi04@gmail.com

K. Vasudevan
e-mail: karthick.vasudevan@reva.edu.in

H. Prabhavathi
Department of Studies & Research in Biotechnology, Tumkur University, Tumakuru, Karnataka, India
e-mail: prabhavathi.h@gmail.com

M. A. Kareem (✉)
Department of Biochemistry, School of Sciences, Indira Gandhi National Open University, New Delhi, India
e-mail: abdul.kareem@ignou.ac.in

K. Raza (ed.), *Nature-Inspired Intelligent Computing Techniques in Bioinformatics*,
Studies in Computational Intelligence 1066,
https://doi.org/10.1007/978-981-19-6379-7_8

present in them, which even now are beyond our comprehension, nevertheless, the inquisitiveness in humans still tries to decode them for a better understanding of them and make use of the same in developing new techniques and creating new tools (Siddique and Adeli 2015). In the age of the digital era, in which we currently live, the problems we confront today are data-driven. The complexity of said problems makes it that much more difficult to not just solve them but also to understand them completely. Managing such a large volume of data has become a challenge in and of itself, let alone analyzing it and attempting to glean useful information from it. It becomes extremely difficult to both analyze and extract information with the existing computational methods as it is time-consuming as well as expensive, which is the very reason why there is a need for computational methods that are intelligence-based where it cannot just compute this bulk data but at the same time do it in a quick cost-effective manner (Kar 2016). Nature always has inspired the human race time and again, where existing problems are compared to that which is already existent in nature, or by mimicking or imitating the behavior of any living organism and applying it to solve real-world problems. Drawing its influence from this, nature-inspired computation came into existence. In this, the development of various algorithms, programs, and hardware was produced. These products replicated various components of nature around us, be it the physical, biological, or chemical aspects of nature. Consequently, these algorithms are developed with the purpose of enabling autonomous entities to operate computing systems. To put it another way, when a problem arises, these computational systems adapt and modify themselves to deliver the best solution feasible (Liu and Tsui 2006).

The biological aspect of nature is the fastest and most diversified of the numerous elements of nature. It is made up of all the living beings on the globe. Even today, the many processes and systems present in these organisms are examined in order to obtain a deeper and more profound understanding of them so as to apply the same notions to the development of new technologies. The name "bio-inspired algorithms (BIA)" comes from the fact that this biological factor has played a significant influence in the development of computational systems. It has been observed that the application of bio-inspired algorithms has proven time and again to be effective in solving complex problems, particularly those with a variety of constraints as well as problems with multiple solutions and a lack of clarity in determining which the best among them is. The ability of bio-inspired algorithms to learn, adapt, and evolve to varied environments, mimicking a living organism, has proven to be a viable option that has helped in dealing with issues of complexity faced by us. This modern age challenge necessitates not just a solution, but the most optimal solution, which is where BIA comes into play (Chiroma et al. 2017).

The neural network (NN), ant colony algorithm, bee colony algorithm, and invasive weed algorithm are a few of the best-known bio-inspired algorithms that have been developed and are applied in numerous domains (Almufti 2019). Among these, the "artificial immune system (AIS)" is one of the most well-known and widely utilized computational systems. As the name suggests artificial immune system has drawn inspiration from the immune system seen in living organisms, more specifically the one seen in human beings. The immune system ensures that the host is protected from any pathogenic element that could harm the host by causing damage or disease. In this regard, the immune system contains a variety of cell types, each of which plays a distinct role in the defense of the host (Kogut et al. 2020). All of these cells are interconnected in diverse biological pathways seen beginning from the production of the immune cells and continuing through the invasion of the pathogen and the eradication of the pathogenic substance. The immunological concepts, pathways, and mechanisms have been studied and used as the cornerstone of the AIS. This chapter offers a compilation of the most well-known and widely used bio-inspired algorithms focusing especially on artificial immune systems, as well as real-world instances of their application.

In the present day, the world is moving toward more machine learning where problem detection and its cure are done through artificial intelligence itself. We sought to describe the role of biology-inspired algorithms along with the applications of the artificial immune system in this chapter. The core objective of this chapter is to sensitize the readers regarding the possibility of novel perspectives toward the analysis as well as diagnosis of complex problems using artificial immune systems. As a result, we have made a sincere effort to assemble the content available online in the form of open educational resources, research publications, and reference books.

2 Nature-Inspired Computing

Nature-inspired computing (NIC) is an interdisciplinary science that incorporates computing science with research from different areas such as chemistry, physics, mathematics, biology, and engineering. It is used for developing algorithms, hardware, and software that can solve problems, synthesize patterns, behaviors, and organisms. In the natural sciences, NIC is a class of meta-heuristic algorithms that imitate or are inspired by certain natural phenomena. Recently, many nature-inspired computing paradigms have been developed. There are three main classes: (1) physics-based algorithms (PBAs), (2) chemistry-based algorithms (CBAs), and (3) biology-based algorithms (BBAs) as shown in Fig. 1.

Three groups of biology-based algorithms: (1) evolutionary algorithms (EAs), (2) bio-inspired algorithms (BIAs), and (3) swarm intelligence-based algorithms (SIAs) are shown in Fig. 2. BIA refers to phenomena observed in some animal species as well as the movement of certain organisms. Particle swarm optimization

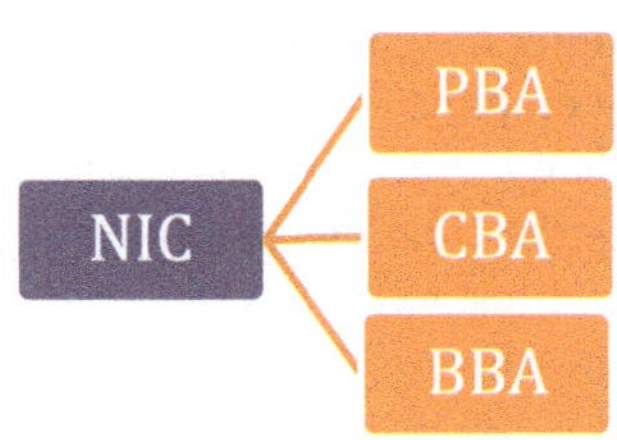

Fig. 1 Classification of nature-inspire computing

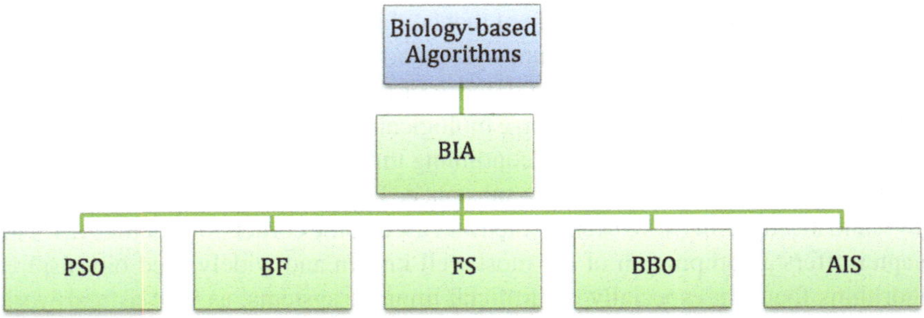

Fig. 2 Classification of biology-based algorithms

(PSO) simulates the social behavior of swarms in nature, such as birds flocking and fish schooling. About 20 different variants of PSO are now available. Bird flocking (BF) is thought of as an element of coordinated movement of multiple individuals for a number of reasons, including protection from predators, habitat search, and social behavior. The fisher school (FS) has very interesting characteristics in its behavior. The majority of fish species are known to form fish schools at some point. FS self-organized systems are observed as complex self-regulating systems composed of individuals with autonomy, and there are many sizes and shapes. In the 1940s and 1950s, MacArthur and Wilson developed mathematical models to describe biogeography and it describes how new species emerge on one island and how others become extinct on another. As a result of the biogeography concept, Simon proposed biogeography-based optimization (BBO). Models based on principles of biological immune systems were proposed by Farmer et al. in the 1980s, which outlined mathematically the interactions between antibodies (Siddique and Adeli 2015).

3 Bio-inspired Algorithms

As human civilization tread steadily toward the future, the problem faced by us tends to get more and more complex. Owing to the rising complexity of day-to-day activities, present technology, particularly in the field of computational sciences, requires constant upgradation to address the issue. The inception of computational sciences is based on the creation of mathematical models and simulations to have a fuller insight into challenging problems and at the same time solve them effectively. It employs mathematical principles to produce algorithms, which are nothing more than a "*set of rules that must be followed by the system, and these algorithms aid in the creation of software applications or computer programs.*" Natural science, in contrast to the advanced areas of computational science, appears to be archaic, as it investigates the nature around us, which is made up of living organisms. However, it is fascinating to find that systems found in living beings like animals and plants

Table 1 Existing algorithms and computer software developed based on them

Algorithms or computer program	Biological counterparts	References
Artificial immune system	Immune system	Fan et al. (2020)
Bio-inspired neural network	Brain	Pérez et al. (2018)
Sensor network	Sensory organs	Nunoo-Mensah et al. (2018)
Membrane computing	Membrane of cells	Theocharopoulou et al. (2019)
Invasive weed optimization algorithm	Growth and spreading of weed	Misaghi and Yaghoobi (2019)
Selfish gene algorithm	Based on Darwinian theory of evolution	Ariff et al. (2016)
Bee colony algorithm, ant colony algorithm, and bacterial foraging algorithm	Based on animal foraging behavior	Luo et al. (2020), Wang et al. (2020)
Bacterial colony optimization	Based on the life cycle model of *E. Coli*	Chen et al. (2020), Niu and Wang 2012)

have contributed to the creation of AI and machine learning. This is known as a bio-inspired algorithm or a biologically inspired algorithm (Yadav and Vishwakarma 2020; Li et al. 2018).

Although bio-inspired algorithms have not been developed in recent years, they have been extensively explored and employed in the development of numerous models, networks, and computational systems that are still in use today and are assisting in the development of far more advanced systems. The extensive research work performed on understanding the mechanisms of how cells and tissues coordinate has led to the design and development of effective computer algorithms (Kar 2016). Table 1 lists the existing algorithms and computer software developed through studies and research done on living beings and their biological systems.

Biological science and computational science have always been two distinct sides of a coin (research), but have been in constant association, with biologists requiring the use of various computational methods used to analyze, interpret, and integrate data generated during experiments. On the other side, various computational algorithms and programs are inspired by the biological systems of different organisms. The association and collaboration between the two fields will increase in the future as progress is made in comprehending various biological systems through research as well as advancements in the field of computational sciences.

Of late many existing and emerging algorithms developed have used various principles which are observed and studied in biology. The evolutionary algorithm and the swarm intelligence algorithm are two significant groups of algorithms that are developed by mimicking diverse biological activities. The former is based on concepts from evolutionary biology and Darwinian principles such as natural selection, commonly known as "survival of the fittest" whereas the latter is based on animal social collective behavior that can be observed easily in organisms such as ants, bees, and other insect colonies. Other than these two groups, there are a variety of algorithms that do not necessarily fit into the aforementioned categories, resulting in less well-known groupings such as ecology-based categories and multi-objective algorithms. The categories are split in such a way that the source of biological inspiration may be understood at the same time the functionality of the algorithm remains tied to the category to which it belongs (Fan et al. 2020). Let us explore some well-known algorithms in bio-inspired algorithms:

3.1 Algorithms Based on Evolution

This algorithm consists of the computational programs which are based on the laws and principles observed in evolutionary biology. The algorithms in this category are created in such a way that the virtual population is forced to live and survive in an environment with a clearly defined challenge that is specific to the objective function. The algorithm then chooses the suitable candidate from a virtual population of candidates in the given environment, taking into account factors that influence evolutionary aspects such as mutation and recombination. Types of algorithms that come under this group are genetic algorithms, artificial neural networks, differential neural networks, evolution strategies to name a few (Lacerda et al. 2021; Falcón-Cardona and Coello 2020).

3.2 Algorithms Based on Swarm Intelligence

Swarm intelligence, also referred to as SI, is loosely based on the concept of collective behavior of organisms (e.g., bees, insects) that live and move together in nature, generally in the form of colonies. The unique feature of swarm intelligence is that it is decentralized, and the colony functions as a self-organized system. This is imitated in the creation of swarm intelligent systems, which typically consist of a population of agents with the ability to interact with other agents in their immediate vicinity as well as their environment. Because there is no central control structure, local interactions are in charge of generating a specific output by combining multiple local interactions between agents. This is especially important in the case of swarm robotics, which is a type of robotics.

Ant colony optimization (ACO) and fish swarm algorithms (FSAs) are a couple of examples of swarm intelligence-based models among others (Rostami et al. 2021; Zedadra et al. 2018).

3.3 Algorithms Based on Ecology

Algorithms based on ecology are a relatively newer type of algorithm. There is a population of individuals in this type of algorithm, each of whom is a potential solution to a specific problem. Here, each population tends to evolve based on the sort of optimization strategy provided. Hence, based on the input parameters different types of populations evolve with every individual in the obtained population representing a viable solution. The algorithm can be modeled in two ways: homogeneous model, in which the entire population is given the same input parameters or optimization strategy, and heterogeneous model, in which different optimization strategies are given to the population. Invasive weed optimization (IWO) and biogeography-based optimization (BBO) are a few examples of ecology-based algorithms (Misaghi and Yaghoobi 2019; Li and Wang 2021).

3.4 Multi-Objective Algorithm

Multi-objective algorithm or optimization, often known as Pareto optimization, deals with optimization issues containing many objective functions that must be optimized at the same time. It is a type of multiple-criteria decision-making optimization system. When there are two or more conflicting objectives to consider and a decision to be made, the algorithm analyzes the situation and makes the best conclusion possible. It is not necessary to obtain a single solution in this scenario; multiple solutions can arise that cater to one, many, or all of the objectives. A few examples of multi-objective algorithms are population-based ACO (P-ACO) and non-dominated sorting genetic algorithm-II (NGSA-II) (Niu and Wang 2012; Li and Wang 2021).

4 Artificial Immune System

Many elements of modern science and technology have been influenced by nature. Bio-inspired algorithms aim to enhance the computational approaches by simulating natural events, with the intention of reaching comparable and favorable qualities observed in the existing natural system. Evolutionary algorithms, neural networks, membrane computing, and swarm intelligence, are just a few of the algorithms inspired by biological systems (Kashani et al. 2021). The artificial immune system (AIS) is also a type of computational system that is inspired by biology, where programs and algorithms have been developed that mimic, the human immune system. To develop AIS concepts in immune networking, clonal selection and negative selection have been studied extensively and used for developing the AIS. These mentioned theories have proved to be very effective in various application areas (Molina et al. 2020).

The immune system being a highly reliable, autonomous, error-tolerant, and adaptive mechanism, these characteristics are extremely valuable in the creation of new computer programs and systems. All AIS algorithms are designed such that they can replicate the behavior and features of the cells playing a pivotal role in the immune system, such as the B and T lymphocytes, along with the dendritic cells (DCs), but the algorithm designed can vary in complexity and can perform a variety of functions. Negative selection, clonal selection, and immunological networks have been the focus of AIS research so far as they happen to be the well-studied aspects in immunology. While a lot of effort has been taken in developing and applying AIS-based algorithms, the same does not hold true for theoretical research in the field, as conceptual literature on immunology has lately begun to emerge, with the increase in realizing the relevance of the subject (Jim et al. 2022; Brabazon et al. 2015).

4.1 Biological Aspects Focused on Developing the Algorithm

The immune system observed in vertebrates consists of a variety of cells as well as other associated molecules which collaborate with various organ systems present in the body, such as the nervous and endocrine systems, to keep the host in a homeostatic state. The immune system's primary concern is to defend the host's body from disease-causing agents such as bacteria, viruses, fungi, and other parasites. These are known as pathogens, which are the invading agents. The recognition of a pathogen or any of its associated molecules, called an antigen, triggers an immune response (Greensmith et al. 2010).

The immune system is categorized into two parts namely innate and adaptive. Innate immunity is geared such that it can fight against all disease-causing agents that enter the body, rather than specific invaders. The innate immune system, present in all vertebrates from birth, is critical for the initiation as well as the modulation of immune responses. Innate immune systems are also known to consist of specialized cells that have developed not just the property to recognize but also to bind to molecular patterns found exclusively in microbes which could be a potentially infectious agent. However, because it is essentially static in nature, the innate immune system does not provide comprehensive body protection (Al-Enezi et al. 2011).

Adaptive immunity, also referred to as acquired immunity, permits the immune system to target any pathogenic agent that the innate system is unable to eliminate. The adaptive system is set up such that it can target specific invaders, of which some could have been encountered before in the host's body whereas others have never been seen before. Lymphocytes, or white blood cells, make up the majority of the adaptive immune system, especially the B and T lymphocytes. These lymphocytes assist in the recognition as well as the destruction of antigenic substances. Antigen-specific adaptive immune responses are those responses that are directed against an antigen that can stimulate the adaptive immune system. The majority of immunological research is focused on the adaptive immune response (Çipe et al. 2022). Most AIS research efforts have concentrated mainly on the immune system's memory and learning mechanisms, which are inherent in the clonal selection and immunological networks, as well as the negative selection principle for the construction of detectors that have the ability to identify changes in self (Iwasaki and Medzhitov 2015).

4.2 Clonal Selection Theory

This theory explicates the working of the adaptive immune system when there is an antigenic stimulation. The immune system works such that it proliferates the immune cells only if they have the ability to recognize the antigenic substrate and do not let the cells multiply which does not have the ability to do so. Both the T and B cells are subjected to clonal selection as they have receptors present on their surface that can recognize the antigenic substrate. B cells activate and differentiate into a slightly modified version of themselves called the plasma or memory cells when their antigen receptors (antibodies) interact with an antigen. Before this occurs, replicas of B cells are created, which are then subjected to somatic hypermutation, resulting in a diverse B cell population. Plasma cells generate a significant quantity of antigen-specific antibodies, which aid in the elimination of the antigen resulting in an effective immune response. Memory cells that are also produced in the process remain in the host which helps in inducing a quick secondary response when exposed to the same (or similar) antigen in the future. Acquired immunity is the term for this phenomenon (Gendreau and Potvin 2010; Müller et al. 2018).

4.3 Immune Networks

Immune networks, though not a widely accepted theory, especially among immunologists, happen to pave the way to creating the AIS. The development of a coherent network of B lymphocytes aids the immune system to achieve immunological memory. This is the ability to respond to a disease that has been encountered before but more quickly and efficiently by the immune system. The capacity of paratopes that are present on the B cells matches to idiotopes present on the other B cells causing this network of B cells. The stimulation or activation of the B cells is due to this interaction observed between the idiotopes and the paratopes.

This occurs as B cell paratopes react to the idiotopes present on the other B cells in the same way that they would to a foreign antigen. Although as a control mechanism to prevent this interaction, there is some inhibition observed between the B cells. This B cell interaction observed in the network is considered to lead to a memory structure that is stable and also contributes to the retention of the memory cell even at times when the antigen is not present (Timmis et al. 2008).

4.4 Negative Selection

A huge proportion of AIS work has been founded on the basis of this process. Lymphocytes which play an important role in immunity can react to antigens from the pathogen as well as in some cases materials of the host's cells as well. If and when this triggers a full immune response, the host organism may be harmed and cause serious health issues. This is known as autoimmunity where the cells of the host are attacked by the host's immune system. Negative selection is a process used to safeguard the organism from self-reactive responses, hence is of extreme importance. During the synthesis of the lymphocytes, there is a process called somatic hypermutation that occurs, and in some cases, the lymphocytes produced have the ability to recognize self-antigens, hence tend to attack the host along with foreign antigens (Fernandez-Leon et al. 2014).

In the thymus, where the T lymphocytes are present, the cells are subjected to negative selection. The thymus forms a barrier which is also known as the blood-thymic barrier and is highly impervious to macromolecules. Thymocytes which are the immature form of the T cells mature in the thymus and undergo selection without coming into contact with foreign antigens because of the blood-thymic barrier. Antigen presentation cells known as APCs present the T cells with a major histocompatibility complex or a peptide belonging to the host itself during the selection phase. The T cells that bind strongly with either the MHC complex or the self-peptide are removed by a regulated cell death procedure better known as apoptosis. This results in the production of only T cells that can recognize foreign antigens and are not self-reactive (Hooper et al. 2012).

4.5 Artificial Immune System Algorithms

Four main AIS algorithms are involved to enable different AIS applications as shown in Fig. 3. The four AIS algorithms are negative selection-based algorithms, clonal selection-based algorithms, artificial immune networks, danger theory, and dendritic cell algorithms.

4.5.1 Negative Selection-Based Algorithms

This algorithm depends on the standard capacity that safeguards the body against self-receptive lymphocytes. Immune systems recognize antigens that are not familiar to them and do not react with "self-cells." T cells are generated through a pseudo-arbitrary hereditary re-plan process in which the receptor is delivered. In the thymus, these receptors pass through unfavorable determination processes. T cells that respond to "self-cells" are discarded, whereas those that do not bind to self-cells are allowed to leave the thymus. White blood cells are allowed to develop freely

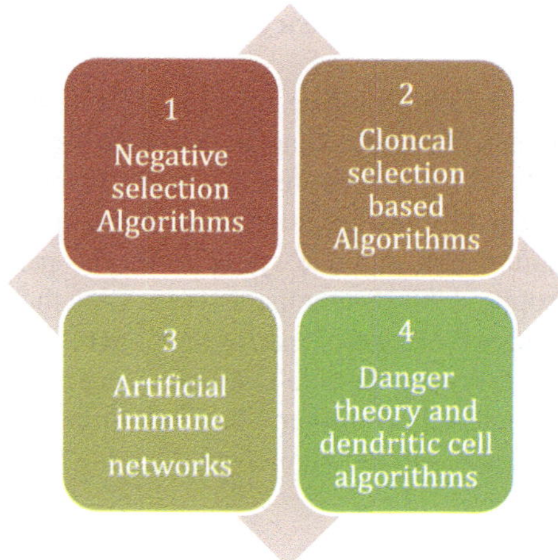

Fig. 3 Four AIS algorithms

throughout the body and contribute to immunological interaction against unfamiliar antigens (Sam-Yellowe and Sam-Yellowe 2021).

4.5.2 Clonal Selection Algorithms (CSAs)

This algorithm is planned based on the mechanism called antigen–antibody identification. Which is referred to as clonal selection theory which has brought about the advancement of a few artificial immune algorithms mentioned as CSA (Somayaji et al. 1998).

4.5.3 Artificial Immune Networks (AINs)

In AISs, artificial immune networks are another useful paradigm. Farmar et al. established the immune network model, which became the foundation for other AINs algorithms, and Ishida proposed the primary immune network based on this model (Castro and Zuben 2000). Timmis and coworkers recharacterized these immune networks, naming them artificial immune network (AINE). According to Knight and Timmis (2001), AINs use an artificial recognition ball (ARB) to identify indistinguishable B cells (Ishida 1990). If the partiality between two ARBs is less than a network affinity threshold (NAT), two B cells are consolidated. In 2001, Timmis and Neal created resource limited artificial immune systems (RLAIS) in view of AINE. In this series of models, we included information about the fixed absolute number of B cells contained in ARBs which acted as a unit, but each ARB also played a distinct role in getting assets from the standard. ARBs that are not ready to receive assets are removed from the organization. Cloning, mutation, and collaboration of B cells all take place at the ARB level (Knight and Timmis 2001).

4.5.4 Danger Theory and Dendritic Cell Algorithms

DCs inside an innate immune system are cells that answer some particular peril signals. The three fundamental kinds of DCs, for example, (1) immature dendritic cells, (2) semi-mature DCs, and (3) mature DCs. The dendritic cell algorithm is organically roused improvement taking motivation from the danger theory of the mammalian immune system with the particular capacity of DCs. Matzinger first proposed this danger theory expressing that the job of the acquired immune system is to answer signals of risk, as opposed to separating self from non-self. The hypothesis expresses that the aide T cells actuate a caution signal giving the co-feeling of antigen-explicit cells to answer. The dendritic cell algorithm (DCA) is aroused by the capacity of the DCs of the human invulnerable framework (Knight and Timmis 2001).

5 Artificial Immune System Applications

Several AIS applications have been designed and developed as a result of the biological immune network. AIS uses the immune system to explore how it can be applied to computing problems. These AIS algorithms apply some of the principles of natural immune systems in order to solve complex problems using advanced techniques like machine learning. A variety of processes are monitored using AIS, including chemical and nuclear plant power transformers, uninterruptible power supply systems, and gearbox systems.

5.1 Fault Detection, Diagnosis, and Recovery Using AIS

Multiple AIS fault detection and recovery applications have been developed and designed using the self/non-self-discrimination principle. The immune cells have receptors on the surface of their cells that distinguish between what belongs to the body, called the self, and what does not belong, called the non-self. Applications to technical components and machinery include counter systems, direct current (DC) motors, induction machines rotorcrafts, and airplanes. Applications to processes include chemical processes and nonlinear systems. The dendritic cell algorithm is used to develop fault detection and isolation method (FDI). In the year 2017, Alizadeh et al. proposed a test model procedure for wind turbines (Alizadeh et al. 2016). Through this technique, it is feasible to distinguish as well as segregate sensor flaws. A measurable correlation is completed to assess the exhibition of the proposed conspire. The biological differentiation of dendritic cells mechanism was proposed by a model. This model isolates dendritic cell combination data, describes the functionalities of outside signals used in wireless sensor networks, and provides the DC mathematical model. The performance of a real-time intrusion detection system was evaluated in terms of robustness, complexity, and scalability, which resulted in greater recognition with lower energy use (Daudi 2015).

5.2 *Immune Network-Based Approaches*

Several AIS applications for monitoring different types of processes have been developed from the biological immune network, including monitoring plant power systems, gearbox systems, and uninterruptible power supply systems. The approaches are generally used either to locate the origin of faults after detecting faulty states, such as in, or to classify certain fault samples without focusing on the detection task (Mohapatra and Khilar 2020).

5.3 *Automobile Hydraulic Brake Fault Diagnosis*

Based on clonal selection theory, the authors have developed a clonal selection algorithm to diagnose the problem with hydraulic brakes in vehicles. For each problem condition in a brake system, nine fault conditions were simulated and tested. Inspired by the clonal selection theory of acquired immunity, the clonal selection classification algorithm (CSCA) has shown success in a broad range of engineering problems. The purpose of this study is primarily to evaluate the clonal selection algorithm in analyzing brake fault development. Therefore, an effort has been made to classify hydraulic brake faults by using the CSCA algorithm (Bayar et al. 2015).

5.4 *Fault Detection in Refrigerator System*

The AIS algorithm is used to spot anomalies in supermarket freezer cabinet temperature data. In-store refrigerators, the researchers are especially looking for early signs of icing. r-bit matching guidelines give an accurate detection for erroneous data in addition to information encoding (Jegadeeshwaran and Sugumaran 2015).

5.5 *Fault Diagnosis, Detection, and Isolation of Wind Turbine*

After being inspired by an immune system, a proposal for fault detection and isolation of wind turbine systems has been developed. The authors created a hierarchical system called the negative selection algorithm (NSA) to recognize and segregate both separately occurring errors and concurrent failures. Various fault conditions were put under a nonparametric statistical comparison test to evaluate the proposed work. Support vector machines and NSA display similar results in simulations, while NSA creates better results under some fault conditions (Daudi 2015).

5.6 Fault Diagnosis in Transformer

Using an artificial immune network, the scientists proposed a new way of diagnosing defects. They integrated it with a NN's radial basis function (RBF), with the same structure. The proposed strategy has fewer invisible layers and a higher diagnostic rate than RBF. Wang and colleagues l presented an artificial immune network in combination with fuzzy c-means clustering to identify the sorts of errors in transformers. Experiments have shown that the suggested method is capable of accurately classifying various types of power transformer defects (Daudi 2015).

6 Conclusion

Despite the fact that there is a vast spectrum of BIA available today, the scientific community is largely unaware of their existence and use. Apart from a few well-known BIA, there is little literature that shows how they can be used to solve real-world problems. The objective of this chapter is to bridge the gap that is present between the ones who develop these algorithms and their peers in the scientific community. Practitioners, particularly those working in the field of information or data science, might benefit from gaining a basic understanding of these algorithms. They may use this as a starting point to explore the aforementioned BIA for their applications in real-life problems in various domains depending on both the nature of the problem being addressed and the complications which are associated with them.

References

Aldhaheri S, Alghazzawi D, Cheng L, Alzahrani B, Al-Barakati A (2020) Deepdca: novel network-based detection of IoT attacks using artificial immune system. Appl Sci 10(6):1909

Al-Enezi JR, Abbod MF, Alsharhan S (2011) Artificial immune systems-models, algorithms and applications

Alizadeh E, Meskin N, Khorasani K (2016) A negative selection immune system inspired methodology for fault diagnosis of wind turbines. IEEE Trans Cybern 47(11):3799–3813

Almufti SM (2019) Historical survey on metaheuristics algorithms. International Journal of Scientific World. 7(1):1

Ariff NM, Khalid NE, Hashim R, Noor NM (2016) Selfish gene algorithm versus genetic algorithm: a review. In: IOP conference series: materials science and engineering, vol 160, no 1. IOP Publishing, p 012098
Bayar N, Darmoul S, Hajri-Gabouj S, Pierreval H (2015) Fault detection, diagnosis and recovery using artificial immune systems: a review. Eng Appl Artif Intell 1(46):43–57
Brabazon A, O'Neill M, McGarraghy S (2015) Artificial immune systems. In: Natural computing algorithms. Springer, Berlin, Heidelberg, pp 301–332
De Castro LN, Von Zuben FJ (2000) The clonal selection algorithm with engineering applications. In: Proceedings of GECCO, vol 2000, pp 36–39
Chen H, Zhang Q, Luo J, Xu Y, Zhang X (2020) An enhanced bacterial foraging optimization and its application for training kernel extreme learning machine. Appl Soft Comput 1(86):105884
Chiroma H, Herawan T, Fister I Jr, Fister I, Abdulkareem S, Shuib L, Hamza MF, Saadi Y, Abubakar A (2017) Bio-inspired computation: Recent development on the modifications of the cuckoo search algorithm. Appl Soft Comput 1(61):149–173
Çipe F, Arısoy ES, Correa AG (2022) Immunological Responses to Infection. In: Pediatric ENT infections. Springer, Cham, pp 3–17
Daudi J (2015) An overview of application of artificial immune system in swarm robotic systems. Adv Robot Autom 4(1)
De Lacerda MG, de Araujo Pessoa LF, de Lima Neto FB, Ludermir TB, Kuchen H (2021) A systematic literature review on general parameter control for evolutionary and swarm-based algorithms. Swarm Evol Comput 1(60):100777
Falcón-Cardona JG, Coello CA (2020) Indicator-based multi-objective evolutionary algorithms: a comprehensive survey. ACM Comput Surveys (CSUR) 53(2):1–35
Fan X, Sayers W, Zhang S, Han Z, Ren L, Chizari H (2020) Review and classification of bio-inspired algorithms and their applications. J Bionic Eng 17(3):611–631
Fernandez-Leon JA, Acosta GG, Rozenfeld A (2014) How simple autonomous decisions evolve into robust behaviours?: a review from neurorobotics, cognitive, self-organized and artificial immune systems fields. Biosystems 1(124):7–20
Gendreau M, Potvin JY (eds) (2010) Handbook of metaheuristics. Springer, New York
Greensmith J, Aickelin U, Tedesco G (2010) Information fusion for anomaly detection with the dendritic cell algorithm. Inf Fusion 11(1):21–34
Hooper LV, Littman DR, Macpherson AJ (2012) Interactions between the microbiota and the immune system. Science 336(6086):1268–1273
Ishida Y (1990) Fully distributed diagnosis by PDP learning algorithm: towards immune network PDP model. In: 1990 IJCNN international joint conference on neural networks. IEEE, pp 777–782
Iwasaki A, Medzhitov R (2015) Control of adaptive immunity by the innate immune system. Nat Immunol 16(4):343–353
Jegadeeshwaran R, Sugumaran V (2015) Brake fault diagnosis using clonal selection classification algorithm (CSCA)—A statistical learning approach. Eng Sci Technol Int J 18(1):14–23
Jim LE, Islam N, Gregory MA (2022) Enhanced MANET security using artificial immune system based danger theory to detect selfish nodes. Comput Secur 1(113):102538
Kar AK (2016) Bio inspired computing—A review of algorithms and scope of applications. Expert Syst Appl 15(59):20–32
Kashani AR, Camp CV, Rostamian M, Azizi K, Gandomi AH (2021) Population-based optimization in structural engineering: a review. Artif Intell Rev 4:1–08
Knight T, Timmis J (2001) AINE: an immunological approach to data mining. In: Proceedings 2001 IEEE international conference on data mining. IEEE Computer Society, pp 297–297
Kogut MH, Lee A, Santin E (2020) Microbiome and pathogen interaction with the immune system. Poult Sci 99(4):1906–1913
Li W, Wang GG (2021) Elephant herding optimization using dynamic topology and biogeography-based optimization based on learning for numerical optimization. Eng Comput 4:1–29

Li G, Jin Y, Akram MW, Chen X, Ji J (2018) Application of bio-inspired algorithms in maximum power point tracking for PV systems under partial shading conditions—A review. Renew Sustain Energy Rev 1(81):840–873
Liu J, Tsui KC (2006) Toward nature-inspired computing. Commun ACM 49(10):59–64
Luo Q, Wang H, Zheng Y, He J (2020) Research on path planning of mobile robot based on improved ant colony algorithm. Neural Comput Appl 32(6):1555–1566
Misaghi M, Yaghoobi M (2019) Improved invasive weed optimization algorithm (IWO) based on chaos theory for optimal design of PID controller. J Comput Des Eng 6(3):284–295
Mohapatra S, Khilar PM (2020) Immune inspired fault diagnosis in wireless sensor network. In: Nature inspired computing for wireless sensor networks. Springer, Singapore, pp 103–116
Molina D, Poyatos J, Ser JD, García S, Hussain A, Herrera F (2020) Comprehensive taxonomies of nature-and bio-inspired optimization: Inspiration versus algorithmic behavior, critical analysis recommendations. Cogn Comput 12(5):897–939
Müller V, De Boer RJ, Bonhoeffer S, Szathmáry`E (2018) An evolutionary perspective on the systems of adaptive immunity. Biol Rev 93(1):505–528
Niu B, Wang H (2012) Bacterial colony optimization. Discrete Dyn Nat Soc
Nunoo-Mensah H, Boateng KO, Gadze JD (2018) The adoption of socio-and bio-inspired algorithms for trust models in wireless sensor networks: a survey. Int J Commun Syst 31(7):e3444
Pérez J, Cabrera JA, Castillo JJ, Velasco JM (2018) Bio-inspired spiking neural network for nonlinear systems control. Neural Netw 1(104):15–25
Rostami M, Berahmand K, Nasiri E, Forouzandeh S (2021) Review of swarm intelligence-based feature selection methods. Eng Appl Artif Intell 1(100):104210
Sam-Yellowe TY, Sam-Yellowe TY (2021) Immunology: overview and laboratory manual. Springer
Siddique N, Adeli H (2015) Nature inspired computing: an overview and some future directions. Cogn Comput 7(6):706–714
Somayaji A, Hofmeyr S, Forrest S (1998) Principles of a computer immune system. In: Proceedings of the 1997 workshop on new security paradigms, pp 75–82
Theocharopoulou G, Giannakis K, Papalitsas C, Fanarioti S, Andronikos T (2019) Elements of game theory in a bio-inspired model of computation. In: 2019 10th International conference on information, intelligence, systems and applications (IISA). IEEE, pp 1–4
Timmis J, Hone A, Stibor T, Clark E (2008) Theoretical advances in artificial immune systems. Theoret Comput Sci 403(1):11–32
Wang H, Wang W, Xiao S, Cui Z, Xu M, Zhou X (2020) Improving artificial bee colony algorithm using a new neighborhood selection mechanism. Inf Sci 1(527):227–240
Yadav A, Vishwakarma DK (2020) A comparative study on bio-inspired algorithms for sentiment analysis. Clust Comput 23(4):2969–2989
Zedadra O, Guerrieri A, Jouandeau N, Spezzano G, Seridi H, Fortino G (2018) Swarm intelligence-based algorithms within IoT-based systems: a review. J Parallel Distrib Comput 1(122):173–187

Nature-Inspired Computing: Bat Echolocation to BAT Algorithm

Astha Singh, Arun Meyyazhagan, and Saurabh Verma

1 Introduction

Bats are a group of wonderful mammals capable of flight. Their adaptations for true flight have mesmerized and motivated the scientists. Bats evolved millions of years ago capable of true flight. Their wings are modified forelimbs with extended digits covered by a thin membrane. This adaptation allowed them to exploit the niche which was dominated by birds and to a certain extent by the insect. Their size varies from a few inches of wing span to more than 5 feet. Their feeding habit is also diverse with some dependent on fruits and nectar while others feed on insects and even blood. They are represented by more than 1400 species and are cosmopolitan in distribution. They are of immense importance to the ecosystem for pollination and seed dispersal with some species completely dependent on them for these maneuvers. The bat guano is a backbone of farming in some countries. Their importance as reservoirs of diseases has been evident several times still many countries have culinary needs for them as well (Altringham 1996). Their foraging behavior is dependent on echolocation as they detect prey by echolocation. Since this technique cannot differentiate between fruits and trees, it does not hold any importance for frugivore bats; hence, they do not echolocate and have lost this ability although some fruit bats can echolocate for communication, etc. This is performed by sending ultrasonic sounds or calls both in flight and while resting which echo back to the calling bat. The call structure of a bat is more dependent on the environment rather than the phylogenetic relation of bats. The analysis and disclosure of modulation in call structure dynamics to prey catching serves as a raw material for developing engineering modules to solve various

A. Singh · S. Verma (✉)
Tumor Biology, ICMR-National Institute of Pathology, New Delhi, India
e-mail: svarmasv1@rediffmail.com; saurabhverma.nip@gov.in

A. Meyyazhagan
Cytogenetic Department, Biomedical Research Centre, Euroespes, Spain

K. Raza (ed.), *Nature-Inspired Intelligent Computing Techniques in Bioinformatics*,
Studies in Computational Intelligence 1066,
https://doi.org/10.1007/978-981-19-6379-7_9

kinds of problems. The sense of ultrasonic hearing still exists in frugivore bats due to evolutionary implications. The calls made by echolocating bats are diverse and change between species but similarity remains within synaptic species. The principle is always similar to the echoes from calls giving trajectories of the prey position speed and velocity (Marimuthu 1996). The initial use of ultrasound was as SONAR (short for sound, navigation, and ranging). Several NIC algorithms have been proposed in recent years based on natural maneuvers including the cuckoo search algorithm (Yang 2010), cuckoo search (CS), artificial bee colony algorithm (ABC) (Kumar and Kumar 2013), particle swarm optimization (Kennedy and Eberhart 1995), ant colony optimization (Dorigo et al. 2006), termite colony optimization (TCO) (Hedayatzadeh and Salmassi 2010), cat swarm optimization (Chu et al. 2006), etc. In the present study, we go through the journey of how BAT a heuristic nature-inspired computing was developed from the bat echolocation applied to solving global optimization problems. The applications and new variants of this algorithm have also been accounted for briefly.

2 Discovery of Echolocation

Lazzaro Spallanzani in 1790 first discovered that the bats could navigate in total darkness even after surgical removal of eyes. Charles Jurine demonstrated that bats could not evade obstacles in their path if their ears were plugged. Both Spallanzani and Jurine concluded that the "bats could see from their ears." In 1920, Hartridge hypothesized that bats could emit ultrasound and listen to them. It was 18 years later that D. R. Griffin demonstrated that the bats could emit ultrasound. He used a microphone sensitive to ultrasound to show that bats could emit pulses of ultrasound. The pulses of ultrasound sounds called "calls" are emitted by bats and after hitting objects in the surrounding it hits back into their ears. Their ears are sensitive to these sounds and help them in making an image of the surroundings as well as the prey and its trajectory (Fig. 1) (Marimuthu 1996).

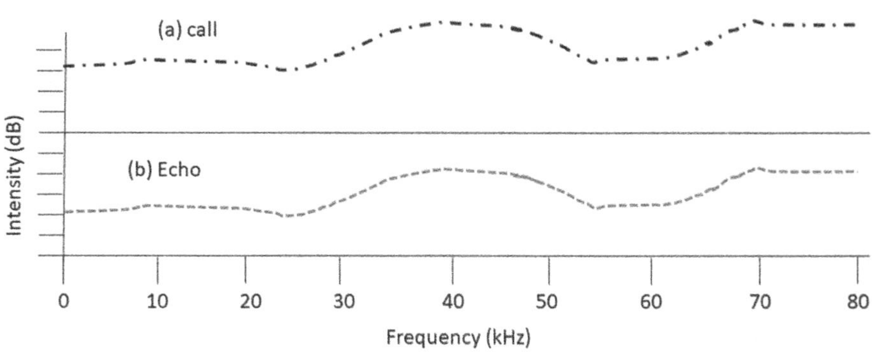

Fig. 1 Diagrammatic representation of a spectrogram of a bat call (**a**) and its echo (**b**)

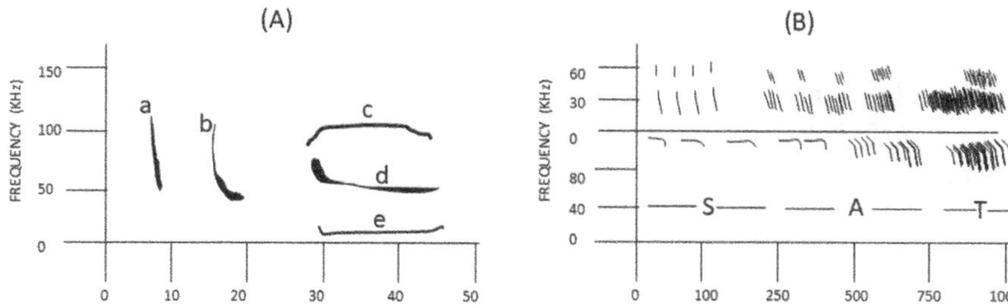

Fig. 2 **a** Various combination of CF and FM calls are shown here. *a* Steep broadband call that starts with sharp FM and ends with low frequency. *b* Steep call that ends with shallow FM. **c** A long CF with increasing and decreasing tails. *d* Call that starts with a shallow FM and long CF. *e* A long CF with decreasing and increasing tail. **b** Typical calls for search (S), approach (A), and (T) terminal or prey catching

3 Vocalization in Bats

The sounds produced are either from the mouth or the nostrils. The noseleaf is a specialized structure used for narrowing and focusing the outgoing beam of sound. The vocalizations are generally of two categories:

- **Broadband signals**: These calls are short and vary from 20 to 140 kHz less than 5 ms in duration. Starting from a high frequency and going down to a very low frequency in a very short span. These are called frequency-modulated or FM calls.
- **Narrowband signals**: These have a constant frequency (CF) with 100 ms duration.

Most bats used both the FM and CF while in flight to gather information in different combinations Fig. 2a. The combination varies according to the habitat. Nowadays, bat detectors having specialized software are used to detect and record bat calls (Marimuthu 1996).

The search calls are made to detect potential targets to prey, and hence longer CF calls are used but in the approaching stage; the calls become shorter with an increase in frequency to increase the details picked by the bats (Marimuthu 1996). In the final stage that bats emit calls that are very short and high in frequency as shown in Fig. 2b.

4 The BAT Algorithm

The echolocation behavior of bat-inspired BAT algorithm was given by Yang in 2010 and is one of the hallmarks of NIC (Yang 2010). It was initially applied for problem-solving but later on, new variants were given which increased the applicability of this algorithm in diverse areas (Fig. 3).

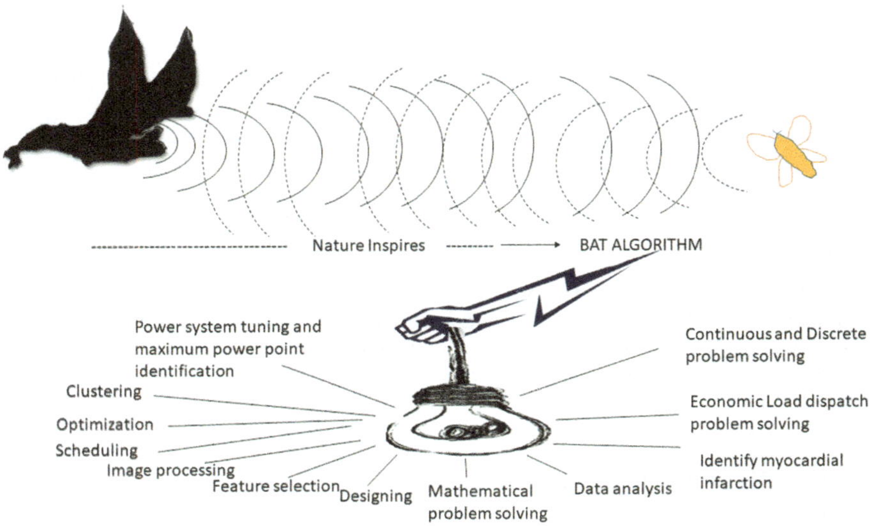

Fig. 3 Inspiration and a few areas of application of the BAT algorithm

The idealized rules of this algorithm include.

1. All bats can sense distance and differentiate between prey and background
2. Bats fly with a velocity v_i at a position x_i with a frequency f and loudness (of call) A_0 to search for prey. They can modulate the frequency and rate of pulse emissions $r2$ [0; 1] of calls according to the distance of the target.
3. The loudness of calls has been assumed to be maximum A_0 to a minimum constant value $A_{\min}$.

The algorithm is as follows.

Each bat B_i is initialized from position X_{i} with velocity v_{i} at frequency fi. The time interval t has a max iteration T. The movement of each bat in virtual space is defined by velocity and frequency in the following equations.

$$f_i = f_{\min} + (f_{\max} - f_{\min})\beta; \tag{1}$$

$$v_i^t = v_i^{t-1} + (x_i^{t-1} - x_*) f_i; \tag{2}$$

$$x_i^t = x_i^{t-1} + v_i^t \tag{3}$$

where $\beta \in [0; 1]$

4.1 Drawbacks

This algorithm is for continuous optimization. The diversity of the algorithm does not increase with population. The exploration phase is described in linear equations of changing loudness and pulse rates, thus limiting its application in different applications. The use of BA is restricted to continuous variables hence to use it for combinatorial problems (Kumar and Kumar 2013).

4.2 New Variants and Application

BA has been used in a wide variety of applications despite its shortcomings. BA has been used in solving problems of optimization, image processing, data analysis, schedule designing, grouping, classification, etc. The new variants have increased the application areas by limiting the shortcomings and increasing their problem-solving capacity (Fig. 3).

After the BA was formulated, dozens of studies were done mostly between 2012 and 2014 (Fig. 4). New variants are still been given which highlights the potential for variation in this algorithm.

The *binary BAT algorithm* (BBA) given in 2012 by Nakamura et al., uses a sigmoid function to convert the continuous position of bats into binary values thus helping in feature selection. Table 1 highlights a list of studies so far involving the BAT algorithms.

These were applied for ergonomic clustering (Umar and Rashid 2021), bi-objective functions (Khan et al. 2011), global engineering optimization (Yang 2011),

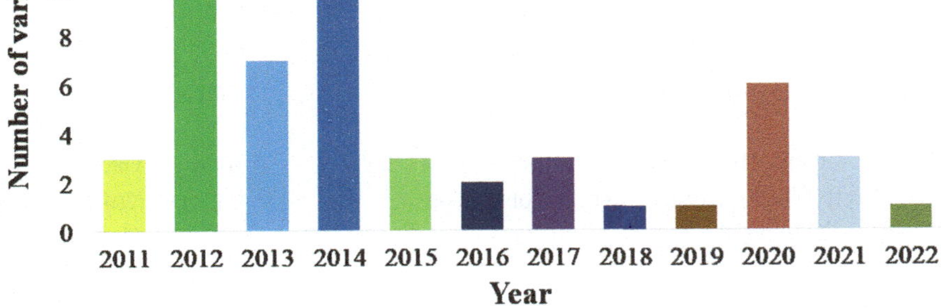

Fig. 4 Year-wise number of variants since the BAT algorithm was given

Table 1 List of variants and application studies on BAT algorithm so far

S. No.	Variant	Application	References
1	Fuzzy logic BAT algorithm	Ergonomic clustering of office work spaces	Umar and Rashid (2021)
2	Multi-objective BAT algorithm	Bi-objective ZDT3 functions	Khan et al. (2011)
3	BAT algorithm	Energy modeling in a gas turbine	Yang (2011)
4	BAT algorithm	Global engineering optimization	Yang and Gandomi (2012)
5	Binary BAT algorithm	Solving combination problems	Nakamura et al. (2013)
6	k-means BAT algorithm	Clustering	Komarasamy and Wahi (2012)
7	Modified BAT algorithm	Phishing website detection	Lemma and Bin Mohd Hashim (2011)
8	BAT algorithm with mutation	Image matching	Damodaram and Valarmathi (2012)
9	Modified BAT algorithm	Record deduplication	Du and Liu (2012)
10	BAT algorithm	Realistic hybrid flow shop scheduling	Faritha Banu and Chandrasekar (2012)
11	Metaheuristic BAT algorithm	Classification of microarray data	Marichelvam and Prabaharam (2012)
12	BAT algorithm	Scheduling problem	Mishra et al. (2012)
13	BAT algorithm	Optimal capacitor placement	Musikapun and Pongcharoen (2012)
14	BAT algorithm	Topology optimization in microelectric applications	Reddy and Manoj (2012)
15	Differential operator and Lévy flights trajectory	Function optimization	Yang et al. (2012)
16	Improved BAT algorithm	Global optimization	Xie et al. (2013)
17	Hybrid BAT algorithm	Problem-solving	Jamil et al. (2013)
18	Binary BAT algorithm	Economic load dispatch	Fister et al. (2013)
19	BAT algorithm	combined economic load and emission dispatch	Mallikarjuna et al. (2013)
20	Naïve BAT algorithm	Feature selection	Ramesh et al. (2013)
21	Hybrid BAT algorithm	global numerical optimization	Taha et al. (2013)
22	Compact BAT algorithm	Data analysis	Wang and Guo (2013)
23	Hybrid self-adaptive BAT algorithm	Problem-solving	Dao et al. (2014)

(continued)

Table 1 (continued)

S. No.	Variant	Application	References
24	Complex valued BAT algorithm	Problem-solving	Fister et al. (2014a)
25	Chaotic BAT algorithm	Problem-solving	Li and Zhou (2014)
26	Chaotic BAT algorithm	Optimization tasks	Gandomi and Yang (2014)
27	BAT algorithm with Gaussian walk	Problem-solving	Afrabandpey et al. (2014)
28	Modified BAT algorithm	Problem-solving	Cai et al. (2014)
29	Cloud model BAT algorithm	Problem-solving	Yılmaz et al. (2014)
30	Improved BAT algorithm	Image processing	Zhou et al. (2014)
31	BAT algorithm	Sports training session planning	Alihodzic and Tuba (2014)
32	BAT algorithm	Fuel arrangement optimization for reactor core	Fister et al. (2014b)
33	Enhanced BAT algorithm	Skeletal structure design	Kashi et al. (2014)
34	modified BAT algorithms	Economic dispatch	Kaveh and Zakian (2014)
35	Discrete binary BAT algorithm	Multidimensional knapsack problem	Latif and Palensky (2014)
36	Metaheuristic BAT algorithm	Power system tuning	Sabba and Chikhi (2014)
37	Improved BAT algorithm	Multiple UCAVs	Sambariya and Prasad (2014)
38	Chaotic bat swarm optimization	Problem solving	Li and Peng (2014)
39	Improved BAT algorithm	Identify myocardial infarction	Jordehi (2015)
40	Self-adaptive BAT algorithm	Problem solving	Kora and Kalva (2015)
41	Chaotic BAT algorithm	Economic dispatching	Fister et al. (2014c)
42	Modified BAT algorithm	Distribution feeder problem	Adarsh et al. (2016)
43	Directional BAT algorithm	Continuous optimization problem	Kavousi-Fard et al. (2016)
44	BAT algorithm	Multi-objective optimization	Chakri et al. (2017)
45	Bat inspired algorithm	Load frequency control of hydrothermal power system	Tharakeshwar et al. (2017)
46	Island BAT algorithm	Optimization tasks	Elsisi et al. (2017)
47	BAT algorithm	Principal component analysis	Al-Betar and Awadallah (2018)

(continued)

Table 1 (continued)

S. No.	Variant	Application	References
48	Scanning bat algorithm	Maximum power point tracker	Cui et al. (2019)
49	BAT algorithm	Problem-solving	Eltamaly et al. (2020)
50	Discrete BAT algorithm	Event planning optimization	Huang and Ma (2020)
51	Modified BAT algorithm	Global optimization	Delalić et al. (2020)
52	Binary BAT Algorithm	Feature Selection	Ahmed et al. (2020)
53	BAT algorithm	Design of work scheduling	Qasim and Algamal (2020)
54	Discrete BAT algorithm	Pairwise biological network alignment	Batubara et al. (2020)
55	Modified BAT algorithm	Large-scale bound-constrained global optimization	Chen et al. (2021)
56	Improved BATt algorithm	Path planning for robot	Mashwani et al. (2021)
57	BAT algorithm	LQR weight optimization for vibration control in the vehicle suspension system	Yuan et al. (2021)

combination problem-solving (Yang and Gandomi 2012), clustering by k-means BA (Nakamura et al. 2013), function optimization (Komarasamy and Wahi 2012), global optimization (Lemma et al. 2011), problem-solving using hybrid BATt algorithm (Damodaram and Valarmathi 2012), economic load dispatch (Du and Liu 2012), feature selection (Faritha Banu and Chandrasekar 2012), global numerical optimization (Marichelvam and Prabaharam 2012), data analysis (Mishra et al. 2012), problem-solving using hybrid self-adaptive BAT algorithm (Musikapun et al. 2012), complex-valued bat algorithm (Reddy and Manoj 2012), chaotic BAT algorithm for problem-solving (Yang et al. 2012), and optimization (Xie et al. 2013). Problem-solving using BA with Gaussian walk (Jamil et al. 2013), modified BA (Fister et al. 2013), cloud model BA (Mallikarjuna et al. 2013), image processing (Ramesh et al. 2013), sports session planning (Taha et al. 2013), Fuel arrangement optimization (Wang and Guo 2013), skeletal design (Dao et al. 2014), economic dispatch (Fister et al. 2014a), solving multidimensional knapsack (Li and Zhou 2014), power system tuning (Gandomi and Yang 2014), multiple ucavs (Afrabandpey et al. 2014), identify myocardial infarction (Cai et al. 2014), problem-solving using self-adaptive BATt algorithm (Yılmaz et al. 2014), phishing website detection (Zhou et al. 2014), image matching (Alihodzic and Tuba 2014), record duplication (Fister et al. 2014b), energy modeling in a gas turbine (Kashi et al. 2014), realistic hybrid flow shop scheduling (Kaveh and Zakian 2014), classification of microarray data (Latif and Palensky 2014), scheduling problem (Sabba and Chikhi 2014), combined economic load and emission dispatch (Sambariya and Prasad 2014), optimal capacitor (Li and Peng 2014), topology optimization in microelectric applications (Jordehi 2015), economic dispatching (Kora and Kalva 2015), problem-solving using chaotic BAT swarm optimization (Fister et al. 2014c), continuous optimization problem (Adarsh et al. 2016), distribution feeder problem (Kavousi-Fard et al. 2016), multiobjective optimization (Chakri et al. 2017), load frequency control (Tharakeshwar et al. 2017), optimization task (Elsisi et al. 2017), principal component-based BA (Al-Betar and Awadallah 2018). Eltamly et. al. applied a scanning BATt algorithm for detecting maximum power point in photovoltaic energy system (Cui et al. 2019). BAT algorithm based on integration strategy and Gaussian distribution was given in 2020 (Eltamaly et al. 2020). A discrete BAT algorithm for event planning was given by Delalic et al. in the same year (Huang and Ma 2020). Huda et al. gave a modified BAT algorithm for global optimization (Delalić et al. 2020).Omar et al. used different transfer functions in the binary BAT algorithm for feature selection (Ahmed et al. 2020). Sumiharni et al. gave a design of work scheduling using BAT algorithm (Zakariya and Algamal 2020). Chen et al. used a discrete BAT algorithm for pairwise network alignment (Batubara and Sari 2020). For solving large-scale bound-constrained global optimization problems modified BAT algorithm was given by Mashwani et al., in 2021 (Mashwani et al. 2021).Yuan et al. gave an improved BAT algorithm for path planning of robots (Yuan et al. 2021).Yuvapriya et al. used the BAT algorithm for LQR weight optimization of vehicle suspension sytem (Yuvapriya et al. 2022).

5 Conclusion

Bat echolocation is one of the evolutionary hallmarks which helps them in nocturnal habits. The development of BA has given a new approach to formulating algorithms and their computation application. It is a metaheuristic model used for global optimization. The development of its new variants has magnified its applicability, thereby allowing for its use in solving multiple problems with global applicability. It seems to be a popular choice among researchers to find new areas of its application and add new variants for gaining precision in its problem-solving capability. There is still scope for increasing its efficacy by incorporating more parameters of bat foraging like catching insects that can echolocate to blind the bat as an escape. Further, the species specificity and environment similarity of bat calls is another aspect that may be incorporated. A combinatorial approach to combining BA with other algorithms can open new realms of problem-solving.

References

Adarsh BR, Raghunathan T, Jayabarathi T, Yang X-S (2016) Economic dispatch using chaotic bat algorithm. Energy 96:666–675

Afrabandpey H, Ghaffari M, Mirzaei A, Safayani M (2014) A novel bat algorithm based on chaos for optimization tasks. In: 2014 Iranian conference on intelligent systems (ICIS). IEEE, pp 1–6

Ahmed HI, Hamed ET, Saeed Chilmeran HTh (2020) A modified bat algorithm with conjugate gradient method for global optimization. Int J Math Math Sci 2020(Article ID 4795793):14. https://doi.org/10.1155/2020/4795793

Al-Betar MA, Awadallah MA (2018) Island bat algorithm for optimization. Expert Syst Appl 107:126–145. ISSN 0957-4174

Alihodzic A, Tuba M (2014) Improved bat algorithm applied to multilevel image thresholding. Sci World J

Altringham JD (1996) Bats: biology and behaviour. Oxford University Press

Batubara S, Sari DK, Wicaksono DA (2020) Design of Job scheduling using BAT algorithm to minimize makespan in hybrid flowshop. In: Proceedings of the international conference on industrial engineering and operations management, Dubai, UAE, 10–12 Mar 2020

Cai X, Wang L, Kang Q, Wu Q (2014) Bat algorithm with Gaussian walk. Int J Bio-Inspired Comput 6(3):166–174

Chakri A, Kehlif R, Benouaret M, Yang X-S (2017) New directional bat algorithm for continuous optimization problems. Expert Syst Appl 69:159–175

Chen J, Zhang Y, Xia J-F (2021) Pairwise biological network alignment based on discrete bat algorithm. Comput Math Methods Med 2021(ID 5548993):12. https://doi.org/10.1155/2021/5548993

Chu S, Tsai P, Pan J (2006) Cat swarm optimization. In: Proceedings of Pacific Rim international conference on artificial intelligence, pp 854–858

Cui Z, Li F, Zhang W (2019) Bat algorithm with principal component analysis. Int J Mach Learn Cyber 10:603–622. https://doi.org/10.1007/s13042-018-0888-4

Damodaram R, Valarmathi ML (2012) Phishing website detection and optimization using modified bat algorithm. Int J Eng Res Appl 2(1):870–876

Dao T-K, Pan J-S, Chu S-C, Shieh C-S et al (2014) Compact bat algorithm. In: Intelligent data analysis and its applications, vol II. Springer, pp 57–68

Delalić S, Alihodžić A, Tuba M, Selmanović E, Hasić D (2020) Discrete bat algorithm for event planning optimization. In: 2020 43rd International convention on information, communication and electronic technology (MIPRO), pp 1085–1090. https://doi.org/10.23919/MIPRO48935.2020.9245276

Dorigo M, Birattari M, Stiitzle T (2006) Ant colony optimization. IEEE Computational Intelligence Magazine, pp 28–39

Du ZY, Liu B (2012) Image matching using a bat algorithm with mutation. Appl Mech Mater 203(1):88–93

Elsisi M, Soliman M, Aboelela MAS, Mansour W (2017) Optimal design of model predictive control with superconducting magnetic energy storage for load frequency control of nonlinear hydrothermal power system using bat inspired algorithm. J Energy Storage 12:311–318

Eltamaly AM, Al-Saud MS, Abokhalil AG (2020) A novel scanning bat algorithm strategy for maximum power point tracker of partially shaded photovoltaic energy systems. Ain Shams Eng J 11(4):1093–1103. https://doi.org/10.1016/J.ASEJ.2020.02.015

Faritha Banu A, Chandrasekar C (2012) An optimized appraoch of modified bat algorithm to record deduplication. Int J Comput Appl 62(1):10–15

Fister I Jr, Fister D, Yang X-S (2013) A hybrid bat algorithm. Elektrotehniskivestnik 80(1–2):1–7

Fister I, Fong S, Brest J, Fister I (2014a) A novel hybrid self-adaptive bat algorithm. Sci World J

Fister I, Rauter S, Yang X-S, Ljubic K, Fister I Jr (2014b) Planning the sports training sessions with the bat algorithm. Neurocomputing

Fister I, Fong S, Brest J, Iztok F (2014c) Towards the self-adaptation in the bat algorithm. In: Proceedings of the 13th IASTED international conference on artificial intelligence and applications

Gandomi AH, Yang X-S (2014) Chaotic bat algorithm. J Comput Sci 5(2):224–232

Hedayatzadeh R, Salmassi A (2010) Termite colony optimization : a novel approach for optimizing continuous problems. In: Proceedings of Iranian conference on electrical engineering (ICEE 2010), pp 553–558

Huang J, Ma Y (2020) Bat algorithm based on an integration strategy and Gaussian distribution. Math Probl Eng 2020(Article ID 9495281):22. https://doi.org/10.1155/2020/9495281

Jamil M, Zepernic H-J, Yang XS (2013) Improved bat algorithm for global optimization. Appl Soft Comput

Jordehi AR (2015) Chaotic bat swarm optimisation (CBSO). Appl Soft Comput 26:523–530

Kashi S, Minuchehr A, Poursalehi N, Zolfaghari A (2014) Bat algorithm for the fuel arrangement optimization of reactor core. Ann Nucl Energy 64:144–151

Kaveh A, Zakian P (2014) Enhanced bat algorithm for optimal design of skeletal structures. Asian J CivialEng 15(2):179–212

Kavousi-Fard A, Niknam T, Fotuhi-Firuzabad M (2016) A novel stochastic framework based on cloud theory and θ-modified bat algorithm to solve the distribution feeder reconfiguration. IEEE Trans Smart Grid 7(2):740–750

Kennedy J, Eberhart R (1995) Particle swarm optimization. In: Proceedings of IEEE international conference on neural networks, pp 1942–1948

Khan K, Nikov A, Sahai A (2011) A fuzzy bat clustering method for ergonomic screening of office workplaces. In: Third international conference on software, services and semantic technologies S3T 2011. Advances in intelligent and soft computing, vol 101

Komarasamy G, Wahi A (2012) An optimized K-means clustering technique using bat algorithm. Eur J Sci Res 84(2):263–273

Kora P, Kalva SR (2015) Improved bat algorithm for the detection of myocardial infarction. SpringerPlus 4:666. https://doi.org/10.1186/s40064-015-1379-7

Kumar B, Kumar D (2013) A Review on artificial bee colony algorithm. Int J Eng Technol 2(3):175–186

Latif A, Palensky P (2014) Economic dispatch using modified bat algorithm. Algorithms 7(3):328–338

Lemma TA., Bin Mohd Hashim F (2011) Use of fuzzy systems and bat algorithm for exergy modelling in a gas turbine generator. In: IEEE colloquium on humanities, science and engineering (CHUSER'2011), 5–6 Dec 2011, pp 305–310

Li L, Zhou Y (2014) A novel complex-valued bat algorithm. Neural Comput Appl 1–13

Li YG, Peng JP (2014) An improved bat algorithm and its application in multiple ucavs. Appl Mech Mater 442:282–286

Mallikarjuna B, Reddy KH, Hemakesavulu O et al (2013) Economic load dispatch problem with valve-point effect using a binary bat algorithm. ACEEE Int J Electr Power Eng 4(3)

Marichelvam MK, Prabaharam T (2012) A bat algorithm for realistichybridflowshopschedulihng problems to minimize makespan and mean flow time. ICTACT J Soft Comput 3(1):428–433

Marimuthu G (1996)The strange way of bats. Resonance 40–48

Mashwani WK, Mehmood I, Bakar MA, Koçcak I (2021) A modified bat algorithm for solving large-scale bound constrained. Glob Optim Prob 2021(Article ID 6636918). https://doi.org/10.1155/2021/6636918

Mishra S, Shaw K, Mishra D (2012) A new meta-heuristic bat inspired classification approach for microarray data, Procedia Technol 4:802–806.8

Musikapun P, Pongcharoen P (2012) Solving multi-stage multi-machine multi-product scheduling problem using bat algorithm. In: 2nd International conference on management and artificial intelligence (IPEDR), vol 35, IACSIT Press, Singapore, pp 98–102

Nakamura RYM, Pereira LAM, Rodrigues D, Costa KAP, Papa JP, Yang X-S (2013) Binary bat algorithm for feature selection 9—Binary bat algorithm for feature selection

Qasim OS, Algamal ZY (2020) Feature selection using different transfer functions for binary bat. Int J Math Eng Manage Sci 5(4):697–706. https://doi.org/10.33889/IJMEMS.2020.5.4.056

Ramesh B, Mohan VCJ, Reddy VCV (2013) Application of bat algorithm for combined economic load and emission dispatch. Int J Electr Eng Telecommun 2(1):1–9

Reddy VU, Manoj A (2012) Optimal capacitor placement for loss reduction in distribution systems using bat algorithm. IOSR J Eng 2(10):23–27

Sabba S, Chikhi S (2014) A discrete binary version of bat algorithm for multidimensional knapsack problem. Int J BioInspired Comput 6(2):140–152

Sambariya D, Prasad R (2014) Robust tuning of power system stabilizer for small signal stability enhancement using metaheuristic bat algorithm. Int J Electr Power Energy Syst 61:229–238

Taha AM, Mustapha A, Chen S-D (2013) Naive Bayes-guided bat algorithm for feature selection. Sci World J

Tharakeshwar TK, Seetharamu KN, Prasad BD (2017) Multi-objective optimization using bat algorithm for shell and tube heat exchangers. Appl Therm Eng 110:1029–1038

Umar SU, Rashid TA (2021) Critical analysis: BAT algorithm-based investigation and application on several domains, World J Eng. https://doi.org/10.1108/WJE-10-2020-0495

Wang G Guo L (2013) A novel hybrid bat algorithm with harmony search for global numerical optimization. J Appl Math

Xie J, Zhou Y, Chen H (2013) A novel bat algorithm based on differential operator and Lévy flights trajectory. Comput Intell Neurosci 2013:453812. https://doi.org/10.1155/2013/453812. PMID:23606827;PMCID:PMC3628216

Yang XS (2010) A new metaheuristic bat-inspired algorithm, In: Nature inspired cooperative strategies for optimization (NISCO 2010). Studies in computational intelligence, vol 284, pp 65–74

Yang X (2010) A new metaheuristic bat-inspired algorithm. In: Proceedings of nature inspired cooperative strategies for optimization (NICSO 2010), pp 65–74

Yang XS (2011) Bat algorithm for multi-objective optimisation. Int J Bio-Inspired Comput 3(5):267. https://doi.org/10.1504/ijbic.2011.042259 (Inderscience Publishers)

Yang XS, Karamanoglu M, Fong S (2012) Bat aglorithm for topology optimization in microelectronic applications. In: IEEE international conference on future generation communication technology (FGCT 2012), British Computer Society, 12–14 Dec 2012, London, pp 150–155

Yang X-S, Gandomi AH (2012) Bat algorithm: a novel approach for global engineering optimization. Eng Comput 29(5):464–483

Yılmaz S, Kucuksille EU, Cengiz Y (2014) Modified bat algorithm. Electron Electr Eng 20(2):71–78

Yuan X, Yuan X, Wang X (2021) Path Planning for mobile robot based on improved bat algorithm. Sensors (basel, Switzerland) 21(13):4389. https://doi.org/10.3390/s21134389

Yuvapriya T, Lakshmi P, Kumar V (2022) Experimental validation of LQR weight optimization using BAT algorithm applied to vibration control of vehicle suspension system. IETE J Res 1–11. https://doi.org/10.1080/03772063.2022.2039079

Zhou Y, Xie J, Li L, Ma M (2014) Cloud model bat algorithm. Sci World J

Social, Emotional and Ethical (SEE) Attributes, Which Configures Our Bioinformatics Systems to Activate the Hidden Forces to Shape Human Decisions

Arif Md. Sattar, Mritunjay Kr. Ranjan, and Gajendra Singh

1 Introduction

Individuals' social, emotional and ethical characteristics such as positive emotional control, self-compassion and interpersonal skills have been noticed and the majority of these characteristics are associated with well-being while making or taking decisions at important stages. The social, emotional and ethical characteristics of a person are also responsible for the development of their understanding and decision-making abilities. One of the most effective decision-making features across nations and cultures, as well as religious and non-religious sects of society, may be achieved via the use of the NSA (Non-Sectarian Approach). According to SE (Secular Ethics), NSA (Non-sectarian Approach) will lead to universal ethics that may be accepted by individuals of any religious belief system is a choice. The development of SEE (Social, Emotional and Ethical) characteristics by the human brain results in the development of a perfect man in terms of his behaviour and well-being towards himself and others. Human brain orientation is changed by our actions and well-being, which allows us to make the best choice possible. In addition, the majority of pioneers and those who have considered the feasibility and promise of SEE (Social, Emotional and Ethical) approach have made substantial contributions to society and

A. Md. Sattar (✉)
Department of Computer Science, Anugrah Memorial College, Gaya, India
e-mail: amsattargaya@gmail.com

M. Kr. Ranjan
Department of Computer Science, Magadh University, Bodh Gaya, India
e-mail: mritunjaykranjan@gamil.com; mritunjay.ranjan@sandipuniversity.edu.in

SOCSE, Sandip University, Nashik, India

G. Singh
Piramal Foundation for Education Leadership, Jhunjhunu, India
e-mail: gajendra@gandhifellowship.org

K. Raza (ed.), *Nature-Inspired Intelligent Computing Techniques in Bioinformatics*,
Studies in Computational Intelligence 1066,
https://doi.org/10.1007/978-981-19-6379-7_10

all sects of the globe, which have been profoundly affected by the SEE learning framework. An ethical and charitable mindset is rooted in fundamental human principles that may be detected via common sense, common experience and science, according to the Dalai Lama Layard et al. (2016). Furthermore, the Dalai Lama has had a long-standing association with Emory University that dates back more than two decades. As a result of this partnership, ground-breaking research in fields such as compassion science has been conducted, paving the way for future collaborations across cultures and disciplines. Its ultimate goal has been to aid in the accomplishment of both his and the university's vision of an education that is both heartfelt and intellectually challenging. The notion that nurturing fundamental human values may be beneficial to oneself and others is gaining momentum fast as research from across the globe illustrates the link between ethical principles and thriving in many aspects of life. As stated in the United Nations' 2016 edition of the World Happiness Report, 'We should judge human development by the degree to which people are enjoying their lives—by the predominance of happiness and, inversely, the absence of sorrow… We should, in all of our transactions, sincerely hope for the happiness of all of those who are within our power to influence and we should nurture in ourselves an attitude of unconditional compassion (Heckman et al. 2012).

Based on common sense, common experience and science derived from the foundations of human interdependence, the SEEL framework leads to the ideal man. SEE Learning takes its cues from psychologists, educators and neuroscientists who are investigating the nature of fundamental human values and prosocial emotions and competencies and how they might be taught and fostered as a basis for their approach. Education should not just concentrate on cognitive skills but also on emotional intelligence, such as emotional awareness and social skills, the capacity to work and collaborate with other people and the ability to cope constructively with conflict. According to Dr. James Heckman, a Nobel laureate in economics who pushes for life skills to be taught in schools, these talents are just as critical to success in life as cognitive aptitude exams ("Sample teaching activities to support core competencies of social and emotional learning," 2021). Increasingly, businesses in a wide range of industries are realizing the value of life skills when it comes to selecting new employees. Because of this, we may conclude that the SEE framework helps companies make recruiting decisions throughout the applicant screening process (Goleman and Senge 2014).

SEE Learning is built on three aspects, each of which encompasses a different form of knowledge or skill that it aims to develop in people: (i) *Compassion*, (ii) *Awareness* and (iii) *Engagement*. What's more, the space beyond the standard three dimensions may be broken down into three areas of study: (i) *Personal*, (ii) *Social* and (iii) *Systems* (Fig. 1).

As Goleman and Senge (2014) recommended in their book: 'The Triple Focus: A Framework for Success', this framework is founded on the evidence-based work of SEL: (i) *a focus on self*, (ii) *a focus on others* and (iii) *a focus on interdependence and systems*. Awareness, Compassion and Engagement (SEE Learning) are portrayed as overlapping because they are intertwined. The various skills may be experienced but are best understood in the context of the total as shown in Fig. 1. An person or a

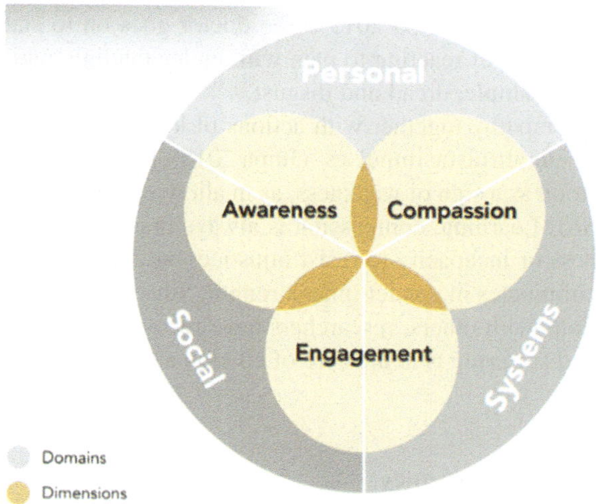

Fig. 1 Various aspects of SEE learning. *Source* The SEE Learning Companion

group must first be aware of the issue or problem in order to take positive action. Second, one must care and feel invested in order to act. Finally, behave wisely. The trinity of awareness, compassion and skillful involvement may benefit oneself and others. This relates to 'head,' 'heart,' and 'hand'. SEE Individually and collectively, learning divides things down into three dimensions.

2 The Three Domains and Three Dimensions

2.1 Compassion

Compassion is at the core of SEE Learning's three dimensions. Compassion is growing compassion, empathy and care for both pleasure and suffering in oneself, others and humanity as a whole. Compassion and kindness cannot be cultivated by simply telling individuals or groups to be empathetic. Compassion requires critical thought. This is a sort of critical thinking that strives to comprehend one's own and others' needs, goals, and values. This involves knowing what will benefit oneself and others in the long run. When this perspective extends beyond oneself, it involves understanding and discerning others' needs, finally acknowledging shared humanity. It is crucial to cultivate self-compassion and compassion for others so that we be aware of our own and others' mental states, particularly their feelings of pleasure and pain. Compassionate global citizens must be aware of their connections and the larger systems in which they live. Similarly, the abilities and behaviours necessary for compassion—whether directed toward oneself, others, or larger communities—must be developed. 'Compassion' is defined by Dr. Thupten Jinpa, a famous compassion scholar and creator of Stanford University's Compassion Cultivation Training programme (Jinpa 2015). The doctor goes on to state that compassion 'offers the potential of reacting to pain with understanding, patience and kindness rather than, for example, dread and disgust…'. 'Compassion is the glue that holds the emotion of empathy together with actions of kindness, generosity and other manifestations of our altruistic impulses' (Jinpa 2015). In certain cultures, compassion is misunderstood as a sign of weakness, as in allowing bullying or other undesirable acts. But in SEE Learning, compassion is always brave compassion. It does not suggest weakness or incapacity to resist injustice, but rather care for others that arises from and culminates in greater inner strength. Since ethics pertains to our actions and interactions with others, researchers have recognized that empathy for others' well-being and suffering is at the core of all ethical thought.

2.2 Awareness

In certain cultures, compassion is misunderstood as a sign of weakness, as in allowing bullying or other undesirable acts. But in SEE Learning, compassion is always brave compassion. It does not suggest weakness or incapacity to resist injustice, but rather care for others that arises from and culminates in greater inner strength. Since ethics pertains to our actions and interactions with others, researchers have recognized that empathy for others' well-being and suffering is at the core of all ethical thought (Goleman and Senge 2014).

2.3 Engagement

One may be aware of a need or opportunity, yet lack the skills to act appropriately and participate successfully. Engagement is the third SEE Learning component. Engagement relates to how one puts into practice the awareness and compassion elements. This component relates to learning and habituating behaviours that promote personal, social and community well-being. This involves self-regulation in the personal domain, social skills and the capacity to connect to others in the social domain and global citizenship awareness and action in the system's domain.

3 The Three Domains Are Three Types of Things

By recognizing compassion as the core of SEE Learning education, the scene is created for individuals/majorities to become more aware of their behaviours and to stop harming themselves and others. This leads to the first SEE Learning domain, Personal, which focuses on self-care. The second category, Social, broadens this emphasis to encompass empathy and compassion for others, as well as good interpersonal abilities for connecting to others. Finally, the Systems domain aims to assist students to acquire global awareness, beliefs and abilities so they may become responsible decision-makers and global citizens. They are *Personal Domain, Social Domain* and *System Domain.*

3.1 The Personal Domain

As global citizens who make responsible choices that benefit themselves and others, SEE Learning is meant to assist in common individually, socially and emotionally. Although all three domains may be tackled separately and in any sequence, the Personal domain greatly supports the Social and Systems domains. Individuals and groups must learn to care for themselves before they can care for others and make ethical decisions. To learn to attention to the needs of individuals, communities and even the whole planet, they must first learn to attend to their own. In SEE Learning, this involves developing 'emotional literacy' and related abilities like attentiveness. Aspects of emotional literacy ability to perceive emotions, relate them to a wider context, including one's own needs, gain discernment about the impact of emotions and manage them effectively. Finally, emotional literacy helps individuals and groups to avoid reactive and impulsive conduct that may hurt themselves or others, while maintaining the mental tranquillity required to make solid long-term judgements. As a result, it is a critical skill for individual success.

3.2 The Social Domain

Emotional literacy and self-regulation are abilities that pupils will need throughout their lives. Because humans are social creatures, the ability to connect successfully to others is also vital. Prosocial qualities may be nurtured via learning, reflection and purposeful practice, according to scientific studies. These practices produce observable changes in the brain, body and behaviour, as well as advantages for physical, mental and social health. Based on accumulating data, an education designed to help children thrive should incorporate not just self-regulation abilities, but also social skills. Social in this sense refers to direct interpersonal relationships.

3.3 The Systems Domain

But we don't only talk to each other one-on-one. Compassion alone is not enough in our increasingly complicated society to achieve effective ethical participation; it must be accompanied by responsible decision-making based on a grasp of the larger systems in which we live. Even acts inspired by compassion might have unforeseen negative repercussions if they are not approached from many viewpoints or evaluated over time. Individuals/majority are growing up in a more sophisticated, global and interconnected environment. The difficulties facing current and future generations are vast and far-reaching, requiring a new collaborative, multidisciplinary and systems-oriented method of thinking and problem-solving.

Nine Domains and Dimensions components. With nine components, SEE Learning's three dimensions and three domains may be matriculated. Systems thinking is an essential new technique of issue solving in today's environment. Systems thinking seeks to understand the causes and conditions that give rise to the problem in context, then address the system's components and processes to solve not just the current problem, but future possible manifestations of it. As per Fig. 2, each domain will study each dimension, yielding nine components. Each component has its own name for clarity. In other circumstances, a linear approach makes logical since some subjects build on and rely on others. To apply emotional literacy to others and larger systems, individuals/majority must first achieve a degree of emotional literacy themselves. Compassion and Engagement need similar abilities learned in the realm of awareness. The emphasis is on increasing capabilities across all nine components, which are strongly interconnected.

4 Literature Review

4.1 How to Take a Decision in a Multi-Solution Environment?

It caught Dan Ariely interest and led him to research human behaviour. Researchers employed physical pain such as heat, cold water, pressure, loud sounds and even the psychological pain of losing money in the stock market to find the answers (Heyman and Ariely 2004). By the time when he was finished, he realized that the burn unit nurses were nice and caring people with a lot of experience applying and removing bandages, but they lacked the proper mentality for making their patients' suffering as comfortable as possible. How could they be so incorrect? He wondered, knowing that their actions were not due to malice, stupidity, or carelessness, given their prior experience and the fact that he knew this nurse personally. He explained to the nurses and doctors that when therapies are performed at a lower intensity and for a longer period of time, patients are less in pain than when the same goal is achieved at a greater intensity and for a shorter period of time. To put it another way, if they had carefully removed the bandage rather than quickly. Behavioural economics is a relatively recent field that combines psychology and economics. They have attempted to understand not just the action but also the decision-making process that goes into such behaviour, whether it's yours, mine, or someone else's. What a bit of work a man is! How Nobel in reason, how infinite in faculty, in form and moving, how

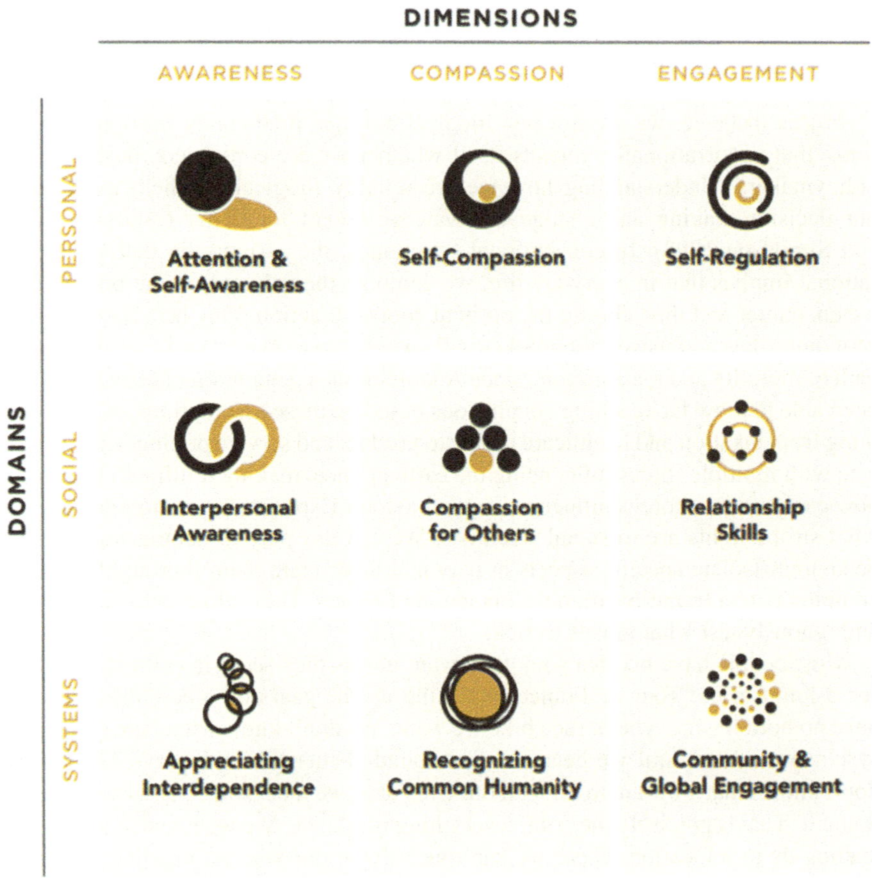

Fig. 2 The three domains and three dimensions. *Source* The SEE Learning Companion

expressive and admirable, in form and moving, how expressive and admirable, in action, how much like Angel's play, in apprehension, how like a god, the beauty of the world, the paragon of animals. This quote expresses a commonly held view of human nature held by economists, politicians, non-professionals and regular people alike. In fact, this opinion is mostly correct. Our bodies and minds can accomplish extraordinary things. We can observe a ball being thrown from a long distance and calculate its trajectory and impact in real-time before moving our bodies and hands to catch it. This assumption about our ability to reason flawlessly has made its way into economics from the field of science. The core concept of rationality serves as the foundation for hypotheses, predictions and recommendations in economics (Tversky and Kahneman 1981). From this perspective and to the extent that we believe in human rationality, we are all economists. I don't mean that we can all instinctively develop complex game theoretical models or understand the generalized axioms of revealed choice; rather, we share the core human nature ideas that economics is built on (Simonson 1989). It is taken aback by surprise. Although human capability is unquestionably justified, there is a huge difference between genuine admiration and preconceived notions about our reasoning capacities.

In this instance, we are not just illogical but also predictably irrational, in the sense that our irrationality repeats itself whether we are consumers, businesses or policymakers. Understanding how we are reliably illogical can help us improve our decision-making and change the way we depart for better results (Tversky and Simonson 1993). In conventional economics, the assumption that we are all rational implies that in everyday life, we compute the value of all the possibilities we encounter and then choose the optimal course of action. Play here too; classical economics has an answer, the market will sweep down owners and return us to the path of morality and reason soon. Since Adam Smith, a generation of economists has been able to draw far-reaching conclusions based on these assumptions about everything from taxation and healthcare policy to product and service pricing. Life is intricate with multiple forces influencing the earth at once, making it difficult to explain how each of these forces influences our behaviour. Experiments are to microscopes what strobe lights are to social scientists. We can use play to demonstrate human behaviour. Isolate specific aspects at play and investigate them thoroughly, adding complexity to a frame-by-frame explanation of events. They allow us to directly and ambiguously test what makes us tick.

Most people have no idea what they want unless they see it in context. Until we see a jump in the Tour de France ratcheting up the gears on a certain model, we have no notion what type of race bike we want. We don't know what type of speaker system we prefer until we hear one that sounds better than the previous one. We don't know what we want to do with our lives until we meet someone who is already doing it. That is precisely the point: everything is relative. We want to flee with lights leading us to a location where we can touch down our wheels, just like an airline pilot landing in the dark.

The central circle, as can be seen in Fig. 3, does not appear to be able to sustain its size. As it is placed within the wider circle, it shrinks. It expands when placed among the smaller circles. The size of the centre circle is the same in both cases, but it appears to fluctuate depending on what is placed adjacent to it. This may appear to be a passing interest, yet it reflects how the mind is the word. We can't help but compare our surroundings to what other people are doing. This is true not only for physical items such as two-star bicycles, puppies, restaurant reservations and dance jobs, but also for experiences such as vacations, educational chances and ethical concerns such as one's emotions, attitude and point of view. We compare jobs to jobs, vacations to vacations, lovers to lovers and wines to wines all the time. All of this relativism makes me think of a line from Crocodile Dundee. Although relativity is straightforward to understand, one aspect of it continues to perplex us. As a result, we not only compare things to one another, but we also compare things that are easy to compare and avoid comparing things that are difficult to compare. To better understand how relativity works, consider the following illustrations (Fig. 4).

On the left side of Fig. 4, we can see two options, each of which is superior in a different way. Option A is superior in every respect. To define it, let's use the word

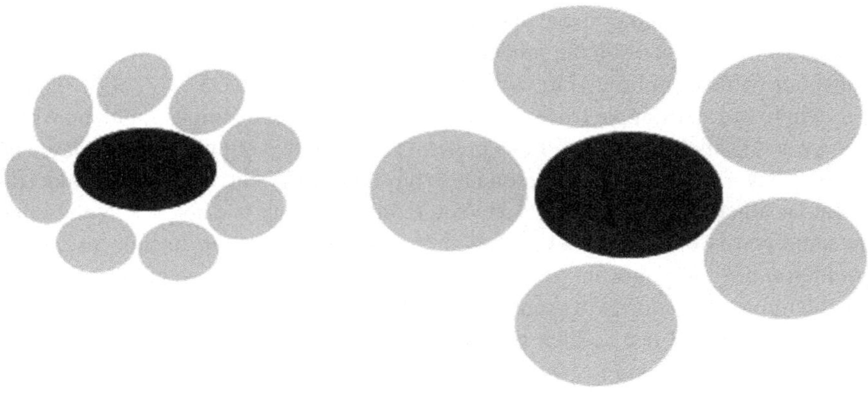

Fig. 3 *Source* Predictably irrational by Dan Ariely

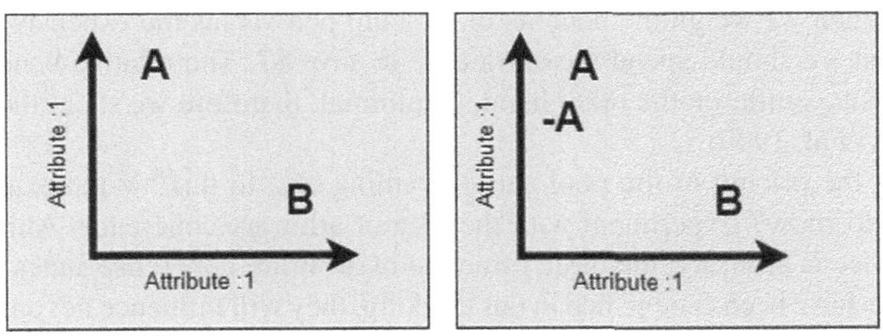

Fig. 4 Example to better understand how relativity works. *Source* Predictably irrational by Dan Ariely

'quality.' Option B is superior in the second attribute, attractiveness. Obviously, these are two quite different options and deciding between them is difficult. Consider what happens if we introduce a new option, '−A.' This option is clearly inferior to option A, but it is also relatively comparable to it, simplifying the comparison and stating that A is not only superior to A-B but also superior to B. In some ways, presenting −A as the decoy establishes an easy relative contrast with A, making A appear superior not only to minus-in the collection but also to people's final choices, even if no one chooses it. Relativity helps us make decisions in life. It has the ability, though, to make us exceedingly sad. Why? When we compare our circumstances to those of others, we get envious and jealous. After all, you are not to seek your neighbour's house, lands, male or female slave, donkey, or anything else that belongs to your neighbour, according to the Ten Commandments. Given our natural need to compare, this may be the most difficult commandment to follow. The good news is that we can sometimes alter the circles that surround us, shrinking them and increasing our happiness. If we are at our class reunion and there is a giant circle in the middle of the room with a drink in his hand marketing his large pay, we can take a few steps apart and talk to someone else. If we're looking for a new home, we may be choosy about which open houses we go to, avoiding those that are out of our price range; if we're looking for a new car, we can focus on the models that we can afford and so on. We might also adjust our focus from limited to wide (Ariely and Wallsten 1995).

Let here use an example from a search undertaken by two famous scholars, Amos Starsky and Daniel Kahneman. Let's pretend you have two errands to run today. The first is to buy a new pen and the second is to buy a business suit. You find a wonderful pen at an office supply store for $25. You're about to buy it when you realize that the same pen is on sale for $18 at another store 15 min away. In this case, what would you do? Do you want to save $7 by driving 15 min? When confronted with this dilemma, the majority of Americans said they would travel to save $7. Now it's time for the second task, which is suit shopping. You find a lovely grey pinstripe suit for $450 and decide to buy it, but a client whispers in your ear that the same suit is on sale for only $443 at a store only 15 min away. In this instance, do you take the second 15-min trip? 'No,' most people say. This is the relativity problem: we evaluate our choices in relation to the alternatives accessible in the immediate vicinity. When we weigh the benefits of the chip pen versus the expensive one, it's evident that we should spend the extra time to save $7. The relative benefit of the less expensive outfit, on the other hand, is minimal, therefore we spent the extra $7 (Sedikides et al. 1999).

Why is the pricing of the pool and everything else in flux? What exactly were we trying to show? Experiment with the idea of arbitrary coherence. Although the original price is arbitrary, the basic principle of arbitrary coherence states that once such values have been established in our thinking, they will influence not only present but also future prices. Dan Ariely decided to conduct 17 interviews to answer these questions.

So, let's take a slow walk through the Economist's deception in Fig. 5. The options were, as you recall:

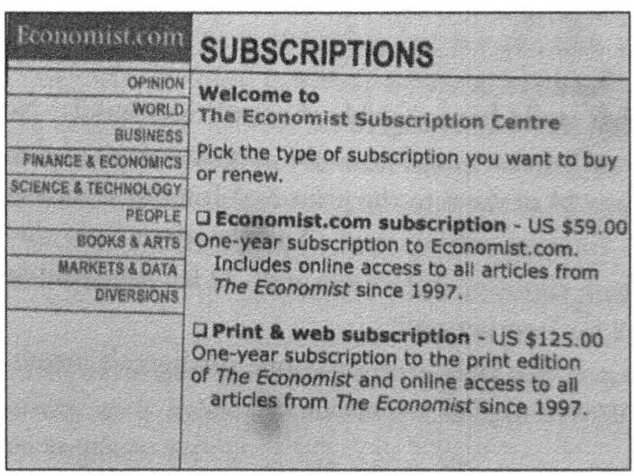

Fig. 5 The economist's deception. *Source* Predictably irrational by Dan Ariely

$59.00 for an internet-only membership.

$125 for a print-only subscription.

$125 for a print-and-internet membership.

When he presented these options to a group of 100 students at MIT's Sloan School of Management, they chose:

Students pay $59-16 for an internet-only membership.

$125 for a print-only subscription–no students

Students pay $125–84 for a print-and-internet subscription.

So far, these Sloan MBAs have proven to be sharp. They could all understand how the print-and-internet deal was better than the print-only deal. But were they convinced by the print's mere presence—the only option? To put it another way, imagine that, would the students reply the same way they did before (i.e. 16 for Internet only and 84 for both) (Fig. 6)? He was sure they'd react the same way, wouldn't they? After all, the choice he removed was one that no one chose, so it shouldn't matter. Is right? Quite the opposite!

This time, 68 students chose the $59 Internet-only option, up from 16 the previous time. Only 32 people chose the $125 combined subscription, down from $84 previously (mentioned in his Book, Predictably Irrational, Harper).

Here we ran an experiment in which we tried to entice individuals away from their previous anchors, toward new opportunities but this was only the start of our adventure. The experiment's next two rounds will allow us to address these concerns. Now that we know how our original decision becomes established in our long-term habits. Consider the case below: As you go past a restaurant, you observe two people waiting to enter. This has to be a fantastic restaurant. You take a place behind them because you believe people are in line. Someone another walks by and notices three individuals in line. He approaches the queue, thinking to himself, 'This must be a

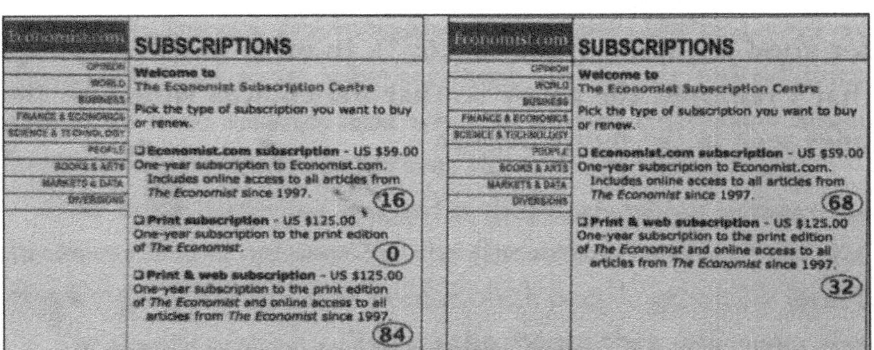

Fig. 6 The economist's deception: subscriptions. *Source* Predictably irrational by Dan Ariely

fantastic restaurant.' This is referred to as 'Harding behaviour'. When we make a decision about anything based on the actions of others, our own actions reflect that decision. Because George Dragon and Dan were so thrilled about the experiment on coherent arbitrariness, they decided to take it a step further. We will demonstrate the many decisions we make in play from minor to important in which anchoring plays a role for one day. We choose the best alternative depending on our fundamental ideas and dislikes, according to economic theory. Market prices, according to traditional economic theory, are decided by the balance of two forces of production at each price, as well as the demand of people with purchasing power at each price. The market price is determined by the point at which these two forces intersect. Consider your current milk and wine usage. The relationships we perceive in the marketplace between demand and supply in the context of arbitrary coherence are based on memories rather than preferences. Consider the likelihood of two new taxes being applied the next day. One will lower the price of wine by 50%, while the other will boost the price of milk by 100%. What do you expect to happen as a result of the price increase? Many people will be happier and have less calcium as a result of this change. Consider what would happen if higher taxes were accompanied by induced forgetfulness about previous wine and milk costs. What if prices fluctuate in the same way, but you don't remember how much you spent on these two goods before? (Ariely et al. 2003).

Another implication of arbitrary coherence is the supposed benefits of the free market and free trade. The fundamental concept of the free market is that if we have something you value more than I do, such as a sofa, trading it will benefit both of us. This means that the mutual advantage of trading is predicated on the assumption that all market players are aware of the worth of what they have and what they desire to gain from the trade (Ariely et al. 2006).

We may need to look elsewhere if we can't rely on supply and demand to identify the right market size and the free-market mechanism to help us maximize our utility. This is especially true when it comes to societal requirements such as healthcare, medicine, water, electricity, education and other critical resources. If you believe that market forces and free markets do not always manage the market in the best interests of consumers, you might agree with those who believe that the government should play a bigger role in regulating some market activities, even if it means restricting free enterprise. A free market based on supply and demand would be ideal if humans were actually sensible. Irrational policies should consider this essential issue even if we are not logical (Sunstein et al. 2002; Simonsohn and Loewenstein 2006).

Why do we regularly overpay for things we don't need? Zero has a long and distinguished history. The Indian astronomer Aryabhata exclaimed 'SatnamSanthanam-Dasagunam' one morning around 498 AD, which translates as 'Place to place in 10 times in value.' The decimal best place value notation was born as a result. Zero had been assigned a task (Simonsohn and Loewenstein 2006).

Why do you wait till an irrational hour to take advantage of the freebie, even if it isn't what you really want? What is it about the freedom that makes it so appealing? Most transactions have advantages and cons, but we tend to overlook the problems when something is free. Polestar 3 produces such an emotional response in us that we mistakenly believe that what is being offered is significantly more valuable than it is. Why do we think it's because people are naturally afraid of losing? This worry is linked to the genuine attractiveness of being free. There is no clear risk of losing when we choose each item. However, if we choose a non-free item, we risk making a poor decision and possibly losing money. As a result, if given the choice, select option 3. Zero isn't just another price for this region in the world of pricing. Although ten cents can have a significant impact on demand, nothing compares to the emotional charge of being free; the zero price effect is in a league of its own. On your PC, play RA (Shampanier et al. 2007).

Several years ago, Amazon.com began offering free shipping on orders over a certain amount. Someone who purchased a single book for $16.95 may have to pay an additional $3.95 for delivery, but if they purchased another book for $31.90, they would receive free shipping. Some buyers didn't need the second book, but the free delivery was so enticing that they were prepared to pay for it. Employees at Amazon were thrilled by the offer, but they soon learned that sales in one country, France, had not grown. Are Amazon consumers more rational than the general public? The French customers, rather than reacting to a different bargain, turned out to be reacting to a different deal (Kahneman and Tversky 1979).

When given an option between two products, we typically choose the one that is free. A free checking account may be preferred over a $5-per-month checking account. However, if the $5 checking account contains free traveller's checks, online billing and other features that the free account does not, we may wind up paying more for this package of services with the free account. Similarly, we may opt for a mortgage with no closing costs but exorbitant interest rates and a free and we may wind up with a product we don't need just because it comes with a free gift. Zero could influence food purchases. On the side of the box, food manufacturers must disclose various types of information, such as calories, fat content and fibre content. Is it possible that we have the same attraction to zero pricing as we do to zero calories, zero trans-fat, zero carbs and so on? If the underlying conditions remain the same, Pepsi will sell more cans if the label reads 'zero calories' rather than 'one calorie.' This isn't simply another price cut. Zero is a one-of-a-kind site. The difference between two cents and one cent is insignificant, but the difference between one cent and zero is enormous. Why do we put up with doing things we aren't paid to do? When enough money is discussed, market norms emerge, but market norms encompass a wide range of behaviours, including self-reliance, assisting and individualism. Kathleen Bose did a series of excellent research to explore if simply thinking about money may encourage people to act differently in this area. Social interactions define one, while market dealings define the other. As indicated by rates, social norms and relational hearts, we apply diverse norms to these two sorts of relationships, as well as incorporate market norms into social interactions. It's difficult to resume social engagement after a mistake like this. After you agreed to pay for the lovely Thanksgiving dinner song, your mother-in-law will remember the incident for years. When a social norm and a commercial norm collide, the social norm disappears for an extended period of time. To put it another way, re-establishing social relationships is difficult. When the rose's bloom has faded. It is unlikely to recover once a market norm has become a social norm. Our personal lives are affected by the fact that we live in both a social and a market world. From time to time, we all need someone to help us move something, watch our kids for a few hours, or collect our mail while we're out of town. What is the most effective method for convincing our friends and neighbours to help us? Isn't it feasible that the play will give it to you as a present? How much or how little societal impact should a song have? As you are surely aware, this social ballet is difficult to perfect, especially when a connection is on the verge of becoming a commercial transaction. The tight balance between social and commercial rules is also seen in the corporate sector. Businesses have tried to present themselves as social companions in recent decades, i.e. they want us to believe that we are family, or at the very least pals who share the same culture. One well-known expression is 'like a kind neighbour'; another is 'Home Depot softly encouraging you to do it.' We can help whoever started the customer service revolution and I had a great suggestion. Customers and a company or family that collaborate benefit the company in a variety of ways. It is impossible to overestimate the value of loyalty. Companies have also attempted to instill social values in their employees. This wasn't always the case. The American workforce used to be more of a market-driven industrial exchange. Working from 9 a.m. to 5 p.m. was standard practice back then. You worked for 40 h and got paid on Friday. Because workers were paid by the hour, they knew when they were working for the boss and when they weren't. When the factory whistled blue, the corporate equivalent, the sale was finalized. This was a simple market transaction that benefited both parties. Companies must understand that treating employees the same way they treat customers demonstrates a long-term commitment. Employees must receive something similar in return if they promised to work harder to meet a critical deadline or if they were asked to board a plane at a moment's notice to attend a meeting, such as support if they become ill or a chance to keep their jobs when the market threatens to take them away. We should think about workplace social standards on a regular basis. Workers' abilities and efforts are becoming increasingly crucial to the country's productivity. Standardized testing and performance-based compensation, in Dan's opinion, will push education away from social standards and toward market norms. The United States has more money bars than any other Western country (Williamson 2008).

5 Mathematical Modulation

We are looking for important elements of human behaviour in the decision-making process in this study. The study also offers a mathematical decision-making model that uses personalization functions and the current state as the key group factors.

6 Model of Decision-Making

In the field of cognitive psychology, the decision-maker model is quite valuable. It has been noticed that decisions correspond to information acquisition, whereas information processing allows an individual to pick among many options and, as a result, act on that decision. This process is crucial and consists of six stages in order. The two models that represent these stages are personalization and instantaneous state (Bezerra et al. 1996).

6.1 The Personalization

The factors of personalization refer to the invariant characteristics of an individual that can be classified into nine categories, each of which corresponds to different personality variables. Ability to cope with reality, knowledge, relation to time, expansiveness, ego, creativity, risk shift, anxiety level and authoritarian attitude are the nine individual categories (Bezerra et al. 1996).

6.2 The Present Situation

The invariant particularities of the circumstance (at the time when the decision is taken) are referred to as the instantaneous state, which is made up of five variables. Fatigue, individual ability to cope with stress at a given time (coping), sleep deprivation, morale and motivation are the variables (Bezerra et al. 1996).

7 Process of Decision-Making

Perception activity, mental representation, data processing, issue solving, solution selection and decision-making to act are the six steps of the decision-making process. Two assumptions must be met in order to determine the relationships between the various variables of the decision-making process in relation to personalization and instantaneous state: (1) the variables are independent of one another; (2) each variable in personalization and instantaneous state influences at most three other stages of the decision-making process, to ensure variable selectivity.

8 Equations for Decision-Making Models

The functions of influence ($F(p_i)$) describing the interactions between the elements of personalization and the elements of DMP have been studied based on these hypotheses and the decision-maker model shown in Fig. 7. With respect to the DMP, a comparable examination of the correlations between the functions of degradation ($I(t_i)$) and the instantaneous state was also carried out.

8.1 Formulation of Human Behaviour into Mathematical Equations

Two functions are used to formulate the concepts, functions F are polynomial functions of one or several variables. The function I is also polynomial. On the assumption that each variable in personalization and instantaneous state is independent, the functional relationships of $F(p_n)$ and $I(t_m)$ are expressed by the following functions:

Assume that personal state parameters are as follows:

$$p_1 = \text{Reality}$$

$$p_2 = \text{Knowledge}$$

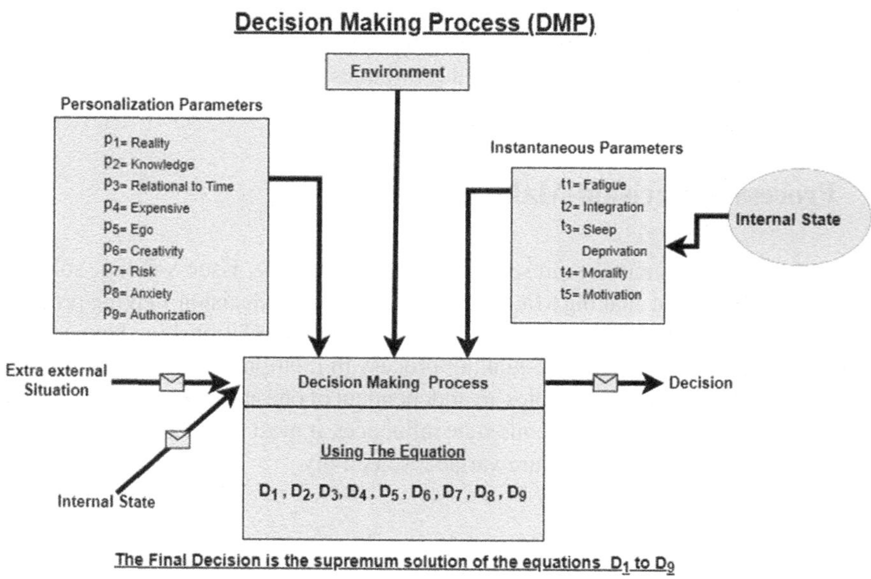

Fig. 7 Flowchart of decision-making process (DMP)

$$p_3 = \text{Relation to time}$$

$$p_4 = \text{Expensivness}$$

$$p_5 = \text{Ego}$$

$$p_6 = \text{Creativity}$$

$$p_7 = \text{Risk}$$

$$p_8 = \text{Anxiety Level}$$

$$p_9 = \text{Authorization}$$

Instantaneous state parameters are as follows:

$$t_1 = \text{Fatigue}$$

$$t_2 = \text{Integration}$$

$$t_3 = \text{Sleep deprivation}$$

$$t_1 = \text{Fatigue}$$

$$t_4 = \text{Morality}$$

$$t_5 = \text{Motivation}$$

8.2 *Predictably Irrational Behaviour*

A group of exceptional Berkeley undergraduate students has participated in a variety of sessions in a variety of locations in less than three months. The participants in the series of sessions evaluated what their sexual and moral choices would be if they were aroused in the following way: However, they anticipated their own selections in the set of sessions that took place when they were in a heated, aroused state—but this time, because they were actually in the grip of passion, it's possible that they were more aware of their own preferences in that condition. We came up with consistent and unequivocal results at the end of the inquiry, which was exceedingly obvious, even dangerously clear. Imagine looking in the mirror one morning and finding that someone else—something alien yet human—has taken over and is controlling your body. Your lips are narrower than before, your incisors are longer, your nails are dirtier and your face is flatter. You're also shorter and hairier, with longer incisors, filthy nails and a flatter face than before. From across the room, two ice reptile eyes stare back at you. You have a strong desire to rape or destroy something. You are not the person you believe you are. Without a doubt, you're a monster. What happens when our illogical self-awakens in an emotional region we think we're familiar with but are actually unfamiliar with? Is it possible to predict how we or others will behave when we are 'out of our heads'—when we are truly angry, hungry, afraid, or sexually aroused? Is there anything that we can do? It's difficult to move from one emotional state to another. It isn't always possible and it isn't always simple. To make informed decisions, we must first comprehend our emotional state following the experience. Learning to bridge this gap is necessary for making important life decisions (Ariely 2010).

The above concepts can be formulated as:

$$D_1 = F(p_1, p_2, p_5, p_8) + I(t_1, t_5) \tag{1}$$

8.3 *Why Can't We Just Force Ourselves to Do What We Want?*

How emotions seize our attention and force us to view the world in new ways. Procrastination is caused by the same problem. We are in a reasonable state when we make a commitment to save money. We restore our composure when we commit to exercising and eating a balanced diet. However, just as we promise to save, we see a new car, a mountain bike, or a pair of shoes that we simply must have. When we want to exercise frequently, we find an excuse to sit in front of the television all day. What about your diet? I'm going to have that slice of chocolate cake and then begin my diet the next day. Dan Ariely and team, procrastination is giving up on our long-term goals in exchange for immediate gratification is there any other way to tackle procrastination in the healthcare field? Consider how we could repackage the majority of our medical and dental treatments to make them more predictable and straightforward. The team tells us a tale to assist us to comprehend this concept. He used to be so convinced that having a 'self-control' credit card would be a great idea that He requested an appointment with one of the major financial organizations. His excitement was rewarded when this prestigious bank responded and offered him to visit their New York headquarters (Ariely 2010).

The above concepts can be formulated as

$$D_2 = F(p_1, p_2, p_3, p_4, p_7, p_9) + I(t_1, t_2, t_5) \quad (2)$$

8.4 Why Do We Place an Excessive Value on Our Possessions?

Dan and his team were given a list of students who had won the lottery as well as a list of those who had not and we began calling them. William, a senior chemistry student, was the first person we chatted with. William had a lot going on. He had a flurry of projects and e-mails to do before school started after spending the previous week camping. Despite having made it to the front of the line, he was dismayed to learn that despite being at the front of the line, he had not been chosen as one of the lottery winners. Ownership has an impact. Our lives, in an odd way, affect many of the decisions we make. As Adam Smith phrased it, 'the community itself matures to be what is truly a commercial society' when 'every man and woman becomes in some degree a trader.' That is, without a doubt, a mind-blowing thought. Much of our life's story can be told via the ups and downs of our material possessions. Clothing, food, vehicles and residences are all items that we buy. We also sell goods such as houses and automobiles, as well as our time. We spend a lot of time and money on our belongings, so why not get the most bang for our buck? Wouldn't it be helpful to know how much we'd adore a new house, a new car, a different couch and an Armani outfit in order to make well-informed purchasing decisions? Unfortunately, this is not always true. The majority of our time is now spent fumbling around in the dark. Why? This is caused by three irrational peculiarities in human nature. Working on something gives you a greater sense of ownership. Think about the last time you constructed something. Knowing which piece goes where and which screw goes where helps you feel more in control. The possession of tangible items is not the only kind of ownership. It can also be used to communicate a variety of viewpoints. Whether in politics or athletics, we must decide what we will do with a concept after we have acquired it. We adore it, perhaps even more than we should. We value it more than it is worth to us. Furthermore, we often find it difficult to let go because we cannot handle the prospect of losing it. So, what options do we have left? a rigid and unyielding ideological framework (Ariely 2010).

The above concepts can be formulated as:

$$D_3 = F(p_1, p_2, p_3, p_4, p_7, p_8, p_9) + I(t_2, t_3, t_4, t_5) \quad (3)$$

8.5 Why Do Options Cause Us to Lose Sight of Our Main Goal?

We had postulated about human behaviour and Albert verified it: given a simple arrangement and a clear aim (in this case, to obtain money), we are all capable of following the source of our delight. In terms of dating, Albert had successfully sampled one date, tried another and even had a fling with another before calling it quits. The best part is that after sampling them all, he didn't go back to the other selections. Throughout the game, he remained with the elite. How can we break free from this unreasonable need to pursue meaningless alternatives? Erich, a philosopher, released Escape from Freedom in 1941. People have too much opportunity in today's democracy, he observed. This is especially true in today's environment. We're told all the time that we can be or do anything we want. The problem is reaching this goal. Before you die, there are 1000 things you must see and we must see at least 999 of them. But are we overstretching ourselves? As their participants hurried from door to door, they witnessed the temptation mentioned. Running from door to door is an unusual enough human behaviour on its own. Their drive to pursue doors of little value—possibilities that are almost dead or don't appeal to us—is perhaps stranger to us. The other side of this tragedy arises when they fail to recognize that some things are in fact closing doors that require immediate care. They may work longer hours without realizing that our children's youth is slipping away. These doors occasionally close too slowly for us to perceive their absence. One of Dan's friends told me that the best years of his marriage were when he was in New York and his wife was in Boston and they only saw each other on weekends. Before they had this arrangement, they used to spend their weekends catching up on work rather than enjoying each other when they lived in Boston. However, as the situation evolved and they discovered they'd only be spending weekends together, their shared time became limited and had a set end (the time of the return train). They decided to spend their weekends together rather than working because they knew the clock was ticking. In research, they found that hurrying frantically to avoid doors closing is a fool's game. It will deplete not only our emotions but also our bank accounts. What we need to do is start closing some of our doors on purpose. Of course, small doors are simple to close. We can cross Kwon do off our daughter's extensive list of extracurricular activities on our holiday card lists. It should be simple to choose between two extremely similar solutions. When picking between MIT and Stanford a few years ago, Dan experienced similar dilemma (I selected MIT). When faced with these two options, he spent weeks weighing their overall appeal. So he did. At this point in his predicament, he required further information and research on the ground. As a result, he attended both schools. He met people and asked how they liked it. He looked into various schools for our children. Wife Sumi and he spoke about how the two scenarios may fit into their ideal lives. As he became more absorbed in this decision, his academic studies and production began to suffer. Ironically, his research was neglected while he searched for the ideal work location.

The above concepts can be formulated as:

$$D_4 = F(p_1, p_2, p_3, p_5, p_7, p_8) + I(t_2, t_4, t_5) \tag{4}$$

8.6 Why Do We Get What We Expect from Our Minds?

It is more likely to be good if we feel something will be good before it occurs; conversely, it is more likely to be bad if we believe something will be awful before it occurs. But how far-reaching are the consequences? What exactly are they doing to change our minds? What effect do they have on the physiology of the experience? To put it another way, can previous information truly influence the brain activity that underpins taste perception, so that when we expect something to taste good (or bad), it actually does? Expectations, as can be shown, have a huge impact on nearly every aspect of our life. Consider the case when you need to hire a caterer for your daughter's wedding. Josephine's Catering boasts about its 'flavourful Greek salad with kalamata olives and feta cheese,' in addition to its 'great Asian–style ginger chicken.' Culinary Sensations, a second caterer, serves a 'succulent organic breast of chicken roasted to perfection and drizzled with a merlot demi-glace, resting on a bed of herbed Israeli couscous,' as well as a 'mousse of the freshest Roma cherry tomatoes and crisp field greens paired with a warm circle of chèvre in a fruity raspberry vinaigrette'. Finally, the importance of presentation cannot be overstated. There's a reason why understanding how to plate food is equally as important as knowing how to prepare it. You'll notice a difference if you remove the Styron foam packaging from your take-out food and serve it on a nice dish with garnishes (especially if you have guests). Stereotypes are also shaped by expectations. After all, a stereotype is a way of categorizing information in order to predict future events. The brain can't start again every time it confronts a new situation. It must build on its previous experiences. As a result, stereotypes aren't necessarily dangerous on their own. They provide shortcuts to aid us in our never-ending search to make sense of our complex world. As a result, we expect an elderly person to require assistance while using a computer and a Harvard freshman to be quite intelligent. While stereotypes give us specific expectations about members of a group, they can also have a negative impact on our attitudes and behaviours. Every one of these tests reminds us that expectations are more than just waiting for a sugary cola to make us feel better. Expectations assist us to understand what is being said in a noisy environment despite the occasional loss of a few syllables. Expectations also help us comprehend text messages received on our cell phones, even when parts of the words are jumbled. And, while expectations can make us laugh at times, they are also incredibly powerful and useful in many instances (Ariely 2010).

The above concepts can be formulated as:

$$D_5 = F(p_3, p_4, p_5, p_7, p_9) + I(t_2, t_4, t_5) \tag{5}$$

8.7 Why Can a 50-Cent Aspirin Perform Functions that a Penny Aspirin Cannot?

PLACEBO is derived from the Latin phrase 'I will please.' Sham mourners were paid to lament and sob for the deceased during funerals in the fourteenth century and the word was coined to describe them. By 1785, it had made its way into the New Medical Dictionary, where it was associated with fringe medical practices. Two mechanisms in general form the expectations that allow placebos to operate. The first is belief, which is our trust or confidence in the medication, technique, or provider. The fact that a doctor or nurse is paying attention to us and comforting us not only helps us feel better, but it also starts our internal healing processes. Even a doctor's excitement for a certain therapy or technique may bias us to a favourable result. From our experiments with our 'pharmaceuticals,' we learned that prices influence the placebo effect. But does pricing have an impact on ordinary consumer goods as well? We discovered the ideal topic in SoBe Adrenaline Rush, a beverage that claims to 'upgrade your game' and provide 'better functioning.' Are we DOOMED, then, to get lower benefits every time we get a discount? If we rely on our irrational instincts, we will If we see a discounted item, we will instinctively assume that its quality is less than that of a full-priced item—and then, in fact, we will make it so. What's the remedy? If we stop and rationally consider the product versus the price, will we be able to break free of the unconscious urge to discount quality along with price? As science researchers, we respect trials that put our views and the effectiveness of various therapies to the test. At the same time, it seems apparent to me that trials, especially those employing medical placebos, present a slew of serious ethical concerns. Indeed, the experiment with mammary ligation, which I discussed at the start of this chapter, posed an ethical issue: there was a backlash against doing fake procedures on patients.

The above concepts can be formulated as:

$$D_6 = F(p_3, p_4, p_7, p_8) + I(t_1, t_4, t_5) \tag{6}$$

8.8 Why Are We Untrustworthy and What Can We Do About It?

For one thing, we could reintroduce the Bible into public discourse. If the only purpose is to eradicate dishonesty, this might not be such a bad concept. Others, on the other hand, may object, stating that the Bible either endorses a particular faith or just integrates religion with the commercial and secular realms. A different form of oath, on the other hand, might work. The experiment with the Ten Commandments hit me the most because pupils who could only recollect one or two of them were just as influenced as those who could recall nearly all ten. This shows that it was the mere thought of any form of moral standard, rather than the Commandments themselves, that prompted honesty. This isn't to imply we shouldn't make an effort. What makes honesty so important? For openers, let us not forget that the United States currently holds economic domination around the world in part because it is (or is thought to be) one of the world's most honest countries in terms of its standards of corporate management (Ariely 2010).

The above concepts can be formulated as:

$$D_7 = F(p_1, p_2, p_3, p_8) + I(t_2, t_4) \tag{7}$$

8.9 *Why Do We Become More Honest When We Deal with Cash?*

It can be difficult to appreciate the amount to which non-monetary items influence our cheating because we are so capable of rationalizing our tiny deception. We might conclude that office supplies are included in our overall salary or that lifting a pencil or two is a common occurrence if we take a pencil. We could argue that stealing a can of Coke from a shared refrigerator is acceptable on occasion, as we've all had cans of Coke taken from us. Perhaps Lay, Skilling and Fastow reasoned that faking Enron's records was acceptable since it was a temporary measure that could be rectified once business conditions improved. We've discovered that if given the chance, people will cheat. What's interesting is that the majority of us are completely unaware of this. In another experiment, we asked students if they thought people would cheat more for tokens than for cash and they said no, that the amount of cheating would be comparable. After all, the tokens represented real money and could be exchanged for real money in a matter of seconds, they continued. They predicted that our participants would see the tokens as legitimate currency as a result. Dishonesty issues, by the way, are not limited to individuals. In recent years, business, in general, has suffered from a decrease in honesty. I'm not talking about large-scale deceptions like those perpetrated by Enron and WorldCom. I'm talking about minor infractions like stealing Cokes from the refrigerator. To put it another way, there are corporations out there that are stealing stuff one step away from cash rather than cash. As a result, we're back to our original query: isn't cash odd? When we deal with money, our brains are wired to think about our acts as though we've just signed an honour code. Indeed, examining a dollar bill reveals that it was designed to conjure up a contract: It clearly says THE UNITED STATES OF AMERICA, with a shadow beneath it that provides the appearance of three-dimensionality. There's also George Washington himself (who, as we all know, is incapable of telling a falsehood). Then, on the other hand, it becomes even more solemn: WE TRUST IN GOD. Then there's that odd pyramid with its unblinking eye above it! And it's looking right at us! Money's sanctity may be aided by the fact that it is a readily identified unit of transaction, in addition to all of this symbolism. It's tough to argue that a dime or a dollar is not the same thing (Ariely 2010).

The above concepts can be formulated as:

$$D_8 = F(p_3, p_4, p_7, p_8) + I(t_4, t_5) \tag{8}$$

8.9.1 Where Are the Free Lunches in Behavioural Economics?

You can see from what we've said so far about this experiment that it delivers some basic life advice—a free lunch. To begin, it's a good idea to plan your order ahead of time and stick to it when dining at a restaurant. Being influenced by others' decisions may lead you to choose a less desirable option. If you're worried about being swayed in any event, making your order known to the table before the waiter arrives is a solid strategy. In this manner, even if another diner orders the same food before you, you've staked a claim on your order and it's less likely that the other diners will think you're not unique. The wisest course of action, however, is to place an order first. Perhaps restaurant owners should require guests to write their orders in private (or discreetly pass them to wait for staff) so that no diner is influenced by the orders of their peers. The joy of eating out costs us a fortune. Increasing the happiness gained from these meetings is perhaps the cheapest and most straightforward way to do it.

The concept of 'free lunches' is one of the most significant distinctions between traditional and behavioural economics. Standard economics implies that all human decisions are logical and well-informed, based on a precise understanding of the worth of all commodities and services, as well as the amount of pleasure (utility) that each option is projected to provide. Everyone in the market is striving to maximize profit and improve their experiences, according to this set of assumptions. As a result, the economic theory asserts that there are no free lunches—if there were, they would have been found and valued long ago. Behavioural economists, on the other hand, believe that humans are subject to irrationality due to irrelevant influences from their immediate environment (which we call context effects), irrelevant emotions, short-sightedness and other forms of irrationality. They believe that these factors impact the development of strategies, tools and approaches to help us make better decisions and improve our overall health. In behavioural economics, free lunches are specifically characterized as such. Isn't it true that there are tools, procedures and rules that can help us all make better decisions, achieve our objectives and improve our overall well-being in the process? Free lunches indicate, from a behavioural economics perspective, that there are tools, tactics and regulations that can help us all make better choices and, as a result, achieve our goals. If I were, to sum up, one significant takeaway from the findings presented in the book 'Predictably Irrational,' it would be that we are puppets in a game whose dynamics we don't fully comprehend. We often consider ourselves in the driver's seat, in perfect control of our decisions and the path our lives take; however, this image is based on our desires and how we want to be perceived, not on reality (Ariely 2010).

The above concepts can be formulated as:

$$D_9 = F(p_5, p_7, p_8) + I(t_1, t_3,) \tag{9}$$

9 START-Action STRATEGY (SAS)

The dynamic operation of the DMP is described in terms of the decider's internal and exterior states (*E*) (Bezerra et al. 1996) as shown in Fig. 7. The Start-Action Strategy (SAS) can be activated using a formula shown in Fig. 8.

State of the Union

Individual and immediate personalization features are considered in the internal state (II) (IS). Previously, these functions were referred to as 'influence' and 'instantaneous states' (Bezerra et al. 1996).

State of the Environment

The word 'external state' refers to the unique characteristics of the current circumstance. The following criteria can influence the recognition of the situation and the decision to be made: a perceived threat due to the environment; internal state constraints (the individual's demands); and an imbalance of the internal state in reaction to the environment (Bezerra et al. 1996).

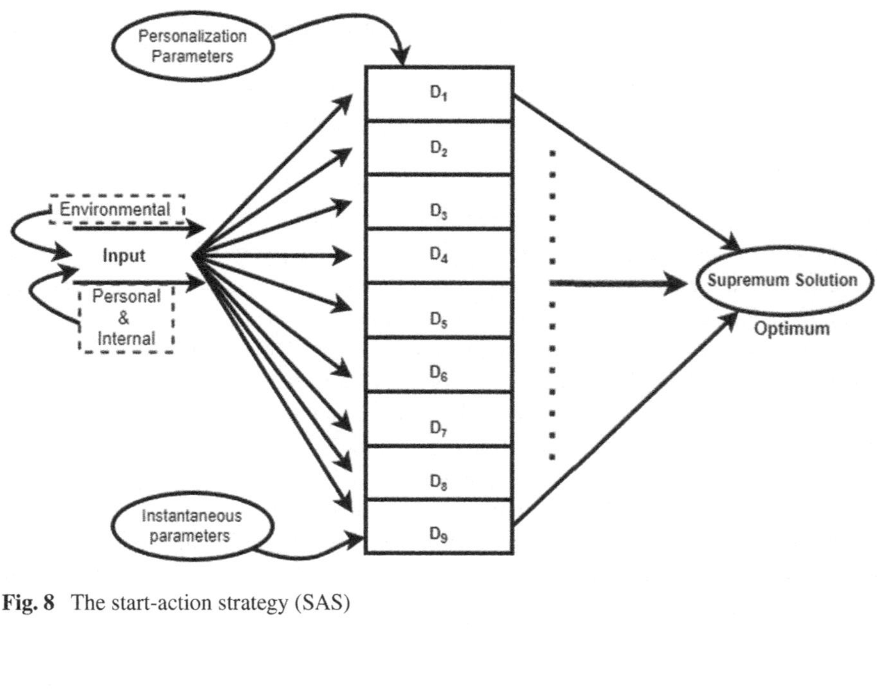

Fig. 8 The start-action strategy (SAS)

The Limitations

The SAS constraints, which lead to a decision, are linked to the process's maturation as well as the information's incubation (the DMP, the internal state, the external state and/or the decider). The following are some of the constraints (Bezerra et al. 1996).

- A satisfactory solution (one that eliminates the perceived threat posed by the environment, the individual's internal needs and/or the individual's internal imbalance in response to the environment); and
- The time limit (this refers to the obligation to react within a specific time frame in order for the decision to be effective)

10 Conclusion and Future Prospects

Compassion is about cultivating compassion, empathy and care for oneself, others and humanity as a whole. Compassion is caring for others that starts from and culminates in greater inner strength, not a lack of ability to combat injustice. Compassion is not a sign of weakness, but rather a desire to help others that begins with and leads to greater inner strength, allowing a human's biological system to make sound decisions. Researchers have discovered that empathy for the well-being and suffering of others sits at the foundation of all ethical ideas since ethics is concerned with our actions and interactions with others. The three SEE Learning domains are Personal Domain, Social Domain and System Domain. SEE Learning is intended to help individuals, socially and emotionally. Individuals and societies must learn to care for themselves before they can care for others or make ethical decisions. According to SEE Learning, well-rounded education for children should include not just self-regulation skills but also social skills. In our increasingly complex environment, compassion alone is insufficient; it must be supplemented with ethical decision-making based on an understanding of broader systems. Systems thinking is a major new problem-solving strategy in today's environment. Learning can be studied in three dimensions and three domains. Each domain will investigate each dimension, for a total of nine components. The emphasis is on improving capabilities across all nine interconnected components. In this chapter, we try to show how these attributes are responsible to configure the biological system of a human being.

Behavioural economics is a new area of study that blends psychology and economics. Researchers employed physical pain such as heat, cold water, pressure, loud sounds and even the psychological pain of losing money in the stock market to find the answers. It is an attempt to comprehend both behaviour and the decision-making process that leads to it. If we believe in human rationality, we must first agree on the fundamental human notions upon which economics is founded. Understanding why we are illogical all the time can help us make better decisions. This has something to do with our inclination to compare our surroundings to what others are doing. We are always comparing jobs to jobs, vacations to vacations, lovers to lovers and wines to wines. Everything is relative, which is exactly the point. When we compare our circumstances to those of others, we become enraged and envious.

This may be the most difficult commandment to obey, given our innate desire to compare. However, there are instances when we can lessen the circles that encircle us, improving our happiness. We compare our alternatives in the relativity issue to those available in our immediate surroundings. We might also adjust our focus from narrow to broad. We select the best alternative based on our core values and dislikes. If supply and demand cannot be used to determine market size, we may need to go elsewhere. Zero is a one-of-a-kind website. The difference between two cents and one cent is substantial. When enough money is traded, market norms emerge; nonetheless, market norms cover a wide range of activities. The social norm is temporarily lost when a social and a business norm collide. When it comes to money, getting anything for free is certainly appealing. We spend a lot of money on the enjoyment of eating out. Customers may be required to write their orders in private by restaurant owners. The best line of action is to place an order as soon as possible. Face-to-face interactions are the most cost-effective strategy to boost happiness. The concept of 'free lunches' is one of the most striking differences between traditional and behavioural economics. We have access to tools, procedures and rules that can help us make better decisions if we get free lunches. In this chapter, we will attempt to demonstrate that human decisions are illogical and frequently predictable. This paper emphasizes the properties of the mathematical model for human decision-making. The DMP is broken down into nine steps. Each one is composed of formal equations called influence functions. The first nine use personalization elements and instantaneous elements. The system input is made up of the DMP phases. A number of DMP phases comprise the system's supremum output. As a result of their acts, SAS is triggered. Finally, we design a mathematical model to make a decision in a multi-solution environment. Figures 7 and 8 graphically demonstrate the model.

From the above model we can also perform the followings:

- Automatic decision-making system using Brain mapping and image processing.
- Nature-Inspired Intelligent systems can be used in human decisions.
- It can also be used in an automated Business decision system.

References

Ariely D (2010) Predictably irrational, revised and expanded edition: the hidden forces that shape our decisions. HarperCollins

Ariely D, Wallsten TS (1995) Seeking subjective dominance in multidimensional space: an explanation of the asymmetric dominance effect. Organ Behav Hum Decis Process 63(3):223–232. https://doi.org/10.1006/obhd.1995.1075

Ariely D, Loewenstein G, Prelec D (2003) "Coherent arbitrariness": stable demand curves without stable preferences. Q J Econ 118(1):73–106. https://doi.org/10.1162/00335530360535153

Ariely D, Loewenstein G, Prelec D (2006) Tom Sawyer and the construction of value. J Econ Behav Organ 60(1):1–10. https://doi.org/10.1016/j.jebo.2004.10.003

Bezerra, S., Cherruault, Y., Fourcade, J., & Veron, G. (1996). A mathematical model for the human decision-making process. *Mathematical Computing Modelling*, *24*(10), 21–26.

Goleman D, Senge PM (2014) The triple focus: a new approach to education

Heckman JJ, Kautz T (2012) Hard evidence on soft skills. Labour Econ 19(4):451–464. https://doi.org/10.1016/j.labeco.2012.05.014

Heyman J, Ariely D (2004) Effort for payment: a tale of two markets. Psychol Sci 15(11):787–793

Jinpa T (2015) A fearless heart: why compassion is the key to greater wellbeing. Hachette UK

Kahneman D, Tversky A (1979) Prospect theory: an analysis of decision under risk. Econometrica 47(2):263. https://doi.org/10.2307/1914185

Layard R (2016) Comment on Sugden and Teng. Policies Happiness, 58–61. https://doi.org/10.1093/acprof:oso/9780198758730.003.0004

Sample teaching activities to support core competencies of social and emotional learning. (2021, October 8). CASEL. https://casel.org/sample-teaching-activities-to-support-core-competencies/

Sedikides C, Ariely D, Olsen N (1999) Contextual and procedural determinants of partner selection: of asymmetric dominance and prominence. Soc Cogn 17(2):118–139. https://doi.org/10.1521/soco.1999.17.2.118

Shampanier K, Mazar N, Ariely D (2007) Zero as a special price: the true value of free products. Mark Sci 26(6):742–757. https://doi.org/10.1287/mksc.1060.0254

Simonsohn, U., & Loewenstein, G. (2006). Mistake: The effect of previously encountered prices on current housing demand. *The Economic Journal, 116*(508), 175–199. https://doi.org/10.1111/j.1468-0297.2006.01052.x

Simonson I (1989) Choice based on reasons: the case of attraction and compromise effects. J Consum Res 16(2):158. https://doi.org/10.1086/209205

Sunstein CR, Kahneman D, Schkade D, Ritov I (2002) Is incoherence outrageous? Stanford Law Rev 54(6):1293. https://doi.org/10.2307/1229622

Tversky A, Kahneman D (1981) The framing of decisions and the psychology of choice. Science 211(4481):453–458. https://doi.org/10.1126/science.7455683

Tversky A, Simonson I (1993) Context-dependent preferences. Manage Sci 39(10):1179–1189. https://doi.org/10.1287/mnsc.39.10.1179

Williamson O (2008) The new Palgrave dictionary of economics. Palgrave Macmillan

Nature-Inspired Computing in Drug Design, Development, and Therapeutics

Applications of Nature-Inspired Computing and Artificial Intelligence Algorithms in Solving Personalized Therapy Complications

Ritu Karwasra, Kushagra Khanna, Nitin Sharma, Rishabh Malhotra, Shivkant Sharma, Surender Singh, Asim Ali, and Saurabh Verma

1 Introduction

It is an astonishing fact that a drug is effective for some people, whereas for others it is less effective or ineffective. Likewise, some people develop dreadful diseases such as cancer but others do not. The question lies in their genomic alterations or in other differential factors like age, disease severity scores, molecular markers, lifestyle, and environmental characteristics that could be the reasons for these tribulations. So currently, each individual patient's illness is treated as unique, and medication is tailored in accordance with the individual's history and biology (Schork 2019). This unique medical practice is termed as *personalized medicine* or *precision medicine* or sometimes *stratified medicine*. It is a promising domain for clinical research and

R. Karwasra · A. Ali
Central Council for Research in Unani Medicine (CCRUM), Ministry of AYUSH, Govt of India, Janakpuri D Block, New Delhi 110058, India

K. Khanna · R. Malhotra
Llyod Institute of Management and Technology, Greater Noida, Uttar Pradesh 201306, India

N. Sharma
Shriram Murti College of Engineering & Technology (Pharmacy), Bareilly Nainital Road, Bhojipura, Bareilly 243202, India

S. Sharma
Department of Pharmacology, Gurugram University, Mayfield Garden, Sector 51, Gurugram 122003, India

S. Singh
Department of Pharmacology, All India Institute of Medical Sciences, New Delhi 110029, India

S. Verma (✉)
Tumor Biology, ICMR-National Institute of Pathology, New Delhi, India
e-mail: svarmasv1@rediffmail.com

K. Raza (ed.), *Nature-Inspired Intelligent Computing Techniques in Bioinformatics*, Studies in Computational Intelligence 1066,
https://doi.org/10.1007/978-981-19-6379-7_11

patient care that integrates multi-modal data from an individual for better understanding and treatment of disease and making patient-tailored decisions (Agarwala et al. 2018). With the involvement of large and complex datasets in personalized medicine approaches, newer techniques and tools will be needed to process and understand these datasets. Artificial intelligence (AI), machine learning (ML), and nature-inspired computing (NIC) progressed rapidly and developed techniques that enabled the processing, storage, and analysis of these intricate datasets (Siddique and Adeli 2015). Traditional statistics methods and early computing techniques could not able to compute these intricate datasets. AI helps in predicting disease, finding accurate diagnostics, and providing treatment to control the occurrence of foreseen disease. NIC is a specific domain of data science that detects, treats, and manages disease in terms of precision and accuracy. NIC techniques are stimulated by independent, developing, adaptive, and autopoiesis behavior of living creatures such as birds, flies, animals, and even humans. All of these are stochastic computing techniques that are widely used in solving numerous real-life problems particularly in diagnostics of diseases and in personalized treatment (Mitra and Shankar 2014). Personalized therapy complications are numerous as the indulgence of varied massive data from individual patient characteristics, and it can be accomplished with the facilitation of data science (Kosorok and Laber 2019).

Presently, data science impacts the era of personalized medicine, and the US-FDA also listed over 160 pharmacogenomic biomarkers related to cancer and neuroscience (Federal Drug Administration 2019). Various—omics data such as genomics, epigenomics, transcriptomics, metabolomics, proteomics, and bioimages (MRT, CT scans), electronic medical records (EMRs), and even the data of health claims and mobile health applications are the sources for personalized medicine (Awwalu et al. 2015). It is imperative that certain diseases such as cancer, immunological, and neurological-related disorders are complex, and identification of a single biomarker is not possible (Mitra and Uma Shankar 2015). Multianalyte signatures derived from binary, high throughput data allow patient characteristics in a holistic manner. Identification of biomarker signatures is an essential requirement of data science. A highly cited example in this domain is MammaPrint (a predictive test for breast cancer) that includes 70-gene signatures. In 2007, FDA approved it because of its effectiveness in predicting the development of distant metastases. Another example is Geno2pheno, a computational tool to estimate drug resistance to HIV and a gene signature termed as S3 score for the prediction of renal cell carcinoma prognosis. Several commercial players such as Google, Amazon, Pfizer, IBM, and Microsoft working on this domain of data science-based solutions with personalized medicine (Papadakis et al. 2019). Improving cancer treatment by integrating with molecular profiling is an emerging practice in the arena of personalized medicine. It reduces the risk of adverse events or treatment failures (Lassen et al. 2021). The ETH Zurich technology platform NEXUS Personalized Health Technologies established a workflow for the diagnosis of cancer (Singer et al. 2017). This workflow is based on high throughput sequencing of tumor samples. Sequencing data is analyzed with an algorithm that combines multiple databases for annotation, and numerous variant callers, and locates individual variant frequencies and gene expression levels into large cohorts (Dhal et al. 2019). The era of personalized medicine is feasible with the advent of data science and artificial learning methods. Nowadays, NIC and AI collectively encompass several steps in personalized therapies, yet acquire certain challenges on the way forward. We will discuss some NIC and AI algorithms that are beneficial in the personalized medicine domain along with the challenges encountered.

2 NIC and AI Algorithms Used in Personalized Medicine

There are a number of algorithms in NIC and AI that are used in specific personalized medicine. NIC algorithms include ant colony optimization, particle swarm optimization, bat algorithm, firefly algorithm, artificial bee colony algorithm, differential evolution, marine predators algorithm, red fox optimization, polar bear optimization, gravity, and mass iterations algorithms (Mitra and Shankar 2014). Together some AI algorithms also help in the personalized medicine domain including artificial neural networks (ANNs), naive Bayesian (NB), and support vector machines (SVM). Figure 1 illustrates the applications of algorithms developed by the human diagnostic performance-based personalized medicine.

3 Nature-Inspired Algorithms

NIC intelligence is a technique that relies on the behavioral response of living organisms of nature by exploring the search space using naturally endowed skills of animals, birds, and insects to find an optimal solution.

Some major types of NIC algorithms are described with some set of examples.

3.1 Genetic Algorithm (GA)

GA is an adaptive heuristic search algorithm, developed by John Holland in 1960. This algorithm is based on the crossover of chromosomes like generating new offspring by uniting genes. It depends upon four factors viz-a-viz crossover rate, mutations, population size, and generations which make it efficient. A study conducted used GA and NB for the diagnosis of heart diseases in patients (Lutimath et al. 2022). In this study, the GA algorithm selects 6 attributes (predictive accuracy is 99.2%). Dhanwani and Wadhe (2013) developed a model of cardiac disease using hybrid GA and ANN with a prediction accuracy of 98.6% (Dhanwani and Wadhe 2013). Similarly, Gupta and Kaur (2014) also used hybrid GA with ANN but the prediction accuracy achieved was 97.8% for early diagnosis of heart diseases (Gupta and Kaur 2014). Aishwarya and Anto (2014) worked on the applications of GA to detect diabetes in 270 patients' datasets (prediction accuracy was 89.5%)

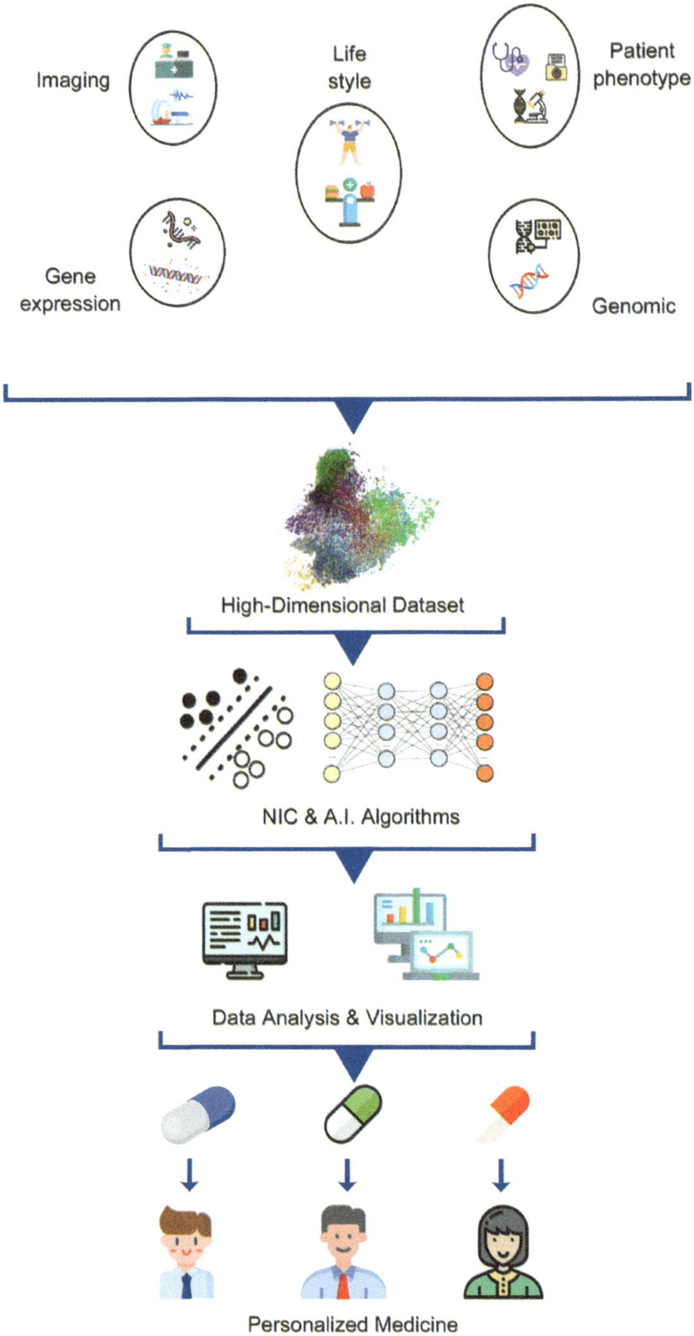

Fig. 1 Role of NIC and AI algorithms in personalized medicine

(Aishwarya and Anto 2014). Bhuvaneshwari et al. used GA with a neural network for lung cancer diagnosis with 90% prediction accuracy (Bhuvaneswari and Therese 2015). A study conducted by Khan Mansoori et al. (2014) proposed GA with SVM to predict the occurrence of ovarian cancer with a prediction accuracy of 96.3% (Khan Mansoori et al. 2014). Numerous researches were conducted by employing the GA method with several other algorithms for disease diagnosis, and their management in patient samples. Coupling of GA with SVM, NB, ANN, backpropagation, distance method, and random forest provides a good predictive accuracy rate. Some of the major research findings are with disease diagnosis tabulated in Table 1.

Table 1 Outcomes of research work in personalized medicine using GA algorithm

S. No.	Methodology	Disease	Sample size	Predictive accuracy (%)	References
1	GA	Heart disease	300 records	73.5	Kaur and Singh (2015)
2	GA and random forest	Cervical cancer	480 records	81.71	Ashok et al. (2016)
3	GA with fuzzy logic	Diabetes	770 records	87	Ephzibah (2011)
4	GA and ANFIS	Breast cancer	290 records	88	Turabieh and Muhanna (2016)
5	GA and ELM	Diabetes	270 records	89.5	Aishwarya and Anto (2014)
6	GA and KNN	Lung cancer	Not mentioned	90	Bhuvaneswari and Therese (2015)
7	GA and SVM	Lung cancer	1400 records	93.19	de Carvalho Filho et al. (2017)
8	GA and SVM	Ovarian cancer	220 records	96.30	Khan Mansoori et al. (2014)
9	GA and ANN	Breast cancer	570 records	96.6	Aalaei et al. (2016)
10	GA and neural networks	Heart diseases	Not Mentioned	97.8	Gupta and Kaur (2014)
11	GA, KNN, and ABC	Heart disease	Not mentioned	98.84	Kumar and Sahoo (2017)
12	GA and naive bayes	Heart disease	Not Mentioned	99.2	Lutimath et al. (2022)

KNN K-nearest neighbor, *ANFIS* Adaptive neuro-fuzzy inference system, *ELM* Extreme learning machine, *NN* Neural networks, *SVM* Support vector machines, *ANN* Artificial neural network, *ABC* Artificial bee colony algorithm

3.2 Ant Colony Optimization (ACO)

This algorithm was developed by Marco Dorigo in 1992 and depends on the behavior of ants in colonies. This algorithm works on the principle of ants searching for their food using a pheromone (the chemical that an animal produces to change the behavior of another animal) trail on a weighted graph system. The algorithm is used for the diagnosis of diseases such as tumors. P. Deoskar et al., developed a model that diagnoses lung cancer with this algorithm (Deoskar et al. 2013). Pheromone was designated by a specific value and each ant built a rule. Further, the effect of pheromone content was evaluated with the help of a heuristic function. Tumors were diagnosed with the highest pheromone value. Garg (2016) focused on the ACO technique to identify white blood cells in leukemia cancer using microscopic images (Garg 2016). Dubey et al. (2014) implemented an algorithm to prevent the risk of heart disease. The solution is based on ant colony optimization and tailored for Bayesian classifiers that ultimately enhance the prediction of cardiovascular disorders (Dubey et al. 2014; Deoskar et al. 2013). Shukla (2014) used ACO to predict the suspects of cancer in the early stage which provides flexibility to the physician to select optimal therapy (Shukla 2014). Patankar et al. (2014) employed ACO for the identification of breast cancer (Patankar et al. 2014). Kavitha and Nithya (2020) developed an ACO-based algorithm for the diagnosis of heart diseases (Kavitha and Nithya 2020). Table 2 summarizes some of the research findings employing the ACO algorithm or its combination for disease diagnosis and treatment.

3.3 Particle Swarm Optimization (PSO)

PSO algorithm is adaptive and was introduced in 1995 by Dr. Russell C. Eberhart and Dr. James Kennedy. This algorithm is based on meta-heuristic swarm intelligence, population-based stochastic optimization method. This technique is inspired by bird clustering to search for the optimal solution. PSO probes probabilistic algorithms that are based on collective intelligence, stimulated by flocking, herding behavior of swarms of fish, or bird flocks to explore better places. PSO assists with time series prediction, and registration of remote sensing images. Recently using the principle of swarm optimization is used to predict heart diseases (Wang et al. 2018). PSO helps to develop a multilayer feed-forward neural network (MLFFNN) to predict heart disease at the initial stage using clinical records. In another work, PSO-optimized neural networks are used for optimizing momentum factors, hidden neurons, and learning rate (Shathesh and Durairaj 2016). PSO is used to generate new patterns and predict complications. It is also used to improve the prediction accuracy rate in the diagnosis of diseases. Sivagowry et al. (2014) imposed a two-stage model for the prediction of heart disease (Durairaj and Sivagowry 2015). Initially, PSO preprocesses the data and later ANN classifies the data. Radha et al. in 2014 used PSO with feature selection to predict the detection of heart-related disorders. Though

Table 2 Outcomes of research work in personalized medicine using ACO algorithm

S. No.	Methodology	Disease	Sample size	Predictive accuracy (%)	References
1	ACO and KNN	Heart disease	1500 records	70.26	Rajathi and Radhamani (2016)
2	ACO	Lung cancer	1830 records	84	Deoskar et al. (2013)
3	ACO and fuzzy logic	Diabetes	Not mentioned	84.24	Ganji and Abadeh (2011)
4	ACO and fuzzy logic	Heart diseases	Not mentioned	91.7	Kavitha and Nithya (2020)
5	ACO, fuzzy logic, and ANN	Diabetes	250 records	92.45	Fiuzy et al. (2013)
6	ACO	Heart disorder	Not mentioned	92.5	F.M R (2013)
7	ACO and fuzzy logic	Breast cancer	700 records	95.72	Anto and Chandramathi (2015)
8	ACO and SVM	Lung cancer	150 records	96.6	Kohad et al. (2014)
9	ACO and ANN	Lung cancer	Not mentioned	98.4	Kohad and Ahire (2015)
10	ACO and SVM	Diabetes	Not mentioned	99.2	Saini (2017)

ACO Ant Colony Optimization, *SVM* Support Vector Machines, *KNN* K-Nearest Neighbor, *ANN* Artificial Neural Network

a hybrid method with FL and extreme learning machine predict disease with more accuracy, sensitivity, and specificity (Radha and Srinivasan 2014). Zhang and Yuan (2015) anticipated kernel-based extreme learning machine (KELM) with PSO for the prediction of various disorders such as diabetes and breast cancer (Zhang and Yuan 2015). A study conducted by Durairaj and Sivagowry in (2015) used PSO to reduce the feature set used in the identification of heart disease with 83% accuracy (Durairaj and Sivagowry 2015). Table 3 summarizes some of the research findings employing the ACO algorithm or its combination for disease diagnosis and treatment.

3.4 Artificial Bee Colony (ABC)

ABC algorithm was established in 2005 by Dervis Karaboga and inspired by the foraging behavior of honey bees to locate food using nectar. This algorithm works on honeybee principles like employee bees exploiting the food sources, onlookers waiting at hives, and those who discover new food sources are scouts (Karaboga et al. 2014). Employees share the food source position to onlookers using the waggle

Table 3 Outcomes of research work in personalized medicine using PSO algorithm

S. No.	Methodology	Disease	Sample size	Predictive accuracy	References
1	PSO with quick reduct algorithm	Heart disease	–	80	Shathesh and Durairaj (2016)
2	PSO	Heart disease	300 records	83	M and S (2015)
3	PSO and NN	Diabetes	–	84.2	Ateeq and Gopinath Ganapathy (2017)
4	PSO	Breast cancer	580 records	85	Mohebian et al. (2017)
5	PSO-BPNN	SRBCT, Blood, and Lymphoma cancer	210 records	89	Vimaladevi and Kalaavathi (2014)
6	PSO and fuzzy logic	Breast cancer	–	91.88	Shahbeig et al. (2017)
7	PSO and ANN	Heart disease	76 records	92.9	Devi and Deepa (2016)
8	PSO, IFCM, and extreme learning machine	Heart disease	–	94.5	Radha and Srinivasan (2014)
9	PSO and LSSVM	Diabetes	185 records	95	Karthikeyan and Vembandasamy (2015)
10	PSO and NN	Heart disorder	–	96.3	Kelwade and Salankar (2017)
11	PSO and KELM	Diabetes, and Breast cancer	770, and 570 records	97.01	Zhang and Yuan (2015)
12	PSO and SVM	Diabetes	770 records	97.83	Soliman AboElhamd et al. (2014)

PSO Particle Swarm Optimization, *SVM* Support Vector Machines, *NN* Neural Network, *KELM* Kernel Extreme Learning Machine, *LSSVM* Least-Squares Support Vector Machine, *IFCM* Interval Fuzzy c-means Clustering, *BPNN* Back Propagation

dance technique and after the collection of food, scouts search for other sources. The location of the food source is a possible solution, and the amount of nectar is the quality of the solution. The ABC algorithm depends on the approach of clustering for complicated data which helps to minimize the time and also improves accuracy. ABC algorithm is used to predict the occurrence of cardiovascular disorders. Subanya and Rajalaxmi evaluated the fitness of this algorithm using the K-nearest neighbor (KNN) method (Subanya and Rajalaxmi 2014b). Coupling of NIC with machine learning algorithms is implemented to predict and diagnose coronary heart diseases.

Table 4 Outcomes of research work in personalized medicine using the ABC algorithm

S. No.	Methodology	Disease	Sample size	Predictive accuracy (%)	References
1	ABC and SVM	Diabetes	770 records	79.29	Uzer et al. (2013)
2	ABC and GA	Diabetes	768 records	84.2	Beloufa and Chikh (2013)
3	ABC and naive bayes	Heart disease	300 records	86.4	Subanya and Rajalaxmi (2014a)
4	ABC	Cancer	450 records	76.8	Alshamlan et al. (2015)
5	ABC and SVM	Colon, Blood lung, Lymphoma	450 records	97.91	Alshamlan et al. (2015)
6	ABC and ANN	Heart disease	76 records	90.5	Devi, K. I., & Deepa (2016)
7	ABC and NN	Blood cancer	–	97.22	Singh et al. (2016)
8	ABC AND ANN	Breast cancer	700 records	95.9	Kumari (2017)
9	ABC	Breast cancer	700 records	99.98	Shah et al. (2019)

In another work, the coupling of fuzzy systems with ABC algorithms was used to predict heart diseases and accuracy computed with the ABC algorithm (Subanya and Rajalaxmi 2014a). ABC approach is used for the detection, diagnosis, and treatment of diverse diseases such as diabetes, immunological, cancer, and heart diseases. Uzer et al. proposed a hybrid approach using ABC for the diagnosis of diabetes with a prediction accuracy of 79% (Uzer et al. 2013). Nazarian et al. (2013) used the ABC method to predict breast cancer, whereas another study conducted by Beloufa et al. in 2013 uses genetic algorithm (GA) with ABC to diagnose diabetes with a prediction accuracy of 84.2% (Nazarian et al. 2013; Beloufa and Chikh 2013). Table 4 summarizes some of the research findings employing the ABC algorithm or its combination for disease diagnosis and treatment.

3.5 *Firefly Algorithm (FA)*

This algorithm was developed in 2008 by Xin-She. It is a global algorithm based on flashing patterns and behavior of fireflies and on light intensity. The flash of fireflies acts as a signal system to attract other fireflies, and it includes three rules. The foremost rule is an attraction of fireflies toward others and this attractiveness is proportional to brightness. Therefore, less bright attracts brighter and as the distance increases, their brightness decreases (Fister et al. 2013). Lastly, they move randomly when no brighter flies. The objective function is the brightness of flies. This algorithm enhanced the accuracy of disease detection. The enhanced accuracy is based on inputs of decision tree accuracy.

4 Artificial Intelligence Tools

Artificial intelligence (AI) is autonomous device capable of perceiving information from the surroundings and subsequently acquired actions. These tools help in effective data handling specifically in case of large heterogeneous data and process immediately (Agbehadji et al. 2020). Learning and training of AI system datasets are necessary before actual deployment in a real-world environment. Types of AI-based functionalities are reactive machines, limited memory, theory of mind, and self-awareness. AI involves machine learning (ML) and deep learning (DL) approaches that also help in diagnosing and managing diseases. AI has a pivotal role in personalized medicine therapy and gained widespread attention in detecting foreseen diseases (Ali and Aittokallio 2019). In the recent COVID-19 pandemic, AI models are useful for ensuring early alert, diagnosis and prognosis, treatment and cures, and social control (Vaishya et al. 2020). The practice of personalized medicine is analogous to the development of drugs and biomedical devices. The initial step involves determining the individual risk factor for disease. The next step comprises the identification of key pathophysiologic processes. AI helps in the identification of targets for therapies; for example, several assays (DNA sequence, molecular pathology imaging) generate massive amounts of data and require statistical tools to identify meaningful patterns and AI aids in identifying such patterns (Agbehadji et al. 2020).

4.1 Artificial Neural Network (ANN)

ANN is widely used in the diagnosis, and imaging of myocardial infarction, arrhythmias, dementia, psychiatric disorders, appendicitis, and pulmonary embolism (Kumari 2017). Neural network models are flexible, dynamic, statistically accurate, and handle noise and incomplete data effectively. ANNs are also termed as black box algorithms. It is able to generalize from a set of examples, with cases of a certain range and if the prediction lies outside the range then the range is completely invalid. ANN accurately diagnoses malignant melanoma, diverse forms of the tumor, and eye-related disorders using analysis of diagnostic criteria and spectral information (Zhang and Szolovits 2008). Research by Amato et al. diagnoses kidney stones using three neural networks (radial basis function, multilayer perceptron, and learning vector quantization) (Amato et al. 2013).

4.2 Naive Bayesian (NB)

This algorithm is based on a probabilistic model and is named after Thomas Bayes (1702–1761). It allows the capture of uncertainty in the model via determining probabilities of dissimilar outcomes. NB is used in medical applications, specifically in personalized medicine. It is robust, requires small training time, and has less input noise. NB algorithm is used on preprocessed text to classify the information based on cure, prevention, and side effects to patients.

4.3 Support Vector Machines (SVM)

SVM produces accurate results on a theoretical basis even in nonlinear separable input data. It is a nonparametric technique and lacks transparency in results. SVM algorithms detect the accuracy of cardiac monitoring devices. Generic systems for monitoring cardiac arrests are unable to differentiate between artifacts and true events, rather SVM provides more accurate patient-specific results (Huang et al. 2018). Srinivasarao (2014) developed a hybrid machine learning-based model fused with SVM classifiers that classify and disseminate healthcare information (Srinivasarao 2014).

4.4 Fuzzy Logic (FL)

Fuzzy cognitive maps are used for the classification and prediction of patient-specific diseases. It acquires a casual relationship with concepts and then analyzes the inference patterns. This is applied in personalized medicine therapies by analyzing the association between different diseases with patient-specific data, for example, blood pressure, blood type, and gene to infer a pattern for the detection of disease (Ghosh et al. 2020). FL was used to detect heart diseases using five input values (blood sugar, blood pressure, chest pain, heart rate, and cholesterol) with two outputs (precautions and result) with twenty-two rule bases. Output detects the presence of heart disease, and the observed result was 92% accurate.

5 NIC and AI in Personalized Medicine: Advantages and Limitations

With the advent of NIC and AI, the treatments are tailored according to patient characteristics, therefore providing better medication effectiveness, reduction in adverse event profiles, lower healthcare costs, disease diagnosed and prevent at early stage.

Another benefit is using sensors, mobile applications and data science, the disease managed at improved rate, fair selection of likely responders at baseline, clinical trials designed with smarter prospective. Schork (2019) NIC and AI proved to be beneficial by many researchers in the identification and management of certain chronic diseases. Dedov (2019) Major benefits of these techniques are combining multi-omics data that includes discrete, binary, continuous and categorical variables and no strict assumptions on normal distribution are made. These tools handle noisy and large variance data to regularization used in NIC and AI methods. Another benefit is that these algorithms are able to recognize multifaceted, nonlinear patterns in training data, which is not possible with human observations. These specialized tools are trained on small datasets (specific for number of features outnumber the number of observations), and thus, the possibility of accurate prediction is high (Radha and Srinivasan 2014). They have the capability of large data search and refining the solutions in a single search process. In addition, they also generate optimum solutions. Another benefit is that it is able to percolate low-level interactions. Generally, the swarm systems are not related to the central control, which makes it a comparable robust system in case of malfunction in any part of the system (Dhal et al. 2019). Conversely, there are numerous limitations in the use of AI and NIC in the development of personalized medicine.

1. Reduced interpretation capabilities of complex, nonlinear patterns. These algorithms sometimes over or under-fit the data in the training and testing sets (Goetz and Schork 2018).
2. NIC computing methods are time-consuming processes, affected by factors such as frequency, the pattern of iterations, and size of the population.
3. Basic swarm intelligence algorithms suffer from a lack of central coordination (Nguyen et al. 2020). Henceforth, the blocking coordination mechanism increases the calculated time.

6 Challenges of NIC and AI in Personalized Medicine

Challenges in personalized medicine comprise of identification or prediction of disease, achieving an accurate diagnosis, and availability of optimized treatment. In the case of data science, the foremost challenge in this domain is the insufficient prediction performance of NIC and AI tools for clinical practice (Chowdhury and Sadek 2012). These techniques capture and describe a signal (mathematically calculated) in the dataset. Signal-to-noise ratio adjustment is the biggest limitation, specifically, in the case of big data analysis (Fan et al. 2014). Linkage of patient-specific characteristics with clinically relevant endpoints is complex, nonlinear (vary with time) and is a multi-analyte dataset. The challenge lies in discriminating relevant from non-relevant patient-specific characteristics. It is also challenging to obtain large patient cohorts with well-defined phenotypes in training models. Another challenge lies in the discrimination of technical and biological noise (highly informative) within a dataset. Another factor that affects the prediction performance of computational tools is the selection of patient samples (under or over-represented specific patient subgroups in the training cohorts) (Ferentinou and Sakellariou 2007). The second most prominent challenge is model interpretation difficulties such as these techniques typically capture statistical dependencies (Shen et al. 2019). The third challenging condition is the lack of insufficient validation and translation of stratification algorithms into clinical practice (Bhattad and Jain 2020).

7 Future Prospective

As we have already discussed, the challenges in AI and NIC model interpretation are imperative bottlenecks for employing personalized medicine. Innovative software solutions that set complex AI and NIC models and their outputs in context to human pathophysiology (Baxi et al. 2022). These map the molecular features onto biological pathways and increased the learning decisions of clinicians. This tool enables clinicians to understand the development of disease over time and with the help of patient history, the rapidity and intensity of disease are also interpreted (Shen et al. 2019). Similar concepts developed in estimating the in-vivo fitness landscape experienced by HIV-1 development of such software and tools facilitate communication between physicians and computational science and construct complex models more interpretable (Ponomariov et al. 2017). There is a need for innovative tools that establish a relationship between knowledge with predictions. Modeling approaches such as hybrid modeling and casual interference techniques are doing well (Dhanwani and Wadhe 2013). Advancement of newer computational modeling approaches that identify critical transitions in patient medical trajectory. These might aid in creating a new simulation model of the disease and prediction of intervention points. It allows the planning of optimum medical intervention at the right dose and time on an individual basis (Hustinx 2019).

8 Conclusion

Personalized medicine is a newer process based on the genetic makeup, lifestyle, genes, and surrounding environment of an individual. It customizes a prevention and treatment strategy that utilizes data from varied sources to understand underlying molecular mechanisms that induce the disease. Generation of massive amounts of data (big data) requires newer approaches to analyze, integrate, and interpret the data. NIC and AI techniques came as imperceptible healthcare technologies. NIC helps to develop a medicare ecosystem that provides clear information about the disease to the clinicians and researchers by incorporating extra primary information from clinical data, including phenotypic details, lifestyle, and non-medical factors that can influence medical resolutions. Applications of combined nature-inspired and artificial intelligence models can accurately screen or detect the relevant details. The use of

NIC tools to identify pertinent features and diminish remaining features. Henceforth, NIC tools and techniques are widely used in imaging prediction (MRI, CT scans) for the diagnosis and management of chronic diseases such as cancer, diabetes, and cardiovascular disorders and are effective in solving personalized therapies and their related complications.

References

Aalaei S, Shahraki H, Rowhanimanesh A, Eslami S (2016) Feature selection using genetic algorithm for breast cancer diagnosis: experiment on three different datasets. Iran J Basic Med Sci 19:476–482

Agarwala V, Khozin S, Singal G et al (2018) Real-world evidence in support of precision medicine: clinico-genomic cancer data as a case study. Health Aff 37:765–772. https://doi.org/10.1377/hlthaff.2017.1579

Agbehadji IE, Awuzie BO, Ngowi AB, Millham RC (2020) Review of big data analytics, artificial intelligence and nature-inspired computing models towards accurate detection of COVID-19 pandemic cases and contact tracing. Int J Environ Res Public Health 17:1–16. https://doi.org/10.3390/ijerph17155330

Aishwarya SS, Anto S (2014) A medical expert system based on genetic algorithm and extreme learning machine for diabetes disease diagnosis. Int J Sci Eng Technol Res 3:1375–1380

Ali M, Aittokallio T (2019) Machine learning and feature selection for drug response prediction in precision oncology applications. Biophys Rev 11:31–39

Alshamlan H, Badr G, Alohali Y (2015) MRMR-ABC: a hybrid gene selection algorithm for cancer classification using microarray gene expression profiling. Biomed Res Int 2015. https://doi.org/10.1155/2015/604910

Amato F, López A, Peña-Méndez EM et al (2013) Artificial neural networks in medical diagnosis. J Appl Biomed 11:47–58

Anto S, Chandramathi S (2015) An expert system for breast cancer diagnosis using fuzzy classifier with ant colony optimization. Aust J Basic Appl Sci 9:172–177

Ateeq K, Gopinath Ganapathy D (2017) The novel hybrid modified particle swarm optimization-neural network (MPSO-NN) algorithm for classifying the diabetes

Awwalu J, Garba AG, Ghazvini A, Atuah R (2015) Artificial intelligence in personalized medicine application of AI algorithms in solving personalized medicine problems. Int J Comput Theory Eng 7:439–443. https://doi.org/10.7763/ijcte.2015.v7.999

Ashok B, Anu S, Nair H, Puviarasan DPA N (2016) Robust diagnosing technique for cervical cancer using random forest classifier

Baxi V, Edwards R, Montalto M, Saha S (2022) Digital pathology and artificial intelligence in translational medicine and clinical practice. Mod Pathol 35:23–32

Beloufa F, Chikh MA (2013) Design of fuzzy classifier for diabetes disease using modified artificial bee colony algorithm. Comput Methods Programs Biomed 112:92–103. https://doi.org/10.1016/j.cmpb.2013.07.009

Bhattad PB, Jain V (2020) Artificial intelligence in modern medicine—the evolving necessity of the present and role in transforming the future of medical care. Cureus. https://doi.org/10.7759/cureus.8041

Bhuvaneswari P, Therese AB (2015) Detection of cancer in lung with K-NN classification using genetic algorithm. Procedia Mater Sci 10:433–440. https://doi.org/10.1016/j.mspro.2015.06.077

Chowdhury M, Sadek AW (2012) Advantages and limitations of artificial intelligence. Artif Intell Appl to Crit Transp Issues 6:360–375 de Carvalho Filho AO, Silva AC, de Paiva AC et al (2017) Computer-aided diagnosis system for lung nodules based on computed tomography using shape analysis, a genetic algorithm, and SVM. Med Biol Eng Comput 55:1129–1146. https://doi.org/10.1007/s11517-016-1577-7

Dedov II (2019) Personalized medicine. Vestn. Ross. Akad. Meditsinskikh Nauk 74:61–70

Deoskar P, Singh D, Singh A (2013) An efficient support based ant colony optimization technique for lung cancer data. Int J Adv Res Comput Commun Eng 2

Devi KI, Deepa SN (2016) Classification of cardiac arrhythmia using artificial neural network with optimization algorithm. Int J Innov Sci Eng Res 3:

Dhal KG, Ray S, Das A, Das S (2019) A survey on nature-inspired optimization algorithms and their application in image enhancement domain. Arch Comput Methods Eng 26:1607–1638. https://doi.org/10.1007/s11831-018-9289-9

Dhanwani D, Wadhe PA (2013) Study of hybrid genetic algorithm using artificial neural network in data mining for the diagnosis of stroke disease. Int J Comput Eng Res 3:95–100

Dubey A, Patel R, Choure K (2014) An efficient data mining and ant colony optimization technique (dmaco) for heart disease prediction. Int J Adv Technol Eng Explor ISSN 2394–7454

Durairaj M, Sivagowry S (2015) A survey on particle swarm optimization and rough set theory in feature selection for heart disease prediction. Int J Comput Sci Mob Comput 4:87–92

Ephzibah E (2011) Cost effective approach on feature selection using genetic algorithms and fuzzy logic for diabetes diagnosis. Int J Soft Comput 2:1–10. https://doi.org/10.5121/ijsc.2011.2101

F.M R (2013) Diagnosis of heart disease based on ant colony algorithm. Int J Comput Sci Inf Secur 11:

Fan J, Han F, Liu H (2014) Challenges of big data analysis. Natl Sci Rev 1:293–314

Federal Drug Administration (2019) Table of pharmacogenomic biomarkers in drug labeling|FDA. United States Gov. pp 1–53

Ferentinou MD, Sakellariou MG (2007) Computational intelligence tools for the prediction of slope performance. Comput Geotech 34:362–384. https://doi.org/10.1016/j.compgeo.2007.06.004

Fister I, Yang XS, Brest J (2013) A comprehensive review of firefly algorithms. Swarm Evol Comput 13:34–46. https://doi.org/10.1016/j.swevo.2013.06.001

Fiuzy M, Qarehkhani A, Haddadnia J et al (2013) Introduction of a method to diabetes diagnosis according to optimum rules in fuzzy systems based on combination of data mining algorithm (d-t), evolutionary algorithms (aco) and artificial neural networks (nn). J Math Comput Sci 06:272–285. https://doi.org/10.22436/jmcs.06.04.03

Ganji MF, Abadeh MS (2011) A fuzzy classification system based on ant colony optimization for diabetes disease diagnosis. Expert Syst Appl 38:14650–14659. https://doi.org/10.1016/j.eswa.2011.05.018

Garg J (2016) Review on implementation of ACO technique for leukaemia detection. 5:859–862. https://doi.org/10.17148/IJARCCE.2016.54211

Ghosh G, Roy S, Merdji A (2020) A proposed health monitoring system using fuzzy inference system. Proc Inst Mech Eng Part H J Eng Med 234:562–569. https://doi.org/10.1177/0954411920908018

Goetz LH, Schork NJ (2018) Personalized medicine: motivation, challenges, and progress. Fertil Steril 109:952–963

Gupta P, Kaur B (2014) Accuracy enhancement of artificial neural network using genetic algorithm. Int J Comput Appl 103:11–15. https://doi.org/10.5120/18133-9258

Huang S, Nianguang CAI, Penzuti Pacheco P et al (2018) Applications of support vector machine (SVM) learning in cancer genomics. Cancer Genomics Proteomics 15:41–51

Hustinx R (2019) Physician centred imaging interpretation is dying out—why should I be a nuclear medicine physician? Eur J Nucl Med Mol Imaging 46:2708–2714

Karaboga D, Gorkemli B, Ozturk C, Karaboga N (2014) A comprehensive survey: artificial bee colony (ABC) algorithm and applications. Artif Intell Rev 42:21–57. https://doi.org/10.1007/s10462-012-9328-0

Karthikeyan T, Vembandasamy K (2015) A novel algorithm to diagnosis Type II diabetes mellitus based on association rule mining using MPSO-LSSVM with outlier detection method. Indian J Sci Technol 8:310. https://doi.org/10.17485/ijst/2015/v8is8/53631

Kaur B, Singh W (2015) Analysis of heart attack prediction system using genetic algorithm. Int J Adv Technol Eng Sci 87–94

Kavitha S, Nithya M (2020) An efficient cardiovascular disease prediction through pheromone based ACO with hybrid fuzzy logic. 1–4

Kelwade JP, Salankar SS (2017) Prediction of heart abnormalities using particle swarm optimization in radial basis function neural network. Int Conf Autom Control Dyn Optim Tech ICACDOT 2016:793–797

Khan Mansoori T, Suman A, Mishra SK (2014) Feature selection by genetic algorithm and svm classification for cancer detection

Kohad R, Ahire V (2015) Application of machine learning techniques for the diagnosis of lung cancer with ant colony optimization. Int J Comput Appl 113:34–41. https://doi.org/10.5120/19928-2069

Kohad R, Ahire V, Kohad MR (2014) Diagnosis of lung cancer using support vector machine with ant colony optimization technique 3:19–25

Kosorok MR, Laber EB (2019) Precision medicine. Annu. Rev. Stat. Its Appl. 6:263–286

Kumar S, Sahoo G (2017) Medical dataset classification using k-NN and genetic algorithm. In: Advances in intelligent systems and computing. pp 813–823

Kumari DJ (2017) Structural redesign of artificial neural network for predicting breast cancer with the aid of artificial bee colony. Indian J Sci Technol 10:1–8. https://doi.org/10.17485/ijst/2017/v10i15/108270

Lassen UN, Makaroff LE, Stenzinger A et al (2021) Precision oncology: a clinical and patient perspective. Futur Oncol 17:3995–4009

Lutimath NM, Ramachandra HV, Raghav S, Sharma N (2022) Prediction of heart disease using genetic algorithm. 49–58. https://doi.org/10.1007/978-981-16-3346-1_4

M D, S S, (2015) Feature diminution by using particle swarm optimization for envisaging the heart syndrome. Int J Inf Technol Comput Sci 7:35–43. https://doi.org/10.5815/ijitcs.2015.02.05

Mitra S, Shankar BU (2014) Integrating radio imaging with gene expressions toward a personalized management of cancer. IEEE Trans Human-Machine Syst 44:664–677. https://doi.org/10.1109/THMS.2014.2325744

Mitra S, Uma Shankar B (2015) Medical image analysis for cancer management in natural computing framework. Inf Sci (ny) 306:111–131. https://doi.org/10.1016/j.ins.2015.02.015

Mohebian MR, Marateb HR, Mansourian M et al (2017) A hybrid computer-aided-diagnosis system for prediction of breast cancer recurrence (HPBCR) using optimized ensemble learning. Comput Struct Biotechnol J 15:75–85. https://doi.org/10.1016/j.csbj.2016.11.004

Nazarian M, Dezfouli MA, Haronabadi A (2013) Classification of breast cancer samples through using the artificial bee colony algorithm. Int J Comput Appl Technol Res 2:522–525. https://doi.org/10.7753/ijcatr0205.1004

Nguyen BH, Xue B, Zhang M (2020) A survey on swarm intelligence approaches to feature selection in data mining. Swarm Evol Comput 54. https://doi.org/10.1016/j.swevo.2020.100663

Papadakis GZ, Karantanas AH, Tsikankis M et al (2019) Deep learning opens new horizons in personalized medicine (Review). Biomed Reports 10:215–217. https://doi.org/10.3892/br.2019.1199

Patankar V, Nawgaje D, Kanphade R (2014) A implementation of ant colony optimization technique for cancer diagnosis 4:2013–2015

Ponomariov V, Chirila L, Apipie F-M, et al (2017) Artificial intelligence versus doctors' intelligence: a glance on machine learning benefaction in electrocardiography. Discoveries 5:e76. https://doi.org/10.15190/d.2017.6

Radha P, Srinivasan B (2014) Feature selection using particle swarm optimization for predicting the risk of cardiovascular disease in type-ii diabetic patients. An Int J Adv Comput Technol 3

Rajathi S, Radhamani G (2016) Prediction and analysis of Rheumatic heart disease using kNN classification with ACO. In: Proceedings of 2016 international conference on data mining and advanced computing, SAPIENCE 2016. pp 68–73

Soliman OS, AboElhamd E (2014) Classification of diabetes mellitus using modified particle swarm optimization and least squares support vector machine. Int J Comput Trends Technol 8:38–44. https://doi.org/10.14445/22312803/ijctt-v8p108

Saini S (2017) Hybrid model using unsupervised filtering based on ant colony optimization and multiclass SVM by considering medical data set. Int Res J Eng Technol 4:2565–2571

Schork NJ (2019) Artificial intelligence and personalized medicine. In: Cancer treatment and research. pp 265–283

Shah H, Chiroma H, Herawan T et al (2019) An efficient bio-inspired bees colony for breast cancer prediction. Lect Notes Electr Eng 520:597–608. https://doi.org/10.1007/978-981-13-1799-6_61

Shahbeig S, Helfroush MS, Rahideh A (2017) A fuzzy multi-objective hybrid TLBO–PSO approach to select the associated genes with breast cancer. Signal Process 131:58–65. https://doi.org/10.1016/j.sigpro.2016.07.035

Shathesh S, Durairaj M (2016) An intelligent hybrid mechanism to predict the risk of cardio vascular disease. Indian J Sci Technol 9. https://doi.org/10.17485/ijst/2016/v9i4/82278

Shen J, Zhang CJP, Jiang B, et al (2019) Artificial intelligence versus clinicians in disease diagnosis: systematic review. JMIR Med. Informatics 7

Shukla R (2014) Cancer detection using frequency pattern ant colony optimization 2:3922–3927

Siddique N, Adeli H (2015) Nature inspired computing: an overview and some future directions. Cognit Comput 7:706–714. https://doi.org/10.1007/s12559-015-9370-8

Singer J, Irmisch A, Ruscheweyh HJ et al (2017) Bioinformatics for precision oncology. Brief Bioinform 20:778–788. https://doi.org/10.1093/bib/bbx143

Singh G, Bathla G, Kaur S (2016) Design of new architecture to detect leukemia cancer from medical images. Int J Appl Eng Res 11:7087–7094

Srinivasarao K (2014) Hybrid machine learning implementations for classifying dis-ease-treatment relations in short texts. Int J Adv Res Comput Sci-Ence Softw Eng 4:63–67

Subanya B, Rajalaxmi RR (2014a) Artificial bee colony based feature selection for effective cardiovascular disease diagnosis. Int J Sci Eng Res 5:606–612

Subanya B, Rajalaxmi RR (2014b) Feature selection using artificial bee colony for cardiovascular disease classification. In: 2014b International conference on electronics and communication systems, ICECS 2014b

Turabieh H, Muhanna M (2016) GA-based feature selection with ANFIS approach to breast cancer recurrence. Int J Comput Sci Issues 13:36–41. https://doi.org/10.20943/ijcsi-201602-3641

Uzer MS, Yilmaz N, Inan O (2013) Feature selection method based on artificial bee colony algorithm and support vector machines for medical datasets classification. Sci World J 2013. https://doi.org/10.1155/2013/419187

Vaishya R, Javaid M, Khan IH, Haleem A (2020) Artificial Intelligence (AI) applications for COVID-19 pandemic. Diabetes Metab Syndr Clin Res Rev 14:337–339. https://doi.org/10.1016/j.dsx.2020.04.012

Vimaladevi M, Kalaavathi DB (2014) Cancer classification using hybrid fast particle swarm optimization with backpropagation neural network. Ijarcce 3:8410–8414. https://doi.org/10.17148/ijarcce.2014.31112

Wang D, Tan D, Liu L (2018) Particle swarm optimization algorithm: an overview. Soft Comput 22:387–408. https://doi.org/10.1007/s00500-016-2474-6

Zhang L, Yuan J (2015) Fault diagnosis of power transformers using kernel based extreme learning machine with particle swarm optimization. Appl Math Inf Sci 9:1003–1010. https://doi.org/10.12785/amis/090251

Zhang Y, Szolovits P (2008) Patient-specific learning in real time for adaptive monitoring in critical care. J Biomed Inform 41:452–460. https://doi.org/10.1016/j.jbi.2008.03.011

Role of Nature-Inspired Intelligence in Genomic Diagnosis of Antimicrobial Resistance

Priyanka Sharma, Geetika Sethi, Manish Kumar Tripathi, Shweta Rana, Harpreet Singh, and Punit Kaur

1 Introduction

Bacterial antimicrobial resistance (AMR), first reported in 1945, continues to emerge as an uncontained global threat (O'neill 2014; Overton et al. 2021). Although all bacterial infections are manageable with proper antibiotic treatment, AMR is leading to an increased mortality rate in infectious diseases (Dahiya et al. 2019). According to recent reports, AMR is prophesied to be responsible for nearly 700,000 deaths per year globally, with an annual estimated cost of 55 billion dollars in the USA (Dadgostar 2019). The six main pathogens, *Escherichia coli* (fluoroquinolone-resistant (Zhao et al. 2021)), *Staphylococcus aureus* (methicillin-resistant (Kuroda et al. 2001)), *Klebsiella pneumonia* (carbapenem-resistant, (Tumbarello et al. 2015)), *Streptococcus pneumonia, Acinetobacter baumannii* (carbapenem-resistant (Lee et al. 2017)), and *Pseudomonas aeruginosa,* were reported to be the leading cause for

P. Sharma · M. K. Tripathi · P. Kaur (✉)
Department of Biophysics, All India Institute of Medical Sciences, New Delhi, India
e-mail: punitkaur1@hotmail.com

P. Sharma
e-mail: priyankap2828@gmail.com

M. K. Tripathi
e-mail: manishtripathi41@gmail.com

G. Sethi
ICMR-AIIMS Computational Genomics Centre, New Delhi, India
e-mail: geetika@bmi.icmr.org.in

P. Sharma · S. Rana · H. Singh
Division of BMI, ICMR, New Delhi, India
e-mail: shweta@bmi.icmr.org.in

H. Singh
e-mail: hsingh@bmi.icmr.org.in

K. Raza (ed.), *Nature-Inspired Intelligent Computing Techniques in Bioinformatics*,
Studies in Computational Intelligence 1066,
https://doi.org/10.1007/978-981-19-6379-7_12

death allied to AMR in 2019 (Dadgostar 2019). Drug resistance to *Salmonella* is a major health challenge in underdeveloped countries and for travelers returning from endemic areas (Dahiya et al. 2017; Sharma et al. 2018). The emerging drug resistance in *Mycobacterium tuberculosis* has raised a panic situation, especially in developing countries (Luz et al. 2022). The decreased drug susceptibility, thus, clearly accentuates the race against antimicrobial resistance in bacterial pathogens and demands coordinated and incessant efforts (Premanandh et al. 2016). Hence, interdisciplinary research and development strategies to improve understanding of the genetic diversities, complemented by data-driven revolutionary computing discoveries may prove to be beneficial in alleviating this looming threat (Katiyar et al. 2020; Sharma et al. 2020).

The underlying reason for the development of AMR is generally the selection pressure posed by the injudicious use of antimicrobial agents. Since the diagnosis of infection and antimicrobial susceptibility test outcomes is a time taking process, the empirical treatment is initially based on the clinical diagnosis which in turn may lead to treatment failure and contribute to the development of resistance (Paterson 2008). International health agencies like Clinical and Laboratory Standard Institute (CLSI), Center for Disease Control (CDC), and European Committees on Antimicrobial Susceptibility Testing (EUCAST) publish recommendations for antibiotic usage on the basis of laboratory parameters, but the delay in diagnosis leads to the empirical use of antibiotics (Paterson 2008). The conventional diagnosis of antimicrobial susceptibility relies on culture-based techniques which require 48–72 h for drug susceptibility results (Opota et al. 2015) (Fig. 1). AMR is measured using disk diffusion or dilution procedures for assessing the growth after the addition of a definite concentration of antimicrobial agents (Khan et al. 2019). There are some novel diagnostic devices based on MALDI-TOF which provide antimicrobial susceptibility testing (AST) results but the sensitivity and the range of tested antimicrobial agents are limiting the factor (Sauget et al. 2018).

Several analytical devices, like BD Phoenix (Becton Dickinson) and VITEK 2 (bioMérieux), are widely employed for point-of-care diagnosis of AMR (Kaur et al. 2019; Layer et al. 2006). With the advancements in technology, *in silico* AST on the basis of genomic characteristics of the causative organism provides a complete list of genomic determinants of AMR are gaining popularity day by day (Katiyar et al. 2020). Whole genome sequencing (WGS) refers to massive parallel and deep sequencing techniques which can provide high throughput sequencing data by sequencing millions of DNA molecules simultaneously (Köser et al. 2014). WGS data can be analyzed with the help of bioinformatics tools in order to annotate the genes, metabolic pathway analysis, identification of pathogenic genes for vaccines, as well as typing methods for phylogenetic and evolutionary relatedness among the strains (Uelze et al. 2020). The analysis of the massive genomic data being produced by sequencing is a computationally demanding task and requires expert bioinformatics skills for the interpretation of results (Köser et al. 2014).

Data-driven discoveries that unleash the true power of data are evidently bolstered by advanced computational systems (Holder et al. 2017; Tripathi et al. 2021a, b). For example, the unprecedented advancements in high-performance computing and

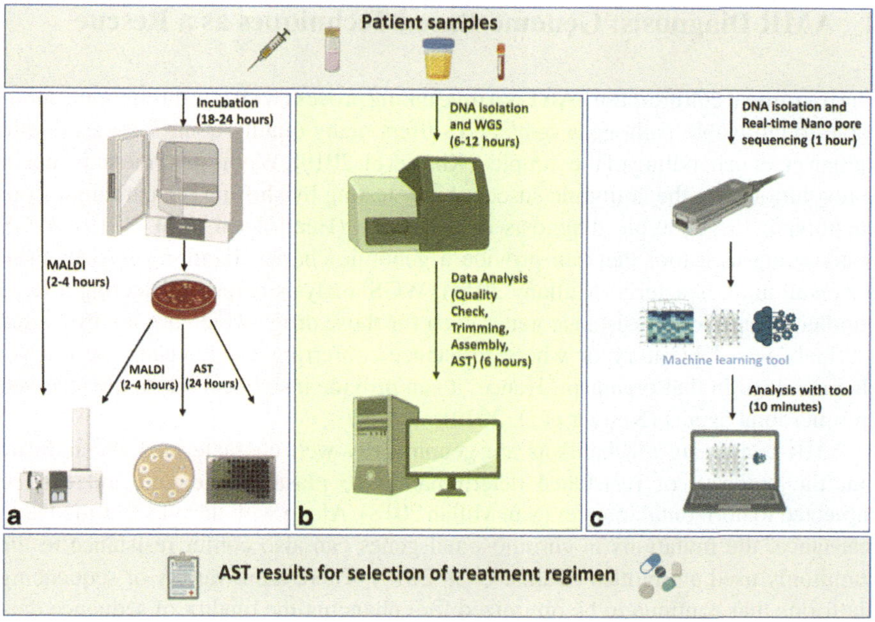

Fig. 1 Diagnosis of antimicrobial resistance using conventional and machine learning methodologies **a** Commonly used AMR detection methods are time-consuming and take 48–72 for diagnosis, although MALDI offers a quick diagnosis, but its sensitivity and specificity are a concern, **b** Genomic diagnosis of AMR markers is an attractive approach but it requires lots of data analysis, **c** Machine learning tools coupled with nanopore sequencers can be used to diagnose AMR in real time. The image has been generated by BioRender (https://biorender.com/) accessed on June 13, 2022

related intelligent computational techniques combined with the deluge of emerging genomics data on bacterial antimicrobial resistance have successfully shifted the paradigms in the field of bacterial antimicrobial resistance (Van Camp et al. 2020). Intelligent computational techniques such as nature-inspired computations, artificial intelligence algorithms, machine learning, deep learning, and metaheuristics have impacted almost every spectrum of human endeavor (Lau et al. 2021). Both nature-inspired intelligent (NII) algorithms and evolutionary artificial intelligence (AI) algorithms have equipped scientists and researchers with better ability to solve complex problems in various interdisciplinary areas including bacterial genomics for AMR (Avershina et al. 2021).

The present chapter comprises the usage of NII in order to identify AMR using WGS data from bacterial pathogens. The basic theories behind the NII algorithms and their use in clinical microbiology for prediction of antibiotic resistance have been enumerated.

2 AMR Diagnosis: Genome-Based Techniques as a Rescue

The traditional culture-based AST is a time taking process with low throughput, applicable to cultivable pathogens only, and suffers many quality control issues due to human error in handling of the samples (Khan et al. 2019). Whole genome sequencing is revolutionizing the antibiotic susceptibility testing by shifting the paradigm from phenotypic to genotypic diagnosis of resistance (Hendriksen et al. 2019). WGS is emerging as a tool that can provide a genomic characterization of AMR even for unculturable bacteria (Mullany 2014). WGS analysis helps in detecting a large number of antibiotic resistance genes even for those drugs which are not evaluated routinely in a laboratory or where resistance-conferring mechanisms are not yet characterized in that organism. Hence, it can provide insight even into the reserved antimicrobial agents (Katiyar et al. 2020).

AMR-conferring mechanisms are genomically well characterized and indicate that the majority of resistance determinants are plasmid-mediated and can be subjected to horizontal transfer (San Millan 2018). Along with the plasmid-mediated resistance, the mutations in chromosomal genes can also confer resistance to the commonly used antibiotics (Kumar et al. 2015). There are a variety of sequencing platforms that continue to be optimized for enhancing the quality of sequence data in a cost-effective manner (Metzker 2010). The Illumina-based sequencers Hi-Seq and Mi-Seq platforms provide a huge amount of short read data with an appreciable accuracy rate but the installation cost and space requirements are limiting factors for clinical implementation of these technologies for their use in routine testing (Allali et al. 2017). Benchtop devices like minion and nanopore sequencers provide an alternative to other sequencing platforms, involve lesser investments, and are fast as well as simple options for genomic sequencing of bacterial pathogens (Kono and Arakawa 2019). These sequencers give the output data in FastQ format which needs to be analyzed to obtain the complete genomic profile of the pathogen (Katiyar et al. 2020). The analysis of WGS data involves multiple steps from quality assessments of data to assembly and annotation of genes, wherein each step requires a number of tools for processing the data (Quainoo et al. 2017). The raw sequences need to be trimmed with a trimming tool like "trimmomatic" to remove the bad quality of data, adapter sequencing, and repeated sequences from the raw data (Bolger et al. 2014). The trimmed data is then assembled to form contigs using *de novo* assembler tools like "spades" or "velvet" tools and also can be aligned to a reference sequence to form the complete genome assembly using bowtie or BWA aligners (Bankevich et al. 2012; Joshi and Srivastava 2011; Langmead and Salzberg 2012; Li and Durbin 2009). For the prediction of AMR, the *de novo* assembly is always preferred over the reference-based assembly as the de novo method provides the best possible assembly results which give us the complete set of resistance genes that could be absent in the reference sequence (Van Camp et al. 2020). The assembled contigs are then tested against the antimicrobial resistance databases like Resfinder, CARD, or ARDB in order to find and annotate the genomic determinants of antimicrobial resistance (Alcock et al. 2020; Bortolaia et al. 2020; Liu and Pop 2009).

The availability of efficient genome sequencers has shifted the main concern from the generation of genome sequencing data to the analysis and interpretation of sequencing results (Van Camp et al. 2020). The genome assembly and annotations are complicated processes and require good bioinformatics skills in order to analyze the genomic data (Hendriksen et al. 2019; Van Camp et al. 2020). This imparts a major challenge for clinicians to interpret WGS data in order to get antimicrobial susceptibility results. Hence, the involvement of ML algorithms has been evidenced as a boon for WGS data analysis to obtain rapid outcomes with minimal computational power (Avershina et al. 2021). Multi-omics data-driven approaches in the field of bacterial antimicrobial resistance (Chernov et al. 2019; Liao et al. 2019) powered by state-of-the-art intelligent computational techniques (Her and Wu 2018) have provided valuable insights, thereby contributing effectively to the race against bacterial antimicrobial resistance (Zhang et al. 2021). Over the years, the field of bacterial genomics in particular (Giulieri et al. 2020; Gorrie et al. 2019; Sánchez-Busó and Harris 2019; Taylor-Brown et al. 2018) has commendably helped in obtaining answers to the pressing questions in the field of bacterial AMR (Giulieri et al. 2020; Hendriksen et al. 2019; Köser et al. 2014). Furthermore, large-scale big data platforms that employ genomic, transcriptomic, and metagenomic platforms have been instrumental in catalyzing advancements related to the fight against bacterial antimicrobial resistance. Moreover, the bacterial genomics footprint has resulted in a data deluge that demands advanced computational systems and tools to derive actionable insights (Saltykova et al. 2019).

3 Nature-Inspired Intelligence

Interestingly, the origin of NII algorithms is very much rooted in the essence of biological systems and their behavior (Wang et al. 2007). Mother nature with its ever-evolving and dynamic landscape is an inspiration to the development of intelligent computing techniques (Yang 2020). Natural systems always had and will continue to have a firm existence ingrained in the structures of computational intelligence and related computational algorithms (Wang et al. 2007).

Nature-inspired computational intelligent techniques refer to the development of algorithms that leverage the way that is adopted by the biological systems to solve real-world challenges (Kang et al. 2012). They mimic the way the real biological systems work when solving complex real-world scenarios (Yang 2020). Tzanetos et al. have effectively compiled a comprehensive database of NII algorithms (Tzanetos et al. 2020). Out of multiple different categories and sub-categories of NII algorithms, four major types, artificial neural networks (ANNs), evolutionary computing, swarm intelligence, and artificial immune systems have been commonly studied (Brownlee 2011; Kang et al. 2012; Siddique and Adeli 2015). The detailed classification of nature-inspired algorithms is represented in Fig. 2.

Mechanistically, an NII algorithm is an optimization technique that seeks an objective within a set of constraints to be gratified and creates an objective function, also

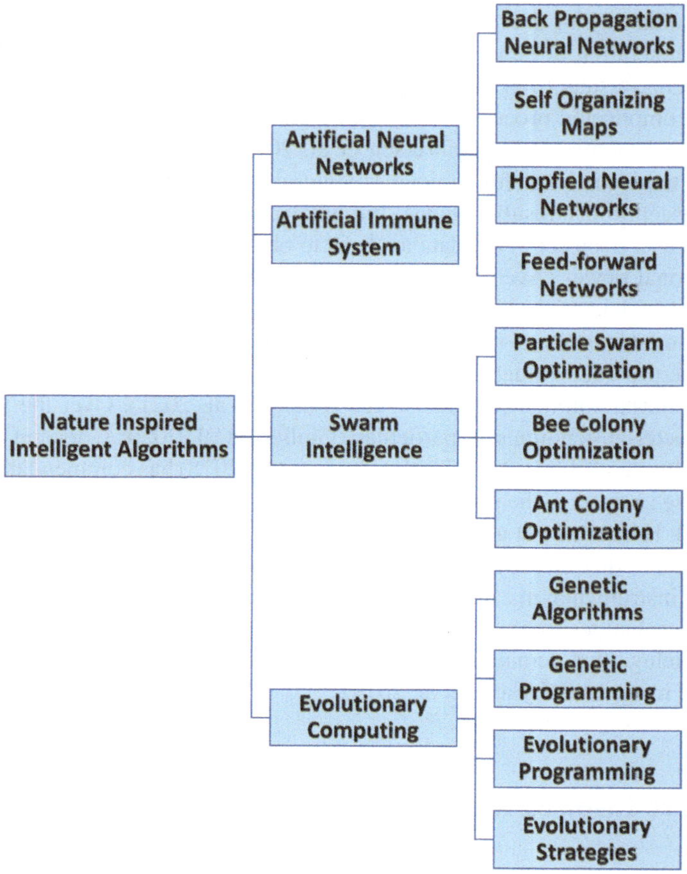

Fig. 2 Classification of nature-inspired intelligence algorithms

known as performance index which is a measure of an optimal solution (Gelbukh et al. 2014). The success of NII algorithms for difficult and complicated optimization problems makes them a considerable choice for exploration. Furthermore, Siddique et al. have summarized the cases, wherein NII algorithms have been advantageously adopted (Siddique and Adeli 2015). Their implementation is considered mainly when it is not possible to solve the addressed issue with conventional methods like complex pattern recognition and classification tasks. Additionally, when the output results contain multidimensional findings, nature-inspired intelligence offers implementable choices (Siddique and Adeli 2015). Recent studies show that NII algorithms are prospectively becoming major contributing factors to foster the rapid evolution of AI tools as hybridization of these NII algorithms with conventional AI tools are becoming increasingly popular (Darwish 2018).

3.1 Artificial Neural Network

Artificial neural network (ANN) incorporates important features like learning and generalizing data. ANN is computational models inspired by human brains. Bernard Widrow first established this model with the inspiration of the human mind. This model encompasses multiple layers of relating nodes analogous to neuronal cells. Like biological neurons, ANN comprises numerous artificial neurons that mimic the functioning of the human mind for the training process. ANN with multiple hidden layers or nodes might be affected autonomously by the noise in data and lead to low-quality data organization, whereas ANN with a limited number of hidden nodes generally fails to discriminate the complex patterns. The complex neural network training process is a highly time-consuming step. Thus, based on its connection, ANN can be classified into the: feed forward, radial basis function, recurrent networks, etc. The ANN network circuit where connections do not make circles is known as the feedforward network, while the network which forms circles is referred to as the recurrent network. Recently, several ANN-based models have been developed to resolve the day-to-day problems that are otherwise complicated to solve (Sheela and Deepa 2013). ANN is widely used in nature-inspired computing simulation to accomplish particular tasks such as clustering, classification, and pattern recognition from biological data. To update the weight bac propagation, these algorithms are employed during the neural network training process. The gradient descent algorithm is employed with the swarm intelligence method or iterative evolutionary method to decrease the time required for training purposes. The classification model is executed for the prediction of the class label of the test set during the classification step (Tripathi et al. 2021a, b. ANN has countless benefits such as good learning ability, parallel execution of tasks, and tolerance to noise (Alaloul 2020). Despite its advantages, it has drawbacks such as time-consuming and expensive costs. The ANN has numerous benefits over statistical approaches as it can distinguish both the linear and nonlinear patterns on the basis of input data. ANN is also termed as a knowledge-based multidimensional model because it can potentially model the association of multidimensional models compared to the linear model. This property empowers ANN to be widely used in genomics studies to mine useful patterns against the desired disease and drug discovery process. Software and packages such as Mathematica, MATLAB, and Neural lab are available which implement ANN algorithms.

3.2 Artificial Immune Systems

The artificial immune systems (AIS) algorithm is inspired by the human immune system. AIS is a sub-branch of biologically inspired computing and natural computation and belongs to the field of artificial intelligence. AIS algorithms are based on immunological principles and are applied to computational engineering problems to correlate the task to be addressed and the immune system. Apart from other bio-inspired techniques, the AIS constitutes a range of algorithms incorporating the different assets of various types of immune cells (Read et al. 2012). All AIS algorithms mimic the nature and properties of various immunological cells such as B cells, T cells, and dendritic cells (DCs).

The artificial immune recognition systems (AIRS) algorithm was initially established by Goodman et al. and verified against Kohenon's learning vector quantization (LVQ) algorithm and the k-nearest neighbors (KNN) algorithm (Rashid et al. 2018). The AIS algorithm is also used for multiclass classification via numerous phases for the training of data points (Pinto et al. 2022). Currently, two generations of AIS algorithms are being used (i) the first-generation algorithm simply relies on the simplified immune model, and (ii) the second generation employs the interdisciplinary relationship, which develops a deeper thought process of the immune system and produces more complex models. These algorithms have been effectively applied against a diversity of problems, including anomaly detection, pattern recognition, optimization, and robotics. This technique is widely used to monitor the network traffic flow.

3.3 Bio-Inspired Swarm Optimization Algorithms

Bio-inspired computing optimization algorithms are the evolving methods that are grounded on the values and motivation from biological evolution to develop robust computing techniques. These methods are documented in the machine learning techniques and address the ideal solution for complex problems. Thus, to tackle the problem of traditional optimization algorithms, recently bio-inspired optimization algorithms are used, representing a promising approach to resolving the complicated optimizations in the engineering and medical science field.

3.4 Particle Swarm Optimization

In the nature-inspired computing methods, particle swarm optimization (PSO) optimizes the data iteratively and assists to solve the problem related to the given amount of data quality. This optimization depends on the population structure and is used for stochastic search in a multidimensional space (Wang et al. 2018). This algorithm successfully solves various optimization problems, including many multi-layered problems. PSO algorithm is different from other optimization algorithms as only the objective function is required and there is no dependence on the gradient function. Thus, this method is an advancement over the popular methods such as steepest descent, gradient descent, conjugate gradient, and Newton method. The approach to designing PSO algorithm is closely correlated to two types of research, one is the evolutionary algorithm, and the other is artificial life, both of which study artificial systems with life features. This method similarly uses the swarm mode, which aids in the search of large regions in the solution space simultaneously (Wang et al. 2018). Swarm intelligence is a branch of AI and is armed with the intelligent behavior of biological swarms for solving real-world problems by simulating biological behaviors.

The principal objective of this algorithm is to regulate the variable which is characterized by vector $X = \{x_1, x_2, x_3, \ldots, x_n\}$, which minimizes or maximizes the suggested optimization formulation of function f_X. Here, X is the position vector which represents variable model in the n-dimensional vector and n denotes the total number of variables. Consider a swarm comprising P particles, with position vector $X_{\text{it}} = \{x_{i1}, x_{i2}, x_{i3}, \ldots, x_{\text{inT}}\}$ and velocity vector $V_{\text{it}} = \{v_{i1}, v_{i2}, v_{i3}, \ldots, v_{\text{inT}}\}$ at a t iteration for every i particle which constitutes it. Further, these vectors are updated via dimension j as per the following equations:

$$V_{ij}(t+1) = wV_{ij}(t) + c_1 r_1 t p\text{best}_{ij} - X_{ij}(t) + c_2 r_2(t) g\text{best}_j - X_{ij}(t) E_1 \quad (1)$$

$$X_{ij}(t+1) = X_{ij}(t) + V_{ij}(t) + E_2 \quad (2)$$

where $i = \{1, 2, \ldots, P\}$ and $j = \{1, 2, \ldots, n\}$, and Eq. (1) represents the three different contributions to a particle's movement in an iteration, whereas Eq. (2) updates the particle's positions.

3.5 Bee Colony Optimization

The genetic bee colony optimization is a metaheuristic optimization algorithm proposed by Karaboga in 2005 (Karaboga and Basturk 2007). This optimization algorithm was designed by assimilating the advantages of genetic algorithm (GA) and artificial bee colony (ABC) for optimizing the numerical problems (Karaboga and Basturk 2007). The ABC algorithm is founded on the model introduced by Tereshko and Loengarov and is motivated by the foraging property of honey bee colonies (Li and Chan 2011). The ABC model contains three important components, namely employed and unemployed foraging bees, and food sources. The employed and unemployed foraging bee components are utilized to search for food sources (Aderhold et al. 2010). To apply the ABC algorithm, the optimization problem is first transformed into the problem of finding the best parameter vector, which helps reduce objective function loss. In further steps, the artificial bees randomly identify the population of the initial solution vector, which iteratively improves the optimization by moving toward a better solution by neighbor search mechanism.

The bee colony optimization algorithm initially creates a randomly disseminated population with SN solution (food source), where SN represents the swarm size. Suppose $Xi = \{xi, 1, xi, 2, \ldots, xi, D\}$ denotes the ith solution of the swarm, and

D represents the dimension size. The employed bee X_i produces a new candidate simulation V_i at the neighborhood of its present position as follows:

$$v_{i,j} = x_{i,j} + \varphi_{i,j}\left(x_{i,j} - x_{k,j}\right) \tag{3}$$

In this, X_k represents the randomly selected candidate solution ($i = k$), j represents random dimension index selected with set {1, 2, ..., D}, and $\phi_{i,j}$ is a random number within [−1, 1]. When the new candidate solution V_i is created, then greedy selection is employed or the approximate problem. Thus, when fitness value of V_i is improved compared to parent X_i, then X_i with V_i will be updated otherwise X_i will be unchangeable.

When all the search processes of the employed bee are complete, then they share the information of the food source with onlooker bees via waggle dances. The onlooker bee then evaluates the nectar information gained from the employed bee and chooses the food source with probability to its nectar amount. This probabilistic selection is a roulette wheel selection mechanism that can be pronounced as follows:

$$p_i = \text{fit}_i \mid \sum \text{j} \, \text{fit}_j \tag{4}$$

where fit_i denotes the fitness value of the ith solution in the swarm. The better the solution i, the higher the probability of selected ith food sources.

3.6 Ant Colony Optimization

Ant colony optimization (ACO) is a population-based metaheuristic search method being employed to solve computational problems for determining the good path through graphs. In this approach, the artificial ant stands for the multi-agent method that mimics real ants' behavior. Here, the pheromone-based communication of biological ants is the predominant paradigm that is used. The optimization problem in ACO is converted into the problem of finding the best path on a weighted graph. The solution is built in a stochastic way by a pheromone model of artificial ants by moving on the graph with a set of parameters linked with graph components. Its value is modified during the optimization process.

In each repetition step, ACO produces global ants and evaluates their fitness, updating the pheromone and edge of the weaker region. If the fitness improves, local ants travel to the better region; otherwise, a new search direction is performed. Thus, the ACO involves both the local and global search. The local ants possess the fitness for moving toward the latent region with an optimized solution having a transition probability of region k,

$$P_k(t) = t_k(t) \sum_{j=1}^{n} t_j(t) \tag{5}$$

where $t_k(t)$ is total pheromone at region k and n is the number of global ants. The pheromone is updated in the successive iterative phase using the following equation:

$$t_i(t+1) = (1-r)t_i(t) \tag{6}$$

where r is the pheromone evaporation rate.

4 Big Data Resources

4.1 Genomic Databases

The availability of reference databases and bioinformatics tools is a prerequisite for the genotypic interpretation of antimicrobial susceptibility (Hendriksen et al. 2019). The development of a machine learning model requires a dataset each for training, testing, and validation of the prediction ability of the designed tool (Avershina et al. 2021). The advancements in sequencing technologies have resulted in the availability of a plethora of genome sequencing data in public data repositories (Lapidus 2019). There exist online repositories for uploading genomic data, where data upload is mandatory before scientific publication. At the time of writing this Chapter, 46, 41, and 112, genomic sequences are available in the Sequence Reads Archive (SRA) of National Centre for Biotechnology Information (NCBI), and hundreds are being added on a weekly basis (Sequence Read Archive 2018). Genomic data represents the most practically attainable details about an organism and can contribute to strain identification, phylogenetic relationships, and pathogenicity of an organism in addition to the antimicrobial susceptibility profile (Quainoo et al. 2017). The SRA and ENA (European Nucleotide Archive) allow the submission of genomic data directly from next-generation sequencing (NGS) platforms (Leinonen et al. 2011; Sequence Read Archive 2018).

NCBI is affiliated with the United States National Institutes of Health and comprises the scientific data deposited from worldwide laboratories (Sequence Read Archive 2018). The data can be submitted from the online submission page and SRA toolkit available on the NCBI website (www.ncbi.nlm.nih.gov). The data identifiers are assigned to each file and provided to the submitters for reference use in their publications. A specific dataset containing the genomic data of pathogenic organisms is available as a sub-domain on the NCBI Pathogens page (https://www.ncbi.nlm.nih.gov/pathogens/). This page contains the metadata regarding the sample source, geographical origin, antimicrobial profile, and date of collection for the biological sample used for data generation (Agarwala et al. 2017). An integrated AMRFinder tool for automatic identification and annotation of AMR-conferring genes in the submitted sequences is available (Feldgarden et al. 2019).

ENA (http://www.ebi.ac.uk/ena) is the prime genomic sequence database of Europe and comprises three sub-domains, namely the SRA, the Trace Archive, and EMBL-Bank (Leinonen et al. 2011). This repository contains the Pathogen Data Portal which can be explored to download the submitted genomic data along with the metadata regarding the bio-samples used for sequencing (https://www.ebi.ac.uk/ena/pathogens/home). It also contains an integrated notebook tool that depicts all the genotypic characteristics along with the antimicrobial susceptibility profiles for real-time monitoring of AMR (Leinonen et al. 2011). ENA also includes metadata regarding phenotypic features of the organisms which can be explored for training, testing, and validation of the machine learning tools (Leinonen et al. 2011). The main shortcoming of sequence databases is the lack of any quality control during sequence submission, and hence, the accountability for data fidelity and quality of the raw sequence data or assembled sequences before deposition lies with the submitters. Moreover, the assembler tools and parameters used for assembly generation affect the outcome and quality of assembled data available in public repositories (Lapidus 2019).

4.2 Antimicrobial Resistance Databases

The identification of genotypic antimicrobial resistance relies on the use of curated reference databases comprising antibiotic resistance markers that should be coordinated and consistent in a way to ensure accurate output results (Mahfouz et al. 2020). A comprehensive catalog of resistance mutations and genes is required to identify the presence of resistance-conferring genomic markers. There are more than 47 freely accessible resources available for antimicrobial resistance determining genes using the nucleotide or amino acid sequence data, for example, ARG-ANNOT, CARD, SRST2, MEGARes, Genefinder, ARIBA, KmerResistance, AMRFinder, and Resfinder (Boolchandani et al. 2019).

AMR data resources can be categorized on the basis of the inclusion of resistance markers. Some databases are only specific to a particular antibiotic class, while some contain all the antimicrobial resistance-conferring mechanisms for all possible antibacterial agents (Bortolaia et al. 2020). The AMR databases like Resfinder harbor only the plasmid-mediated resistance, while CARD includes the information about single nucleotide variations conferring AMR (Alcock et al. 2020; Bortolaia et al. 2020). These databases either have integrated search tools or provide downloadable data which can be entrenched in the bioinformatics pipeline adopted by the user (Mahfouz et al. 2020).

The databases also differ from each other in the requirement of input sequence format (FastQ or assembled contigs). Although the use of assembly-based methodology is computationally demanding, it has an additional value that it denotes the genomic position of the resistance markers and provides information about their horizontal transfer on the basis of flanking transposon sequences. The read-based methods use the raw FastQ sequences and align them against the AMR databases using the aligners like Bowtie2, BWA, and KMA (Mahfouz et al. 2020). The GUI-based databases do not depend on computational skills and are easy to use by clinicians and microbiologists. All the databases have a different set of features, and hence, the AMR prediction can vary from one database to another (Hendriksen et al. 2019). These databases predict AMR based on the determination of known resistance genes. However, to identify novel resistance mechanisms, the complete genome sequences need to be analyzed. An iterative *k*-mer-based analytic procedure is required to identify novel resistance-conferring genomic features but this approach requires complete reference genomes along with the AST data for the training of models (Lupolova et al. 2019; Nguyen et al. 2019).

5 Implementation of ML for Genomic Prediction of AMR

The accurate prediction of antimicrobial resistance from genomic data holds great significance in AMR bioinformatics research (Quainoo et al. 2017; Jamal et al. 2020; Liu et al. 2020). Machine learning has been successfully employed in designing AMR screening tools using deep learning models (Jha et al. 2017; Shi et al. 2019). Various studies have implemented NII algorithms for genomic datasets for bacterial antimicrobial resistance. However, the neural network is the most commonly adopted NII algorithm, followed by the genetic algorithm (Arango-Argoty et al. 2018; Weiss et al., 2016). On the other hand, multiple AI algorithms, for example, support vector machines (SVM), logistic regression, etc., have been leveraged for genomics datasets to help gain insights into bacterial antimicrobial resistance (Lv et al. 2021).

There is an upsurge in the usage of natural metaphors and metaheuristic approaches to address large-scale nonlinear optimizations. Deep neural networks inspired by the function of neuronal cells to provide a set of methods for the analysis of complicated data with multiple result outputs are being widely used in bioinformatics research. The whole genome sequence data of bacterial strains has been utilized for developing machine learning tools to predict the AMR in *K. neumonia, E. coli, P. aeruginosa, Mycobacterium tuberculosis,* and *Staphylococcus aureus* (Jha et al. 2017).

ANN belongs to the group of algorithms commonly employed for the characterization of genomic data using a machine learning approach. Nature-inspired algorithms are proven to possess better proficiency than the traditional algorithms for providing the best output (Bharti et al. 2020; Kang et al. 2012). The machine learning tools adopt different computational approaches for AMR prediction (Lau et al. 2021). The simplest one is the mapping of reads with the database of resistance genes and scoring them on the basis of their alignment to the reference genes. The use of raw sequence reads decreases the chance of false prediction due to loss of information at the time of assembly (Lau et al. 2021). The second approach is matching *k*-mers between the reads and a resistance database. The second approach is the fastest option as it bypasses the enormous computational power and time required for assembly generation (Bradley et al. 2015). The third approach is the matching of assembled contigs against the resistance databases (Lau et al. 2021). The machine learning-based tool "Mykrobe predictor" analyzes the De Bruijn graph containing the information about all reads used for assembly to subsequently evaluate it by comparing it with the De Bruijn graph generated by the AMR database (Bradley et al. 2015). This tool is designed as ready-to-use software that can be downloaded and used offline with WGS data as input to retrieve the complete antibiotic susceptibility profile. The usage of this tool is restricted to only *Staphylococcus aureus* and *Mycobacterium tuberculosis* (Bradley et al. 2015). DeepARG is another tool that predicts AMR using full-length ORF as well as short-sequence reads (Arango-Argoty et al. 2018). DeepARG offers high sensitivity and specificity but fails to identify the unique resistance markers to uncommon drugs like triclosan and sulphonamide (Arango-Argoty et al. 2018). Davis et al. transformed the *k*-mer to a binary matrix employing rapid annotation using subsystem technology (RAST) wherein this matrix was used to predict the presence of AMR-conferring *k*-mers (Davis et al. 2016). PATRIC is another online tool built for the collection of genomic sequences and employs a specific ARG annotation pipeline using NDARO and CARD databases along with data from the scientific literature. This was re-annotated by experts; to build a collection of assembled contigs paired with AST metadata (Davis et al. 2016).

AI is thus emerging as an important tool for AMR predictions. However, sufficient modifications and improvements are required to implement this at the clinical level. The challenges regarding the availability of large training dataset, reliable data resources, and AST metadata need to be resolved to design efficient tools for AMR prediction.

6 Development of Machine Learning Model

Machine learning is the branch of artificial intelligence, where computational machines are trained to perform the functions that humans are supposed to achieve. Machine learning models can be developed using a dataset and can be auto-trained to analyze data on the basis of training algorithms (Davis et al. 2016). Deep learning allows the data presentation and analysis in a hierarchical manner at multiple levels. The deep learning models comprise multiple "deep" layers of neurons that apply nonlinear operations and can be learned automatically using advanced computational algorithms (Jha et al. 2017).

Machine learning can be developed to predict the AMR markers from genomic data, in two manners: the supervised and unsupervised models. The supervised method requires a complete list of genomic features extracted on the basis of characterized data for the training process to create a predictive model which then performs on the basis of features assigned to it at the time of design. A supervised approach

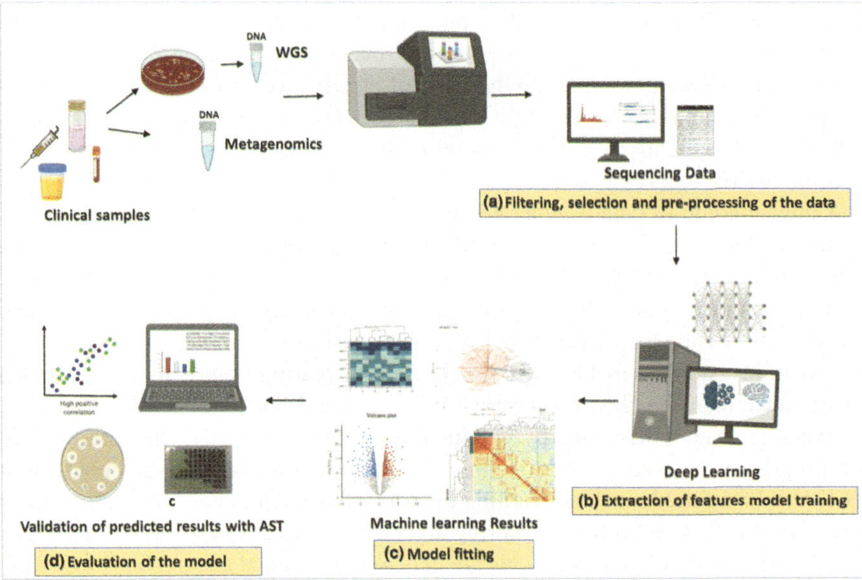

Fig. 3 Workflow for the development of machine learning tool for AMR prediction with genomic data: **a** Filtering, selection, and preprocessing of the data, **b** Extraction of features and model training, **c** Model fitting, and **d** Evaluation of the model. The image has been generated by BioRender (https://biorender.com/) accessed on June 13, 2022

can be used if well-characterized data is available for model development. An unsupervised approach is generally implemented to identify the compiled set of unlabeled sequences that group together on the basis of similarities (Waring et al. 2020).

The main workflow to develop an artificial intelligence tool comprises four main steps (Fig. 3):

(i) Filtering, selection, and preprocessing of the data
(ii) Extraction of features and model training
(iii) Model fitting
(iv) Evaluation of the model

The machine learning models can be trained on the reference AMR databases like CARD and Resfinder or can be trained using the pathogen WGS data (Bradley et al. 2015; Feldgarden et al. 2019). The most commonly used NII algorithm comprises neural networks which have considerably improved the efficiency in analyzing complicated data like the whole genome sequences (Avershina et al. 2021).

One of the utmost renowned and efficient algorithms being adopted to train neural networks is termed an error retro-propagation algorithm which analytically modifies the weight of neuronal connections to achieve the results (Pereira 2013). This algorithm is used in two phases, the first is forward propagation which involves the calculation of neuronal activation from the very first layer and then proceeds to the last layer. The second phase is backward propagation, where the findings of the network are compared to the actual output, for the estimation of a network error. The observed network error is then propagated in a reverse manner to reach the first phase. The weights are then modified on the basis of errors in order to improve the model accuracy (Schmidhuber 2015). The whole process of forward and backward propagation is subsequently reiterated several times to achieve the best efficient model (Pereira 2013).

A huge amount of data is generated from the training set which may hamper the training of learning algorithms by utilizing computational power and affecting the overall predictive accuracy (Waring et al. 2020). Hence, a feature reduction algorithm like a wrapper and filter methods may be implemented to mitigate the effects of higher dimensionality. Deep networks do not perform feature extraction from the input data instead they make a criterion during the learning phase to achieve better performance than the classic neural networks. Several machine learning platforms are available even with a friendly graphical user interface (GUI) which permits the non-programmers to design basic regression learning models. One such platform is WEKA, a Java-based machine learning platform that encompasses multiple, regression, classification, as well as clustering algorithms (Arganda-Carreras et al. 2017). The precision and efficacy of models are generally evaluated using a binary classification output like susceptible or resistant, to calculate the sensitivity and specificity using model evaluation matrices as follows:

Sensitivity: correctly predicted positive samples percentage

$$\text{Sensitivity} = \frac{\text{True Positive}}{\text{True Positive} + \text{False Negative}} \times 100 \tag{7}$$

Specificity: correctly predicted negative samples percentage

$$\text{Specificity} = \frac{\text{True Negative}}{\text{True Negative} + \text{False Positive}} \times 100 \tag{8}$$

Accuracy: correctly predicted positive and negative samples

$$\text{Accuracy} = \frac{\text{True Postive} + \text{True Negative}}{\text{True Positive} + \text{False Positive} + \text{True Negative} + \text{False Negative}} \times 100 \tag{9}$$

7 Challenges and Future Perspectives

The use of NII in the genomic-based prediction of AMR has emerged as a boon for the diagnosis of resistant pathogens in medical microbiology. However, there still exist challenges in their practical use in clinical microbiology. Technological advancements in near future can lead to further development of novel methods that can overcome these issues. Some of the significant challenges combined with the future prospects for successful implementation of NII for genomic prediction of AMR include:

(i) One of the major constraints for use of genomic data is the quality control in the submitted sequences in public repositories. The genome sequences are generated on different sequencing platforms. Some of these platforms are based on different technologies. As a result, the genomic data varies in read lengths, depth, and coverage. As a result, normalization of the data and the training of these reads itself can prove to be a challenging process.

(ii) The genomic sequence available across databases does not generally include phenotypic information. Thus, the insufficiency of phenotypic metadata submitted along with the whole genome sequences restricts AMR tool development.

(iii) The existence of multiple resistance mechanisms for a single antimicrobial agent limits the description of the extent of resistance (minimum inhibitory value) while implementing the clinical breakpoints.

(iv) In addition to the presence of AMR-conferring genes in the genome, their expression also plays an important role in phenotypic resistance and this can be misleading for some pathogens.

(v) The continuously evolving nature of bacterial pathogens contributes to the development of novel mutations which might affect the accuracy of genome-based predictive models. Consequently, the models need to be trained from time to time with the acquisition of the latest data.

(vi) A lack of reproducibility in prediction while using different datasets for the same algorithms.

(vii) Training and testing of machine learning algorithms require tremendous computational programming and bioinformatics skills. The futuristic approaches can include the development of GUI-based programs so that non-bioinformaticians can easily incorporate these algorithms into clinical practice.

(viii) Although deep learning algorithms can be trained on complicated genomic data as the model depth is increased to improve accuracy, the learning process becomes time-consuming and computationally demanding. The advancement in technology and availability of faster processors can facilitate AMR prediction at a faster rate.

8 Conclusion

Machine learning is an attractive approach to envisage the complete antibiograms of bacterial pathogens and can predict the resistance to even those antibiotics which are not in routine testing for clinical samples. Multiple studies have indicated the in-depth potential of nature-inspired intelligent computational systems for genomics-based endeavors, thereby, contributing to alleviating the grave situation arising due to bacterial antimicrobial resistance existing globally.

Although a large amount of genomic data are available on public repositories, unfortunately, these are not uniformly distributed across phenotypes and also represent a skewed geographical distribution. There is a need to deposit the phenotypic parameters along with the genomic data which can be implemented to design novel AMR prediction tools based on machine learning models.

References

Aderhold A, Diwold K, Scheidler A, Middendorf M (2010) Artificial bee colony optimization: a new selection scheme and its performance BT. In: González JR, Pelta DA, Cruz C, Terrazas G, Krasnogor N (eds) Nature inspired cooperative strategies for optimization (NICSO 2010). Springer, Berlin Heidelberg, pp 283–294. https://doi.org/10.1007/978-3-642-12538-6_24

Agarwala R, Barrett T, Beck J, Benson DA, Bollin C, Bolton E, Bourexis D, Brister JR, Bryant SH, Canese K, Charowhas C, Clark K, DiCuccio M, Dondoshansky I, Feolo M, Funk K, Geer LY, Gorelenkov V, Hlavina W, ... Zbicz,K (2017) Database resources of the national center for biotechnology information. Nucleic Acids Res 45(D1). https://doi.org/10.1093/nar/gkw1071

Alaloul WS (2020) Data processing using artificial neural networks In: Qureshi AH, Harkut DG (eds), p ch 6. IntechOpen. https://doi.org/10.5772/intechopen.91935

Alcock BP, Raphenya AR, Lau TTY, Tsang KK, Bouchard M, Edalatmand A, Huynh W, Nguyen ALV, Cheng AA, Liu S, Min SY, Miroshnichenko A, Tran HK, Werfalli RE, Nasir JA, Oloni M, Speicher DJ, Florescu A, Singh B, ... McArthur AG (2020) CARD 2020: antibiotic resistome surveillance with the comprehensive antibiotic resistance database. Nucleic Acids Res 48(D1). https://doi.org/10.1093/nar/gkz935

Allali I, Arnold JW, Roach J, Cadenas MB, Butz N, Hassan HM, Koci M, Ballou A, Mendoza M, Ali R, Azcarate-Peril MA (2017) A comparison of sequencing platforms and bioinformatics pipelines for compositional analysis of the gut microbiome. BMC Microbiology 17(1). https://doi.org/10.1186/s12866-017-1101-8

Arango-Argoty G, Garner E, Pruden A, Heath LS, Vikesland P, Zhang L (2018) DeepARG: a deep learning approach for predicting antibiotic resistance genes from metagenomic data. Microbiome 6(1). https://doi.org/10.1186/s40168-018-0401-z

Arganda-Carreras I, Kaynig V, Rueden C, Eliceiri KW, Schindelin J, Cardona A, Seung HS (2017) Trainable Weka Segmentation: a machine learning tool for microscopy pixel classification. Bioinformatics 33(15). https://doi.org/10.1093/bioinformatics/btx180

Avershina E, Sharma P, Taxt AM, Singh H, Frye SA, Paul K, Kapil A, Naseer U, Kaur P, Ahmad R (2021) AMR-Diag: neural network based genotype-to-phenotype prediction of resistance towards β-lactams in *Escherichia coli* and *Klebsiella pneumoniae*. Comput Struct Biotechnol J, 19. https://doi.org/10.1016/j.csbj.2021.03.027

Bankevich A, Nurk S, Antipov D, Gurevich AA, Dvorkin M, Kulikov AS, Lesin VM, Nikolenko SI, Pham S, Prjibelski AD, Pyshkin AV, Sirotkin AV, Vyahhi N, Tesler G, Alekseyev MA, Pevzner PA (2012) SPAdes: a new genome assembly algorithm and its applications to single-cell sequencing. J Comput Biol 19(5). https://doi.org/10.1089/cmb.2012.0021

Bharti V, Biswas B, Shukla KK (2020) Recent trends in nature inspired computation with applications to deep learning. In: Proceedings of the confluence 2020—10th international conference on cloud computing, data science and engineering. https://doi.org/10.1109/Confluence47617.2020.9057841

Bolger AM, Lohse M, Usadel B (2014) Trimmomatic: a flexible trimmer for illumina sequence data. Bioinformatics 30(15). https://doi.org/10.1093/bioinformatics/btu170

Boolchandani M, D'Souza AW, Dantas G (2019) Sequencing-based methods and resources to study antimicrobial resistance. Nat Rev Genet 20(6). https://doi.org/10.1038/s41576-019-0108-4

Bortolaia V, Kaas RS, Ruppe E, Roberts MC, Schwarz S, Cattoir V, Philippon A, Allesoe RL, Rebelo AR, Florensa AF, Fagelhauer L, Chakraborty T, Neumann B, Werner G, Bender JK, Stingl K, Nguyen M, Coppens J, Xavier BB, ... Aarestrup FM (2020) ResFinder 4.0 for predictions of phenotypes from genotypes. J Antimicrob Chemother 75(12). https://doi.org/10.1093/jac/dkaa345

Bradley P, Gordon NC, Walker TM, Dunn L, Heys S, Huang B, Earle S, Pankhurst LJ, Anson L, De Cesare M, Piazza P, Votintseva AA, Golubchik T, Wilson DJ, Wyllie DH, Diel R, Niemann S, Feuerriegel S, Kohl TA, ... Iqbal Z (2015) Rapid antibiotic-resistance predictions from genome sequence data for *Staphylococcus aureus* and *Mycobacterium tuberculosis*. Nat Commun 6. https://doi.org/10.1038/ncomms10063

Brownlee J (2011) Clever algorithms: nature-inspired programming recipes. Jason Brownlee

Chernov VM, Chernova OA, Mouzykantov AA, Lopukhov LL, Aminov RI (2019) Omics of antimicrobials and antimicrobial resistance. Expert Opin Drug Discov 14(5):455–468. https://doi.org/10.1080/17460441.2019.1588880

Dadgostar P (2019) Antimicrobial resistance: implications and costs. Infect Drug Resist 12. https://doi.org/10.2147/IDR.S234610

Dahiya S, Malik R, Sharma P, Sashi A, Lodha R, Kabra SK, Sood S, Das BK, Walia K, Ohri VC, Kapil A (2019) Current antibiotic use in the treatment of enteric fever in children. Indian J Med Res 149(2). https://doi.org/10.4103/ijmr.IJMR_199_18

Dahiya S, Sharma P, Kumari B, Pandey S, Malik R, Manral N, Veeraraghavan B, Pragasam AK, Ray P, Gautam V, Sistla S, Parija SC, Walia K, Ohri V, Das BK, Sood S, Kapil A (2017) Characterisation of antimicrobial resistance in Salmonellae during 2014–2015 from four centres across India: An ICMR antimicrobial resistance surveillance network report. Indian J Med Microbiol 35(1). https://doi.org/10.4103/ijmm.IJMM_16_382

Darwish A (2018) Bio-inspired computing: algorithms review, deep analysis, and the scope of applications. Future Comput Inf 3(2). https://doi.org/10.1016/j.fcij.2018.06.001

Davis JJ, Boisvert S, Brettin T, Kenyon RW, Mao C, Olson R, Overbeek R, Santerre J, Shukla M, Wattam AR, Will R, Xia F, Stevens R (2016) Antimicrobial resistance prediction in PATRIC and RAST. Sci Rep 6. https://doi.org/10.1038/srep27930

Feldgarden M, Brover V, Haft DH, Prasad AB, Slotta DJ, Tolstoy I, Tyson GH, Zhao S, Hsu C-H, McDermott PF, Tadesse DA, Morales C, Simmons M, Tillman G, Wasilenko J, Folster JP, Klimke W (2019) Using the NCBI AMRFinder tool to determine antimicrobial resistance genotype-phenotype correlations within a collection of NARMS isolates. BioRxiv, 550707. https://doi.org/10.1101/550707

Gelbukh A, Espinoza FC, Galicia-Haro SN (2014) Nature-inspired computation and machine learning. In: Lecture notes in computer science (including subseries lecture notes in artificial intelligence and lecture notes in bioinformatics), 8857(September).

Giulieri SG, Tong SYC, Williamson DA (2020) Using genomics to understand meticillin- and vancomycin-resistant Staphylococcus aureus infections. Microb Genomics 6(1). https://doi.org/10.1099/mgen.0.000324

Gorrie C, Higgs C, Carter G, Stinear TP, Howden B (2019) Genomics of vancomycin-resistant *Enterococcus faecium*. Microb Genomics 5(7). https://doi.org/10.1099/mgen.0.000283

Hendriksen RS, Bortolaia V, Tate H, Tyson GH, Aarestrup FM, McDermott PF (2019) Using genomics to track global antimicrobial resistance. Front Public Health 7. https://doi.org/10.3389/fpubh.2019.00242

Her H-L, Wu Y-W (2018) A pan-genome-based machine learning approach for predicting antimicrobial resistance activities of the *Escherichia coli* strains. Bioinformatics 34(13):i89–i95. https://doi.org/10.1093/bioinformatics/bty276

Holder LB, Haque MM, Skinner MK (2017) Machine learning for epigenetics and future medical applications. Epigenetics 12(7). https://doi.org/10.1080/15592294.2017.1329068

Jamal S, Khubaib M, Gangwar R, Grover S, Grover A, Hasnain SE (2020) Artificial intelligence and machine learning based prediction of resistant and susceptible mutations in mycobacterium tuberculosis. Sci Rep 10(1):5487. https://doi.org/10.1038/s41598-020-62368-2

Jha M, Kawale A, Verma CK (2017) Interpretable model for antibiotic resistance prediction in bacteria using deep learning. Biomed Pharmacol J 10(4). https://doi.org/10.13005/bpj/1316

Joshi N, Srivastava SS (2011) Parallelization of Velvet," a de novo genome sequence assembler. Sci Technol

Kang Q, An J, Wang L, Wu QDi (2012) Nature-inspired computation: a survey. Tien Tzu Hsueh Pao/Acta Electronica Sinica 40(3). https://doi.org/10.3969/j.issn.0372-2112.2012.03.023

Karaboga D, Basturk B (2007) A powerful and efficient algorithm for numerical function optimization: Artificial bee colony (ABC) algorithm. J Global Optim 39(3). https://doi.org/10.1007/s10898-007-9149-x

Katiyar A, Sharma P, Dahiya S, Singh H, Kapil A, Kaur P (2020) Genomic profiling of antimicrobial resistance genes in clinical isolates of *Salmonella Typhi* from patients infected with Typhoid fever in India. Sci Rep 10(1). https://doi.org/10.1038/s41598-020-64934-0

Kaur A, Ruhela A, Sharma P, Khariwal H, Seth S, Kumar A, Kapil A, Elangovan R, Kalyanasundaram D (2019) Simultaneous and high sensitive detection of *Salmonella typhi* and *Salmonella paratyphi* a in human clinical blood samples using an affordable and portable device. Biomed Microdevices 21(4). https://doi.org/10.1007/s10544-019-0441-6

Khan ZA, Siddiqui MF, Park S (2019) Current and emerging methods of antibiotic susceptibility testing. Diagnostics 9 (2). https://doi.org/10.3390/diagnostics9020049

Kono N, Arakawa K (2019) Nanopore sequencing: review of potential applications in functional genomics. Develop Growth Differ 61(5). https://doi.org/10.1111/dgd.12608

Köser CU, Ellington MJ, Peacock SJ (2014) Whole-genome sequencing to control antimicrobial resistance. Trends Genet 30(9). https://doi.org/10.1016/j.tig.2014.07.003

Kumar M, Dahiya S, Sharma P, Sharma S, Singh TP, Kapil A, Kaur P (2015) Structure based in silico analysis of quinolone resistance in clinical isolates of *Salmonella Typhi* from India. PLoS ONE 10(5). https://doi.org/10.1371/journal.pone.0126560

Kumar Tripathi M, Kumar Shrivastava S, Karthikeyan S, Sinha D, Nath A (2021) Application of machine learning and molecular modeling in drug discovery and cheminformatics. Adv AI Tech Appl Bioinform. https://doi.org/10.1201/9781003126164-10

Kuroda M, Ohta T, Uchiyama I, Baba T, Yuzawa H, Kobayashi I, Cui L, Oguchi A, Aoki K, Nagai Y, Lian J, Ito T, Kanamori M, Matsumaru H, Maruyama A, Murakami H, Hosoyama A, Mizutani-Ui Y, Takahashi NK, … Hiramatsu K (2001) Whole genome sequencing of meticillin-resistant *Staphylococcus aureus*. Lancet (London, England) 357(9264):1225–1240. https://doi.org/10.1016/s0140-6736(00)04403-2

Langmead B, Salzberg SL (2012) Fast gapped-read alignment with Bowtie 2. Nat Meth 9(4). https://doi.org/10.1038/nmeth.1923

Lapidus AL (2019) Genome sequence databases: sequencing and assembly. Encyclopedia Microbiol. https://doi.org/10.1016/B978-0-12-801238-3.02495-8

Lau HJ, Lim CH, Foo SC, Tan HS (2021) The role of artificial intelligence in the battle against antimicrobial-resistant bacteria. Curr Genet 67(3):421–429. https://doi.org/10.1007/s00294-021-01156-5

Layer F, Ghebremedhin B, Moder KA, König W, König B (2006) Comparative study using various methods for identification of Staphylococcus species in clinical specimens. J Clin Microbiol 44(8). https://doi.org/10.1128/JCM.00226-06

Lee C-R, Lee JH, Park KS, Jeon JH, Kim YB, Cha C-J, Jeong BC, Lee SH (2017) Antimicrobial resistance of Hypervirulent *Klebsiella pneumoniae*: epidemiology, Hypervirulence-associated determinants, and resistance mechanisms. Front Cell Infect Microbiol 7:483. https://doi.org/10.3389/fcimb.2017.00483

Leinonen R, Akhtar R, Birney E, Bower L, Cerdeno-Tárraga A, Cheng Y, Cleland I, Faruque N, Goodgame N, Gibson R, Hoad G, Jang M, Pakseresht N, Plaister S, Radhakrishnan R, Reddy K, Sobhany S, Hoopen PT, Vaughan R, … Cochrane G (2011) The European nucleotide archive. Nucleic Acids Res 39(SUPPL. 1). https://doi.org/10.1093/nar/gkq967

Li C, Chan F (2011) Complex-fuzzy adaptive image restoration—an artificial-bee-colony-based learning approach. In: Lecture notes in computer science (including subseries lecture notes in artificial intelligence and lecture notes in bioinformatics) 6592 LNAI (PART 2). https://doi.org/10.1007/978-3-642-20042-7_10

Li H, Durbin R (2009) Fast and accurate short read alignment with burrows-wheeler transform. Bioinformatics, 25(14). https://doi.org/10.1093/bioinformatics/btp324

Liao M, Tong T, Zong Y, Zhou X, Cheng L, Huang R, Ren B, Alterovitz G (2019) Application of omics and bioinformatics tools in streptococcus research. Curr Issues Mol Biol 32:327–376. https://doi.org/10.21775/cimb.032.327

Liu B, Pop M (2009) ARDB—antibiotic resistance genes database. Nucleic Acids Res 37(SUPPL. 1). https://doi.org/10.1093/nar/gkn656

Liu Z, Deng D, Lu H, Sun J, Lv L, Li S, Peng G, Ma X, Li J, Li Z (2020) Evaluation of machine learning models for predicting antimicrobial resistance of *Actinobacillus pleuropneumoniae* from whole genome sequences. Front Microbiol 11:48

Lupolova N, Lycett SJ, Gally DL (2019) A guide to machine learning for bacterial host attribution using genome sequence data. Microb Genomics 5(12)

Luz CF, van Niekerk JM, Keizer J, Beerlage-de Jong N, Braakman-Jansen LMA, Stein A, Sinha B, van Gemert-Pijnen JEWC, Glasner C (2022) Mapping twenty years of antimicrobial resistance research trends. Artif Intell Med 123. https://doi.org/10.1016/j.artmed.2021.102216

Lv J, Deng S, Zhang L, Ji L, Senyi D, Le Z (2021) A review of artificial intelligence applications for antimicrobial resistance. Biosaf Health 3(1):22–31. https://doi.org/10.1016/j.bsheal.2020.08.003

Mahfouz N, Ferreira I, Beisken S, von Haeseler A, Posch AE (2020) Large-scale assessment of antimicrobial resistance marker databases for genetic phenotype prediction: a systematic review. J Antimicrob Chemother 75(11). https://doi.org/10.1093/jac/dkaa257

Metzker ML (2010) Sequencing technologies the next generation. Nat Rev Genet 11(1). https://doi.org/10.1038/nrg2626

Mullany P (2014) Functional metagenomics for the investigation of antibiotic resistance. Virulence 5(3). https://doi.org/10.4161/viru.28196

Nguyen M, Long SW, McDermott PF, Olsen RJ, Olson R, Stevens RL, Tyson GH, Zhao S, Davis JJ (2019) Using machine learning to predict antimicrobial MICs and associated genomic features for nontyphoidal Salmonella. J Clin Microbiol 57(2):e01260-e1318

O'neill J (2014) Antimicrobial resistance. Tackling a crisis for the health and wealth of nations

Opota O, Croxatto A, Prod'hom G, Greub G (2015) Blood culture-based diagnosis of bacteraemia: state of the art. Clin Microbiol Infect 21(4). https://doi.org/10.1016/j.cmi.2015.01.003

Overton K, Fortané N, Broom A, Raymond S, Gradmann C, Orubu ESF, Podolsky SH, Rogers Van Katwyk S, Zaman MH, Kirchhelle C (2021) Waves of attention: patterns and themes of international antimicrobial resistance reports, 1945–2020. BMJ Glob Health 6(11). https://doi.org/10.1136/bmjgh-2021-006909

Paterson DL (2008) Impact of antibiotic resistance in gram-negative bacilli on empirical and definitive antibiotic therapy. Clin Infect Dis 47(SUPPL.1). https://doi.org/10.1086/590062

Pereira H (2013) Applicability of data mining technique using Bayesians network in diagnosis of genetic diseases. Int J Adv Comput Sci Appl 4(1). https://doi.org/10.14569/ijacsa.2013.040107

Pinto C, Pinto R, Gonçalves G (2022) Towards bio-inspired anomaly detection using the cursory dendritic cell algorithm. Algorithms 15(1). https://doi.org/10.3390/a15010001

Premanandh J, Samara BS, Mazen AN (2016) Race against antimicrobial resistance requires coordinated action—an overview. Front Microbiol 6:1536. https://doi.org/10.3389/fmicb.2015.01536

Quainoo S, Coolen JPM, van Hijum SAFT, Huynen MA, Melchers WJG, van Schaik W, Wertheim HFL (2017) Whole-genome sequencing of bacterial pathogens: the future of nosocomial outbreak analysis. Clin Microbiol Rev 30(4). https://doi.org/10.1128/CMR.00016-17

Rashid N, Iqbal J, Mahmood F, Abid A, Khan US, Tiwana MI (2018) Artificial immune system–negative selection classification algorithm (NSCA) for four class electroencephalogram (EEG) signals. Front Hum Neurosci 12. https://doi.org/10.3389/fnhum.2018.00439

Read M, Andrews PS, Timmis J (2012) An introduction to artificial immune systems BT. In: Rozenberg G, Bäck T, Kok JN (eds) Handbook of natural computing. Springer, Berlin, Heidelberg, pp 1575–1597. https://doi.org/10.1007/978-3-540-92910-9_47

Saltykova A, Mattheus W, Bertrand S, Roosens NHC, Marchal K, De Keersmaecker SCJ (2019) Detailed evaluation of data analysis tools for subtyping of bacterial isolates based on whole genome sequencing: *Neisseria Meningitidis* as a proof of concept. Front Microbiol 10. https://doi.org/10.3389/fmicb.2019.02897

San Millan A (2018) Evolution of plasmid-mediated antibiotic resistance in the clinical context. Trends Microbiol 26(12). https://doi.org/10.1016/j.tim.2018.06.007

Sánchez-Busó L, Harris SR (2019) Using genomics to understand antimicrobial resistance and transmission in *Neisseria gonorrhoeae*. Microb Genomics 5(2). https://doi.org/10.1099/mgen.0.000239

Sauget M, Bertrand X, Hocquet D (2018) Rapid antibiotic susceptibility testing onblood cultures using MALDI-TOF MS. PLoS ONE 13(10).https://doi.org/10.1371/journal.pone.0205603

Schmidhuber J (2015) Deep learning in neural networks: an overview. Neural Networks 61. https://doi.org/10.1016/j.neunet.2014.09.003

Sequence read archive (2018) The sequence read archive (SRA) overview. NCBI

Sharma P, Dahiya S, Manral N, Kumari B, Kumar S, Pandey S, Sood S, Das B, Kapil A (2018) Changing trends of culture-positive typhoid fever and antimicrobial susceptibility in a tertiary care North Indian Hospital over the last decade. Indian J Med Microbiol 36(1). https://doi.org/10.4103/ijmm.IJMM_17_412

Sharma P, Kumar M, Dahiya S, Sood S, Das BK, Kaur P, Kapil A (2020) Structure based drug discovery and in vitro activity testing for DNA gyrase inhibitors of *Salmonella Enterica Serovar Typhi*. Bioorg Chem 104. https://doi.org/10.1016/j.bioorg.2020.104244

Sheela KG, Deepa SN (2013) Review on methods to fix number of hidden neurons in neural networks. Math Probl Eng 2013:425740. https://doi.org/10.1155/2013/425740

Shi J, Yan Y, Links MG, Li L, Dillon JAR, Horsch M, Kusalik A (2019) Antimicrobial resistance genetic factor identification from whole-genome sequence data using deep feature selection. BMC Bioinformatics 20. https://doi.org/10.1186/s12859-019-3054-4

Siddique N, Adeli H (2015) Nature inspired computing: an overview and some future directions. Cogn Comput 7(6):706–714. https://doi.org/10.1007/s12559-015-9370-8

Taylor-Brown A, Madden D, Polkinghorne A (2018) Culture-independent approaches to chlamydial genomics. Microb Genomics 4(2). https://doi.org/10.1099/mgen.0.000145

Tripathi MK, Nath A, Singh TP, Ethayathulla AS, Kaur P (2021a) Evolving scenario of big data and Artificial Intelligence (AI) in drug discovery. Mol Diversity 25(3):1439–1460. https://doi.org/10.1007/s11030-021-10256-w

Tripathi MK, Sharma S, Singh TP, Ethayathulla AS, Kaur P (2021b) Computational intelligence in drug repurposing for COVID-19. In: Raza K (ed) Computational intelligence methods in COVID-19: surveillance, prevention, prediction and diagnosis. Springer, Singapore, pp 273–294. https://doi.org/10.1007/978-981-15-8534-0_14

Tumbarello M, Trecarichi EM, De Rosa FG, Giannella M, Giacobbe DR, Bassetti M, Losito AR, Bartoletti M, Del Bono V, Corcione S, Maiuro G, Tedeschi S, Celani L, Cardellino CS, Spanu T, Marchese A, Ambretti S, Cauda R, Viscoli C, Viale P (2015) Infections caused by KPC-producing *Klebsiella pneumoniae*: differences in therapy and mortality in a multicentre study. J Antimicrob Chemother 70(7). https://doi.org/10.1093/jac/dkv086

Tzanetos A, Fister I, Dounias G (2020) A comprehensive database of Nature-Inspired Algorithms. Data Brief 31:105792. https://doi.org/10.1016/j.dib.2020.105792

Uelze L, Grützke J, Borowiak M, Hammerl JA, Juraschek K, Deneke C, Tausch SH, Malorny B (2020) Typing methods based on whole genome sequencing data. One Health Outlook 2(1). https://doi.org/10.1186/s42522-020-0010-1

Van Camp, PJ, Haslam DB, Porollo A (2020) Bioinformatics approaches to the understanding of molecular mechanisms in antimicrobial resistance. Int J Mol Sci 21(4). https://doi.org/10.3390/ijms21041363

Wang D, Tan D, Liu L (2018) Particle swarm optimization algorithm: an overview. Soft Comput 22(2):387–408. https://doi.org/10.1007/s00500-016-2474-6

Wang L, Kang Q, Wu QDi (2007) Nature-inspired computation-effective realization of artificial intelligence. Xitong Gongcheng Lilun Yu Shijian/Syst Eng Theory Pract 27(5). https://doi.org/10.1016/s1874-8651(08)60034-4

Waring J, Lindvall C, Umeton R (2020) Automated machine learning: review of the state-of-the-art and opportunities for healthcare. Artif Intell Med 104. https://doi.org/10.1016/j.artmed.2020.101822

Weiss SJ, Mansell TJ, Mortazavi P, Knight R, Gill RT (2016) Parallel mapping of antibiotic resistance alleles in *Escherichia coli*. PLoS ONE 11(1). https://doi.org/10.1371/journal.pone.0146916

Yang X. S. (2020). Nature-inspired optimization algorithms: challenges and open problems. J Comput Sci 46. https://doi.org/10.1016/j.jocs.2020.101104

Zhang A-N, Gaston JM, Dai CL, Zhao S, Poyet M, Groussin M, Yin X, Li L-G, van Loosdrecht MCM, Topp E, Gillings MR, Hanage WP, Tiedje JM, Moniz K, Alm EJ, Zhang T (2021) An omics-based framework for assessing the health risk of antimicrobial resistance genes. Nat Commun 12(1):4765. https://doi.org/10.1038/s41467-021-25096-3

Zhao Q, Shen Y, Chen G, Luo Y, Cui S, Tian Y (2021) Prevalence and molecular characterization of fluoroquinolone-resistant *Escherichia coli* in healthy children. Front Cell Infect Microbiol 11:743390. https://doi.org/10.3389/fcimb.2021.743390

Consequential Innovations in Nature-Inspired Intelligent Computing Techniques for Biomarkers and Potential Therapeutics Identification

Kayenat Sheikh, Salwa Sayeed, Aisha Asif, Mohd Faizan Siddiqui, Misbahuddin M. Rafeeq, Ankita Sahu, and Shaban Ahmad

1 Introduction

Nature is phenomenal. Behind every occurrence, there are invisible dynamic processes and causes. These causes have been intensely studied, analyzed, interpreted, and inferred by scientists across the globe for the past centuries. As a result, steps are taken to explain, understand, duplicate, and implicate nature-inspired artificial systems for improving the standard of living (Siddique and Adeli 2015). Nature-inspired computing (NIC) is an emerging field that has made significant changes in the interdisciplinary fields of science. The term "nature-inspired computing" clearly indicates the scientific approach and algorithms focusing on biology and physics. The

K. Sheikh (✉)
CSIR-Institute of Genomics and Integrative Biology, New Delhi, India
e-mail: kayenat1903945@st.jmi.ac.in

S. Sayeed · A. Asif · S. Ahmad
Department of Computer Science, Jamia Millia Islamia, New Delhi, India
e-mail: Salwa2001129@st.jmi.ac.in

A. Asif
e-mail: Aisha2004009@st.jmi.ac.in

S. Ahmad
e-mail: Shaban184343@st.jmi.ac.in

M. F. Siddiqui
International Medical Faculty, Osh State University, Osh 723510, Kyrgyzstan

M. M. Rafeeq
Department of Pharmacology, Faculty of Medicine, King Abdulaziz University, Rabigh, Jeddah, Kingdom of Saudi Arabia

A. Sahu
Tumor Biology, ICMR-National Institute of Pathology, New Delhi, India

K. Raza (ed.), *Nature-Inspired Intelligent Computing Techniques in Bioinformatics*, Studies in Computational Intelligence 1066,
https://doi.org/10.1007/978-981-19-6379-7_13

past two decades have witnessed a massive amalgamation of computational intelligence and nature-based fields of study (Siddique and Adeli 2015). Several biologically inspired computing algorithms are implemented to solve complex incomprehensible problems. Much known artificial intelligence (AI), machine learning, and deep learning algorithms are extensively used in this area. Techniques based on artificial neural networks (ANNs), multi-layer perceptron (MLP), deep neural networks, and more are employed. Biomarkers play a key role in the detection and diagnosis of diseases. Early detection of severe disorders is critical in saving patients' lives and quick treatment. At present times, biologically intelligent computing techniques have been deployed to serve this purpose at an advanced level (Manning et al. 2014).

The recent examples of nature-inspired intelligent computing (NIIC) techniques applications show its robust future for large-scale biotechnological industrial growth. The COVID-19 pandemic was a major global challenge at different levels. It has claimed the lives of around 6 million people from more than 200 countries since its spread (Qazi et al. 2020). The highly contagious nature of the virus led to several burdens on health care, education, transportation, and more. Healthcare facilities were worst affected due to unpreparedness and suddenness of the chain of events. However, computational biology and nature-inspired computing techniques revealed several unknown facts about the notorious SARS-CoV-2. Until the vaccine discovery, crucial measures were necessary to prevent further contamination and infection (Qazi et al. 2021). A NIIC model is inspired by natural phenomena that include evolutionary and swarm intelligence-based algorithms. The former is based on a genetic algorithm (Sheikh et al. 2022), and the latter mimics the ant colony algorithm (ACO) (Ruhr et al. 2021), dung beetle, particle swarm optimization (PSO), wolf, bat, cat swarm optimization (CSO), and more. NIIC models were employed for the prediction and diagnosis of COVID-19 and effectively used for X-ray image segmentation for detecting the COVID-19 features. NIIC models can be employed to improve performance since AI algorithms pose performance issues (Agbehadji et al. 2020). The outbreak of viral diseases such as COVID-19, MERS, Ebola, HIV, Zika, and more has created a new space for research and applications of computing techniques. Viroinformatics tools applied in viral disease prognosis and analysis are based on NIIC algorithms. In-depth genome studies are possible due to highly compatible genome sequencing and prediction tools (Sheikh and Raza 2021).

Besides viral infections, oncological research has gained pace due to core computational techniques. Due to NIIC methods, the prediction of early stage cancer biomarkers, signaling proteins, genes, and networks has become feasible. Innumerable web servers and tools are available at the current time. Advances are made in AI methods for better image segmentation, prediction, and detection (Sheikh et al. 2022). The therapeutic implications of drugs and plant-based molecules have been simplified using various bioinformatical approaches. In which, NIIC techniques have played a significant role. It allows image processing, classification, clustering, disease prediction, and more (Alam et al. 2022).

2 Nature-Inspired Intelligent Computing Techniques in Biomarkers Prediction

Technology has played a vital role in improving medical facilities over the past few years. The medical advances and their impact enable better healthcare facilities. Researchers have sincere efforts to apply NIIC-based algorithms and techniques in diagnosis, prognosis, treatment, and care (Thimbleby 2013). The biomarker prediction in different diseases enables early diagnosis. Some significant applications are given below.

2.1 Skin Cancer

Skin cancer is a common human malignancy, initially diagnosed visually and with clinical screening. It is then followed by a biopsy, histopathological examination, and skin analysis. Classification of skin lesions through images is challenging since the fine-grained variation in the skin appearance. Deep convolution neural networks (CNNs) have been employed to classify skin lesions. A dataset of 129,450 clinical images was used for training. It allowed the identification of common and deadliest skin cancers—exceptional use cases with low cost and universally accessible to diagnostic centers (Esteva et al. 2017).

2.2 Breast Cancer

Large complex datasets are trained using AI for better mammography screening. Efficient image interpretation with specificity is made possible. A breast cancer diagnosis is possible through deep learning algorithms. In an equal amount of time, metastatic breast cancer can be diagnosed better than radiology methods. Mammography results for breast cancer allow improved treatment and early detection. Efforts have been made to improve the interpretation of screening through AI. Imaging practices have been intensified and enhanced, including more frequent screening, two readings, and supplemental imaging (Bejnordi et al. 2017).

2.3 Colon Cancer

Machine learning-based hybrid model detects both breast cancer and colon cancer. The results of the study are accurate too. The mega-trend diffusion-support vector machine (MTD-SVM) hybrid method is impressive in cutting the cost of computations and enhancing cancer prediction performance (Majid et al. 2014).

2.4 Alzheimer's and Heart Diseases

Degenerative diseases such as Alzheimer's can be detected through precision diagnostics. Most genetic and coronary heart diseases can be detected early through imaging biomarkers (Yu et al. 2018). An 80% more accurate and optimized heart disease diagnosis system based on AI was developed at John Radcliff Hospital in England (Hippisley-Cox et al. 2008). Another team of researchers at Oxford University has developed a highly efficient echocardiography program for detecting coronary heart disease. Up to 90% of the time, the results are accurate due to a decade of imaging data upload (Weng et al. 2017). The Firat University of Turkey designed a heart disease detection system based on neural networks. The heart disease detection results were 80.95% accurate (Muhammad et al. 2020).

2.5 Blood Infections

Harvard University developed a system that predicted infection in blood with 93% accuracy. No human intervention was needed for the same. The system integrated AI and microscopic visualization of blood samples (Smith et al. 2018).

3 Classification of Nature-Inspired Computation

One of the most reviewed tasks is classification. In recent years, many classification techniques for solving numerical and combinatorial optimization problems have been devised. These algorithms mimic the behavior of natural biological systems. Based on this, six significant classes of nature-inspired algorithms are discussed as follows (Fig. 1).

3.1 Swarm Intelligence

A swarm is a collection of homogeneous, simple agents functioning regionally with one another and with their surroundings, with no centralized control to allow for the emergence of interesting global behavior. Swarm-based algorithms are a new type of population-based algorithm that is inspired by nature. These can generate low cost, quick, and reliable solutions to various complicated challenges (Ahmad et al. 2021). As a result, swarm intelligence is defined as the study of computing systems driven by collective behavior. It is also one of AI's most essential concepts. Swarm intelligence is further divided into the following categories.

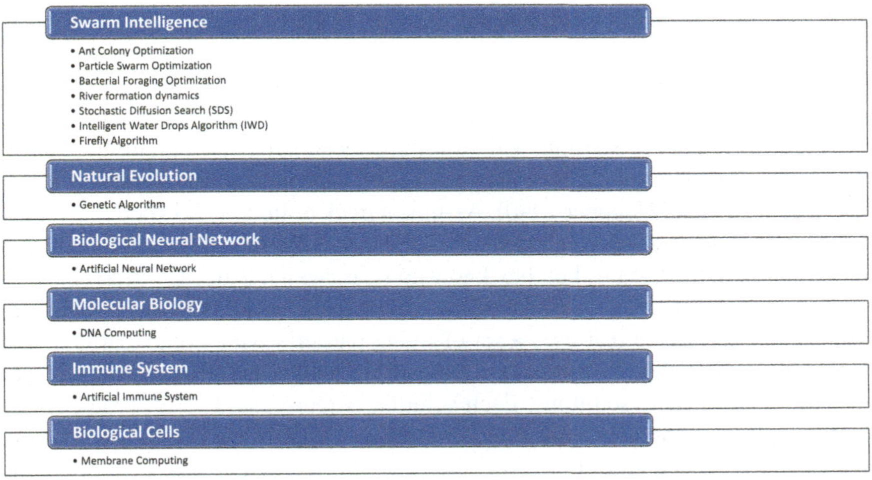

Fig. 1 Classification of nature-inspired computation (Gupta et al. 2021)

3.1.1 Ant Colony Optimization

Marco Dorigo was the first to propose the ant colony optimization (ACO) approach in 1992 and called it ant system. Dorigo, Di Caro, and Gambardella renamed it the ACO metaheuristic in 1999 (Rahmatizadeh and Torkaman 2009). ACO (Ruhr et al. 2021) is a methodology for building metaheuristic algorithms and tackling complex combinatorial optimization problems inspired by actual ant's indirect communication. The key to ants' exceptional ecological success may be sociality (Keller and Gordon 2009). We consider the example of ants, as they have shown great social organization in various ways. They prefer to live in well-organized civilizations where individuals team up, communicate, and divide everyday tasks. Ants have exceptional navigational skills and the ability to build nests and obtain food. They are not just efficient but also hardworking and frugal organisms that can adapt to various environments and climatic extremes. In the actual world, ants leave their colony at random, searching for food, leaving a pheromone trail behind them (Dorigo 1992). For example, Brueckner proposes an infrastructure for ant-based pheromone operations (Odell et al. 2002). The task is turned into a problem of finding the optimum path on a weighted graph to use ACO. By traversing the graph, the artificial ants iteratively construct solutions. A pheromone is a set of parameters associated with graph components (nodes or edges) that the ants alter during runtime. It solves combinatorial issues, network routing challenges, and transportation problems.

3.1.2 Particle Swarm Optimization

Kennedy and Eberhart proposed particle swarm optimization (PSO) in 1995 as a heuristic optimization technique (Kennedy and Eberhart 1995). It is based on birds' intelligence, experience-sharing, and swarming social behavior, initially replicated on a computer by Craig Reynolds (Reynolds 1987) and researched further by biologists (Grenander and Heppner 1990). Assume a flock of birds seeks food in a region at random. In the area being searched, there is only one piece of food. The birds do not know where the food is, but they know how far away it is in each iteration. As a result, the ideal tactic is to follow the bird closest to the meal. PSO is a population-based search technique that uses a set of flying particles with dynamically altered velocities based on their historical performance, as well as their search space neighbors, to locate optimal solutions. Each solution is represented by a particle in the search space. All particles have fitness values assessed by the fitness function to be optimized and velocities that direct the particles' flight. The particles follow the usual optimum particles through the problem space. If the most recent update of the location of any particle in the entire swarm happened to be better than the present position of the swarm's global best particle, the index of the swarm's global best particle is updated before the end of each iteration of PSO.

3.1.3 Bacterial Foraging Optimization

Bacteria foraging optimization algorithm (BFOA) was proposed by Passino, who followed the same swarm-based algorithmic approach. The social foraging behavior of *Escherichia coli* inspired BFOA. Bacteria look for nutrients in a way that maximizes the amount of energy they get per unit of time. Individual bacteria transmit signals to communicate with one another. After considering two primary factors, a bacterium makes foraging decisions. Chemotaxis is the mechanism by which a bacterium moves by taking small steps while searching for food, and a fundamental concept in BFOA is to simulate the chemotactic movement of virtual bacteria in the issue search space.

To understand this in-depth, let us take the example of E. coli and salmonella. They are motile bacteria that push themselves by rotating their flagella. When the flagella rotates counterclockwise, the organism "swims" forward, and when the flagellum rotates clockwise, the bacteria "tumbles" itself in a new direction and swims again (Adler 1966). The bacterium can look for nutrients in obscure routes by switching between "swim" and "tumble". As the bacterium approaches a nutritional gradient, it begins to swim more frequently. As the bacterium moves away from some food in search of more, tumbling and direction shifts become more common. Bacterial chemotaxis is a complicated combination of swimming and tumbling that keeps bacteria in regions with higher nutritional availability. Because of its effectiveness in tackling real-world optimization issues in various application fields, BFOA has already attracted the attention of researchers.

3.1.4 River Formation Dynamics

River formation dynamics (RFD) is a heuristic optimization approach recently developed by Rabanal Basalo et al. that is based on how water drops construct the riverbed (Rabanal Basalo et al. 2017). The approach mimics the movement of drops on the borders between given nodes, allowing for an analysis based on their height, which is altered by soil erosion and silt deposition. When water transforms in the environment, it causes dynamic changes in heights and the formation of lowering gradients. Gradients that follow consecutive drops produce new gradients, strengthening the best ones. As a result, after being created, diminishing gradients are followed by additional drops to create new gradients and reinforce the ones already there. As a result of this approach, researchers could develop optimum solutions to various problems.

3.1.5 Stochastic Diffusion Search

Stochastic diffusion search (SDS) is a multi-agent population-based global search and optimization algorithm that Mark Bishop first developed as a pattern identification technique (Bishop 1989). It is a distributed style of computation that uses the interaction between essential agents. It is better suited to cases in which an objective function is composed of numerous independent partial functions. Each agent has a hypothesis that is iteratively tested by evaluating a partially objective function randomly chosen and parameterized by the agent's current hypothesis. Inter-agent communication allocates hypotheses information throughout the community. SDS agents communicate hypotheses using a one-to-one technique similar to the tandem running procedure observed in some ant species, rather than the stigmergic communication utilized in ACO (Bishop 1989).

Unlike many other nature-inspired search algorithms, SDS provides a solid mathematical framework that characterizes the algorithm's behavior by looking at the allocation of resources, convergence to the global optimum, sturdiness and limited convergence requirements, and linear time ambiguity (Bishop 2007; De Meyer et al. 2000; De et al. 2006).

3.1.6 Intelligent Water Drops Algorithm (IWD)

Shah-Hosseini in 2007 first introduced the intelligent water drops (IWD) algorithm. It is a new swarm-based optimization technique inspired by natural river water drops and is a population-based constructive optimization algorithm (Hosseini 2007). The IWD algorithm is based on the dynamics of river systems and the actions and reactions that occur between river water drops (Rahmatizadeh and Torkaman 2009). IWDs are established using natural water droplets, and IWDs work together to find a preferable and feasible solution.

IWDs are made up of two key characteristics: velocity and soil. These features may vary throughout an IWD's existence. An IWD connects two points: a source and a destination. It then starts its journey with a zero-velocity beginning velocity and zero dirt. It passes through an environment where it eliminates some soil and may gain some speed along its journey. The IWD velocity increases nonlinearly proportional to the inverse of the soil between the two sites from its current location to its following location. As a result, a path with less dirt allows the IWD to accelerate more quickly than a path with more soil. During its journey through the environment, an IWD collects soil. The soil from the pathway connecting the two locations is removed. The amount of soil added to the IWD is nonlinearly related to the inverse of the time; it takes the IWD to travel from one point to the next. The elementary physics equations for linear motion are used to compute this time interval. As a result, the time required is proportional to the IWD's velocity and inversely proportional to the distance between the two points. Furthermore, areas of the ecosystem with more IWDs will have less soil.

3.1.7 Firefly Algorithm

Firefly algorithm (FA) is a population-based optimization approach that resembles the attraction of a firefly to a flashing light. Yang from the University of Cambridge formulated this technique. Gandomi et al. recently used chaotic maps to optimize FA. The FA is a metaheuristic algorithm inspired by fireflies' flashing activity and the concept of bioluminescent communication (Yang 2009).

Winged beetles or insects that produce light and blink at night are fireflies. Bioluminescence is a light with no infrared or ultraviolet frequency that is chemically emitted from the lower abdomen. They employ the flashlight to lure mates or prey in particular. The flashlight also serves as a warning system for the fireflies, reminding them of prospective predators.

FA uses the following three idealized rules (Yang 2009):

1. Since fireflies are unisexual, they will be attracted to each other regardless of their sex.
2. Attractiveness is proportional to brightness, with the brighter firefly attracting the less brilliant firefly. When the distance between the two fireflies was increased, the attractiveness of the two fireflies was reduced.
3. If the fireflies' brightness is the same, the fireflies will relocate at random.

The fireflies' brightness should be linked to the problem's objective function. Due to their desirability, they can divide into smaller groups, each of which swarms around the local models. As a result, FA is intuitive, versatile, and straightforward (Table 1).

Table 1 Types of swarm intelligence-based algorithms and their applications

Types of swarm intelligence-based algorithm	Applications
Ant colony optimization	Sequential ordering (Gambardella and Dorigo 2000), scheduling, DNA sequencing (Blum et al. 2008), 2D-HP protein folding (Blum 2005), protein–ligand docking (Korb et al. 2007), DNA sequencing by hybridization
Particle swarm optimization	RNA secondary structure prediction, gene clustering, phylogenetic tree construction, energy minimization, and protein modeling, face detection (Ramadan and Abdel-Kader 2009), heartbeat classification (Khazaee 2013), object detection, image classification, operon prediction, identification of issue binding web sites in DNA, biomarker choice drug format, radiotherapy making plans, biometrics
Bacterial foraging optimization	Facility layout, feature selection, training kernel extreme learning machine (Chen et al. 2020), secondary and tertiary structure prediction, power loss reduction problem and in the area of PID applications, ligand docking
River formation dynamics	Solving NP-complete problems, such as finding a minimum distances tree and a minimum spanning tree in a variable-cost graph (Gambardella and Dorigo 2000). RFD can also be used to solve difficulties like tree coverage
Stochastic diffusion search	Bone scan, tracking facial features like eyes, sequence detection, cellular automata
Intelligent water drops algorithm	Robot path planning in robotics (Schulman et al. 2014), detection of breast cancer in early stage, biclustering or block clustering, graph coloring
Firefly algorithm	Human retinal images, detection of brain tumor, health monitoring, brain tissue segmentation, DNA binding protein identification, gene selection, iris recognition, jaw fracture classification, diabetes diagnosis, pharmaceutical ingredient detection

3.2 Natural Evolution

Charles Darwin, a nineteenth century naturalist, studied the principle of natural selection. According to this idea, if a member of a species gains a functional advantage (for example, growing wings and learning to fly), its descendants will inherit that advantage and pass it on to their progeny. Only the superior (favored) members of the species survive, whereas the inferior (disadvantaged) members of the same species die off over time. Natural selection preserves a functional advantage that allows a species to compete more effectively than other species. This insight is used to create genetic algorithms.

3.2.1 Genetic Algorithm

Genetic algorithms (Sheikh et al. 2022) are adaptive heuristic search algorithms. As a computational similarity of adaptive systems, it was introduced. They are formed based on natural selection in evolution. Inheritance, mutation, crossover, and selection are the major components of natural evolution that constitute a GA. They are called genetic operators. The optimization process begins with an initial population of randomly generated or constructed solutions using a heuristic. Predetermined rates are used to apply genetic operators. New members are added to the population, and the replacement mechanism chooses the solutions that will survive to the next generation to keep the population number at the desired level.

A genetic algorithm, in short, follows a basic three-stage process (Gupta et al. 2021):

- Creating a string "population,"
- Evaluating each string,
- String selection and genetic modification to establish a new string population.

3.2.2 Biological Neural Network

The human intellect is one of the most inspirational examples of natural intelligence. There are numerous hypotheses about how minds function. It is a considerable debate if it will ever be able to create a machine with quite a mind. When we look at the general anatomy of the human brain, we can see that the components are nerve cells or neurons. Neurons process electrical signals that compose all brain operations and are linked together in complex networks. The brain's immense sophistication and computing capability are based on the simultaneous cooperative behavior of many basic processing units. An artificial neural network can be used to simulate this behavior.

3.3 Artificial Neural Network

A neural network is a parallel computing system that can solve problems that linear computing cannot. It uses statistical signal processing as its foundation. Because the function f is defined as a composite of other functions g (Hippisley-Cox et al. 2008), which can also be defined as a composition of other functions, the name *artificial neural network* (ANN) comes from the word "network." Neurons and weights are the two primary components of an ANN. Neurons are structures made up of simple processing components that are densely interconnected. Each piece is connected to its neighbors via weights, which vary in strength (Shanthi et al. 2009).

ANNs are frequently employed in medicine for various purposes, particularly in cardiology. Diagnoses, electronic signal analysis, medical picture analysis, and radiography have benefited from using ANNs. ANN is also adapted for text, audio, video, speech, facial expression prediction, analysis, forecasting, and pattern identification and recognition.

3.4 Molecular Biology

Biology is also concerned with the information contained in DNA. Mechanism, information, and gene have all played essential roles in the history of molecular biology. Molecular biologists identify and elucidate mechanisms such as DNA replication, protein synthesis, and various gene expression methods to discover and explain. Watson and Crick found the unique double-helix structure, which comprises two strands of DNA twisted around each other. A chain of sugar molecules and phosphate groups runs across each strand. Each sugar group is connected to one of the four "bases" that constitute the genetic alphabet of DNA: guanine (G), cytosine (Agbehadji et al. 2020), adenine (A), and thymine (T). Their arrangement or "sequence" along the molecule makes up the genetic code. It is concerned with the information transformations that occur within the cell.

3.5 DNA Computing

DNA computing is a subset of biomolecular computing that uses DNA as a data transport to perform arithmetic and logic functions. It is a new type of computing that relies on DNA, biochemistry, and molecular biology hardware rather than standard computer hardware. DNA computation technology allows for the creation of single strands that can be utilized to represent binary data bits. Individual strands can be replicated until enough strands can tackle large computational tasks. Parallel processing allows DNA computing to conduct millions of tasks simultaneously. As a result, substantial parallel searches are possible, and a complete set of probable solutions is generated. DNA can store more information per cubic centimeter than a trillion CDs, allowing it to effortlessly handle massive amounts of working memory (Gupta et al. 2021).

DNA sequencing, fingerprinting, mutation detection, and the development and miniaturization of biosensors, which could allow communication between molecular sensory computers and conventional electronic computers, are all applications of natural DNA converted into a computable form (Gupta et al. 2021).

3.5.1 Immune System

The human immune system has several characteristics, including resilience and fault tolerance, that make it a good fit for various computing challenges (Gupta et al. 2021).

3.5.2 Artificial Immune System

Artificial immune systems (AIS) are intelligent algorithms inspired by the human immune system's principles. It is a computational paradigm that emerged through exchanges and interactions between computer scientists and immunologists in the development of algorithms inspired by the immune system of vertebrates. There are two generations of the artificial immune system. The first generation relied on simplified immune models, whereas the second relied on interdisciplinary collaboration to get a deeper understanding of the immune system and, as a result, create more complicated models (Gupta et al. 2021). Anomaly detection, pattern recognition, optimization, and robotics have been effectively implemented using the AIS generations outlined above.

3.6 Biological Cells

Life is closely intertwined with cells; everything alive is made up of cells or has something to do with them. The cell is life's fundamental structural and functional unit. It has a minimal size and a highly complex structure and operation. The cell membrane, nucleus, and cytoplasm are the three primary components of a cell. The cell membrane encloses the cell and regulates the molecules entering and exiting it. The nucleus is a cell structure that houses the nucleolus and most of the cell's DNA. It is also where the majority of the RNA is produced. As known, the cell has a complex internal activity as well as an exquisite connection with its neighbors and the environment in general S. Marcus summarizes life in the equation below:

Life = DNA software + membrane hardware (Marcus 2002)

Most plant and animal cells can only be seen under a light microscope, due to their small size. Electron microscopy has a far better resolution, allowing for a much more comprehensive examination of cell structure. Organisms are either unicellular (comprising a single cell, such as bacteria) or multicellular (containing several cells, such as that plants and animals). Single-celled organisms can live alone, but most cells are organized into tissues, organs, organisms, and groups of organisms. Together with the cell's internal structure and organization, it suggests many ideas that are both mathematically intriguing and possibly valuable in terms of computability. Membrane computing can be used to analyze cell processing.

Membrane Computing: Membrane computing is a subset of natural computing that abstracts computing models from the design and function of living cells and the arrangement of cells in tissues, organs (including the brain), and other higher-order structures such as colonies of cells. P systems are a computational paradigm that abstracts from the behavior and structure of biological cells (Ibarra 2009; Păun 2000). Membrane systems (P systems) are parallel and distributed computing models that process multisets of symbols in cell-like variational frameworks.

A P system progresses from one configuration to the next by enforcing rules following a predetermined strategy (maximally parallel manner). Regulations can modify the shape of objects, move them around, and even alter the structure of the membrane (creation/division/dissolution/movement) (Gupta et al. 2021). Membrane computing has applications in biology, computer science, computer graphics, and linguistics, such as static sorting and public key protocol analysis.

4 Mathematical Methods in Biological Processes

A mathematical model is commonly considered an equation, a group of equations, or any other relationship that describes an event in science, engineering, economics, or another field and allows us to forecast observations quantitatively. By mathematical modeling, we imply the process of formulating and analyzing model equations and comparing observations to the models' predictions (Moskowitz et al. 2009).

High-throughput genomics and proteomics research progressively uses mathematical and computational models to comprehend biomedical data effectively. Hypotheses and experiments are generated with high-performance computer models that allow the simulation of complicated biological processes. Models that are adequately interfaced with biomedical databases are required for rapid access to and sharing of knowledge via data mining and discovery techniques (Ramlal et al. 2021; Motta and Pappalardo 2013).

4.1 Kolmogorov Equations

In 1931, Andrei Kolmogorov attempted to construct a theory of continuous-time Markov processes by extending the idea of Markov processes, which is expressed by the Chapman–Kolmogorov equation (Feller 1949). He observed that depending on the expected behavior over short periods, there are two types of continuous-time Markov processes, namely:

A. Jump processes
   Jump processes occur when you assume that "in a limited time interval, there is an overwhelming probability that the state will remain intact; yet if it changes, the change may be drastic."

B. Diffusion processes
In diffusion processes, "it is certain that some change will occur in any period, however little; only, here, it is certain that the changes over short time intervals will also be minor."

4.1.1 Example of a Kolmogorov Equation

The equation mentioned below is often used to apply to model population growth with birth (Moskowitz et al. 2009).

$$p_n'(t) = (n-1)\beta p_{n-1}(t) - n\beta p_n(t) \tag{1}$$

Here, n is the population index concerning the initial population, β is the birth rate, and finally, $p_n(t) = Pr(N(t) = n)$, i.e., the probability of attaining a specific population size.

Following is the analytical solution (Moskowitz et al. 2009):

$$p_n(t) = (n-1)\beta e^{-n\beta t} \int_0^t p_{n-1}(s) e^{n\beta s} \mathrm{d}s \tag{2}$$

This is a formula for the density $p_n(t)$ in terms of the preceding ones, i.e., $p_{(n-1)}(t)$.

5 Current Trends in Nature-Inspired Intelligent Computing Techniques for Biomarker and Drug Target Identification

The process of drug discovery, from finding the potential target to getting a stable drug against the disease, is a very long and hectic procedure. Currently, various approaches are in progress to get the therapeutic drug target. Nature-inspired computing techniques like neural networks and deep neural networks are being used by researchers in the process of target identification and validation. Machine learning, a domain of artificial intelligence, is based on making the machine learn to act and think like a human, but of course, with more efficiency. The development of high-throughput in biology has given both challenges and advantages to the pharmaceutical industries. Many combinatorial approaches to biology with machine learning (Sheikh et al. 2022; Yadav et al. 2021) have provided essential tools that can help in target validation, identification of biomarkers, and use of ML in the data analysis of clinical trials conducted on patients (Vamathevan et al. 2019). Developments in high-throughput molecular profiling technologies are taking the medical industry from traditional medicine to

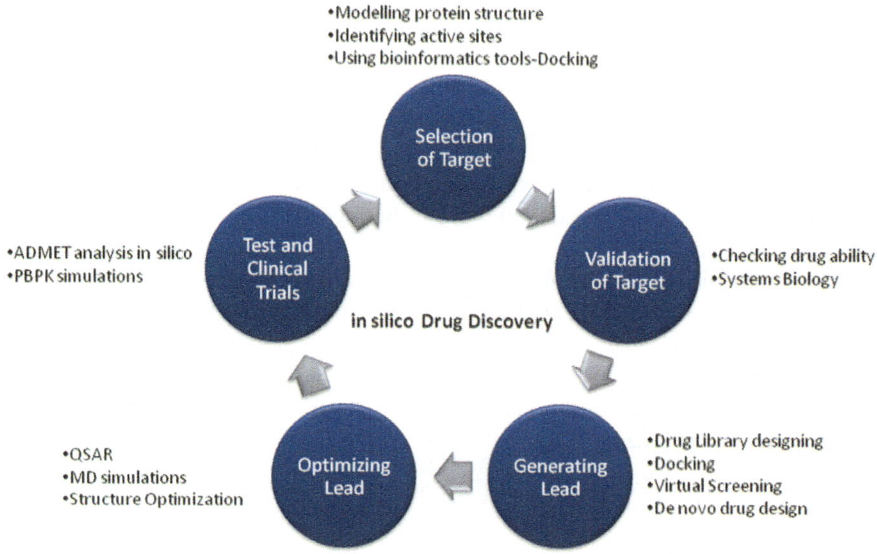

Fig. 2 Computer-based drug discovery and designing

precision medicine to diagnose and treat the disease in individuals. Biomarker signatures can be identified from high-dimensional technologies like gene expression and polymorphisms (Liu et al. 2014) (Fig. 2).

ANNs present the idea of comparing machines' computation and the human nervous system. The field of neural computation's purpose is to make the machines understand how the human brain works and then designs the algorithms and software according to how the human brain processes the information. Deep learning is the new form of ANN, which function is to do feature detection from a colossal number of unlabeled and labeled data by utilizing multi-level deep neural networks (DNNs). Previously scientists have used neural networks to classify coding (exons) and non-coding (introns) regions of DNA. Currently, there are various developments in the application of DNN in drug discovery and drug development, like de novo molecular design, biological-image analysis, and bioactivity prediction (Gawehn et al. 2016). Studying the interaction between protein and ligand is a very crucial part of the research area of molecular docking; various scoring functions are based on the protein–ligand complex structures (Pagadala et al. 2017). Considering the multiple application of convolutional neural networks (CNNs) in image analysis, various studies are conducted in which CNN is applied in scoring protein–ligand interactions (Ragoza et al. 2017). Various current studies have shown that machine learning applications are used to predict drug effects specific to cancer. One such study was conducted by Iorio et al. The researchers screened the 990 cancer cell lines against 265 cancer-effective drugs by applying ANOVA, logic models, and machine learning algorithms like random forest and elastic net regression (Iorio et al. 2016).

The elaboration of nature as a computational mechanism gave the idea of nature as a network of networks where information processing is done on various organizational levels, thus relating natural processes as an information processing system. The main works focused on systems biology, synthetic biology, and cellular computing. Computational systems biology aims to study the complex interactions in biological systems, applying genetic algorithms, and using genetic programming to find a solution to the problems. Recent studies on the medicinal applications of systems biology by studying the host–pathogen interactions are conducted to find the putative biomarkers for improving the field of diagnostic and treatments (Dix et al. 2016). McMillan et al. first time presented a chemistry-based de novo identification of druggable targets in their paper, screening 222 chemical compounds by modeling against more than 100 cells models of various somatic lung cancer lesions, also applying regularized machine learning (elastic net), and probability-based metrics in their study (McMillan et al. 2018).

6 Software and Tools

6.1 BioDiscML

Biomarkers play a very significant role in identifying the natural state of any disease, which is why it is one of the most exciting topics in computational biology (Liu et al. 2014; Beerenwinkel et al. 2016). *Bio*marker *disc*overy by *m*achine *l*earning (BioDiscML) is a biomarker discovery software that takes advantage of the complexity and application of the ML, a software developed by a group of researchers that, by automating various machine learning approaches, can identify the biomarker signature from high-dimensional data like gene expression (Leclercq et al. 2019). This software is written in JAVA 8 language (Fischer 2015) and uses the Weka 3.8 ML library (Holmes and Nevill-Manning 1995; Hall et al. 2009) to support the classification and regression problems, automating many ML steps to identify predictive models. Also, all generated models are then cross-validated, where each step has editable default parameters, and advanced parameters can also be used depending on the user's need. The main requirements are input datasets, class label names, and problem type (classification or regression) to start the program. Thus, BioDiscMl performs data preprocessing, features dimension reduction, and model selection strategy then identifies the best model and searches for correlated features (Leclercq et al. 2019).

6.2 *Psipred 4.0*

The PSIPRED application function uses two trained artificial neural network stages to predict protein secondary structure from the primary structure (Jones 1999). It identifies how the preserved elements from the given protein sequence structure are related to the homologs identified from the database. It is done by matching the "sequence profile" of the protein structure as they are highly preserved, so these regions are more relevant than the whole sequence. However, matching the sequence profile with a similar sequence from the database can be coincidental, so PSIPRED uses the position-specific scoring matrixes (PSSMs) produced by PSI-BLAST to transfer this data to the first neural network (Jones 1999).

A list of NIIC-based software used for biomarkers and potential therapeutics identification is listed in Table 2.

Table 2 Nature-inspired intelligent computing-based software

S. No.	Software	Description
1	**LIGBUILDER**	LigBuilder is software for the multi-purpose of the structure-based drug design platform (Yuan et al. 2020). The scoring functions of this software are based on the three-dimensional structure of the protein/molecule or the ligand-binding site. This software develops ligands piece by piece by using the library of organic filaments, being in the protein's structural constraints
2	**COLIBREE**	COLIBREE stands for combinatorial library breeding is a stochastic optimization algorithm and is based on the particle swarm optimization (PSO) (Hartenfeller et al. 2008). Just like LigBuilder, it also generates ligands from the scratch by adding side chains to the molecular scaffolds, and these side chains are made of molecular building blocks which are linked together through specific linker fragments
3	**DEAP**	DEAP is open-source software written in Python based on the evolutionary algorithm, used for fast prototyping and testing the problems. It is freely available at http://deap.gel.ulaval.ca, under LGPL license (Haynes et al. 2013)
4	**CGP4Matlab**	It is a freely available strong toolbox that is used to solve the problems related to signal processing and image processing. Here, the user can run the Cartesian Genetic Programming in MATLAB
5	**PonyGE2**	It is open-source software developed by the UCD natural computing research and applications group, where they applied the grammatical evolution (GE) in Python (Vu 2021) Grammatical evolution or GE is an evolutionary algorithm that is used to implement grammar in the mapping process from genotype to phenotype

7 Advantages of Nature-Inspired Intelligent Computing Techniques

There are several advantages of innovative NIIC techniques which are successfully reducing the computational complexity and speeding up the data mining process through adaptability, robustness, and scalability (Aličković and Subasi 2017; Schranz 2020; Sharma et al. 2019). Metaheuristic nature-inspired algorithms act as a strong tool that emerges as a problem-solving paradigm related to the biomedical field (Yang and Karamanoglu 2020). The calculations of nature-inspired intelligent computing have turned out to be definitive and well systematized for elucidating tribulations using several techniques including swarm intelligence (SI), genetic algorithm (GA), natural evolution, ANN, deep learning, etc. These techniques are inspired by the nature based on the algorithms that provide a broader perspective and satisfactory solution that can profoundly augment the prevailing technique for drug target identification, identification of biomarkers for expressing the informative gene, drug response biomarkers, data-driven target identification, identifying novel pathways, medical imaging, tracking features, initial disease diagnosis, and many more that brings the new treatments to patients much faster (Best and Sol 2017). Many combinatorial approaches to biology with nature-inspired intelligent computing have provided essential tools to show the strongest signal in the earliest time-point of the disease stages (normal, pre-disease, and disease state) for tracking the progression of the complex disease and provide the finest correlates with the disease groups.

Identified targets/tools/lead molecules through NIIC techniques have gained much attention and shown success. Researchers are interested in the new way of achieving a form of collective NIIC that is also advanced as a nature-inspired computing technique based on the study of self-organization behavior systems. Several collective behaviors (like social insects, colonies of ants, and a flock of birds) are able to handle extremely diverse and dynamic environments that have been extended with weight value to optimize the accuracy of distance measures (Zhang et al. 2019; Derangula and Edara 2021). The major impact of various algorithms to tackle the misclassification errors on disease-based data and also provides valuable insights for opening up new opportunities for the drug discovery process, detection of dynamic network biomarkers signature, and interaction between the genetic variants, producing human-competitive results for natural system or process (Pal and Meher 2013; Coleto-Alcudia and Vega-Rodríguez 2021).

Innovative NIIC techniques have several advantages, which are successfully reducing the computational complexity and speeding up the data mining process through adaptability, robustness, and scalability (Aličković and Subasi 2017; Schranz 2020; Sharma et al. 2019). Metaheuristic nature-inspired algorithms act as a robust tool that emerges as a problem-solving paradigm related to the biomedical field (Yang and Karamanoglu 2020). NIIC techniques are inspired by nature based on the algorithms that provide a broader perspective and satisfactory solution that can profoundly augment the prevailing technique for drug development, drug development, *t*, and medical imaging, initial disease diagnosis, and many more (Best and Sol 2017).

Identified targets/tools/lead molecules through NIIC techniques have gained much attention and success. Several collective behaviors, such as social insects, colonies of ants, and the flock of birds, can handle highly diverse and dynamic environments that have extended with weight value to optimize the accuracy of distance measure (Zhang et al. 2019; Derangula and Edara 2021). The significant impact of various algorithms to tackle the misclassification errors on disease-based data also provides valuable insights for opening up new opportunities for the drug discovery process, detection of dynamic network biomarkers signature, and interaction between the genetic variants, producing human-competitive results for natural systems or process (Pal and Meher 2013; Coleto-Alcudia and Vega-Rodríguez 2021).

8 Limitations and Future Directions

Besides these advantages of successful drug discovery and computational biomarkers, several drawbacks of nature-inspired intelligent computing techniques which not been reached with 100% accuracy.

8.1 Limitations

NIIC techniques have limitations because of having individual and static nature. There is a lack of comprehensive computational mathematical framework and understanding of NIIC techniques as to why and how they work for therapeutic identifications (Sharma et al. 2019; Yang and Karamanoglu 2020). Many studies have been reported with different NIIC algorithms in search space that can solve critical optimization problems in different ways. More exhaustive studies with rigorous mathematics are highly desirable. The biased performance measures may lead to inaccurate distortion of a measurement outcome for identifying the biomarkers and therapeutic targets. It is unclear what measures are fair. The success rates, computing time, accuracy, number of function evaluations, tolerance, means, most acceptable values, and standard derivations are used to compare NIIC techniques, but it is hard to say if these are realistic performance measures (Yang and Karamanoglu 2020).

Advances in digital healthcare technologies, such as AI, VR/AR, 3D printing, robots, and nanotechnology, are transforming the future of health care right in front of our very eyes. We need to keep up with the latest advancements if we want to regulate technology. To be relevant in the coming years, healthcare workers must accept new technologies in the field and work hand-in-hand with technology (Wang et al. 2021). There are many ways in which digital technology can help transform unsustainable healthcare systems, equalize the relationship between doctors and patients, speed up the discovery of new treatments for diseases like cancer and AIDS, and even lead to healthier individuals living in healthier communities (Siddique and Adeli 2015; Tzanetos et al. 2020).

Many diseases can be diagnosed based on the expression levels of a particular gene in an organism (de Jong et al. 2019). The most frequently used microarray technology in the prognosis of many types of cancer can be used to classify and analyze numerous genetically connected diseases (Latha et al. 2020). Due to increased data collection and storage, there has been an increase in interest in applying big data techniques to a wide range of physical challenges (Hermeking 2003). It is, therefore, necessary to first reduce dimensions before classifying them. However, due to the many gene expression levels in the human body, it is difficult to diagnose cancer using microarrays. The quality of performance of classification algorithms is mainly determined by the quality of the input sets of features (Kumar and Greiner 2019). There will be an excellent correlation between features and class labels if they are relevant to each other. As a result, in most cases, the relevance of characteristics is unclear, and the input datasets typically include flaws such as irrelevance and redundancy that are not beneficial during the knowledge discovery process. Thus, this may make it harder to develop a favorable categorization. Reducing the quantity of data for machine learning algorithms requires knowledge of key characteristics, which may be achieved by eliminating unneeded and duplicate features (Yu et al. 2021). This remains a major challenge for the future.

8.2 Future Directions

8.2.1 Artificial Immune System

To execute complex immunological tasks, the artificial immune system, or the AIS method, mimics the basic features and functions of the human immune system. Distribution, flexibility, and decentralization are some of the advantages of the immune system. Many different organs, tissues, and immune cells make up the human immune system, which is dispersed throughout the body. When these components identify, destroy, and remember alien macromolecules and cells, they respond highly coordinated and specialized (Schultz and Grieder 1987).

The primary mechanism of the innate immune system is to identify self from non-self to protect the human body by recognizing and defending against foreign antigens (viruses, bacteria). Lymphocytes are the primary antigen killers in the immune system, out of all immune cells. They have unique binding regions (receptors) that may structurally identify and respond to specific foreign antigens (Akram and Raza 2018). When an antigen attaches to the surface receptors of an immune cell, the contact sensitizes the immune cell population's proliferation and differentiation to that particular antigen. Following antigen elimination, some immune cells become memory cells for future action in the event of a reoccurrence of the same antigen. Specificity, inducibility, variety, memory, differentiating self from non-self, and self-regulation are six fundamental properties of the human immune system that may be determined via the research of immunity processes. Distribution via multi-agent systems applications, flexibility through autonomous agent applications, totally decentralized organization, and other properties of the human immune system's capacity to tackle challenging high-dimension issues are some of the AIS's core qualities (Tay et al. 2015).

Various complicated systems, such as distributed multi-agent systems and autonomous agents, have acquired these enticing qualities. Computational biology includes AIS, DNA RNA, and membrane computing or systems biology (Miho et al. 2018). In order to address tough challenges, this sector makes use of many human body networks and systems. These techniques' fundamental ideas vary significantly from the ACO and PSO algorithms. AIS, in this instance, acts like that of the human immune system. Hepatitis illness is diagnosed using an artificial immune recognition system with fuzzy resource allocation, and lung cancer is diagnosed using an artificial immune recognition system hybridized with principal component analysis and fuzzy weighting preprocessing (Subramanian et al. 2020).

8.2.2 Membrane Computing

Gh. Paun started the membrane computing field at the end of 1998, and it is all about finding computer models that are inspired by the structure and function of live cells (Pérez-Jiménez 2014). Membrane computing has recently been a valuable platform for applications in various fields, including biology and biomedicine. Membrane computing is a novel field of natural computing whose goal is to abstract computing principles from live-cell structure and function and cell cooperation in tissues, populations of cells, and the brain. The created models are referred to as P systems (Păun and Pérez-Jiménez 2006). There was a tremendous amount of progress in this area both in terms of theory, such as Turing computability, language, and automata theory (such as characterizations of Turing completeness), and in terms of computational efficiency, like finding polynomial solutions to NP-complete problems (in system biology, biomedicine, computer graphics, approximate optimization, robot control, cryptography, ecology, and economics modeling) (Schott-Verdugo and Gohlke 2019).

It is possible to divide P systems into three categories: (1) systems that are inspired by live cells; (2) systems that are inspired by tissues; and (3) nervous system-inspired systems. A type of neural-like P system, referred to as spiking neural P systems, is discussed in depth in this study after a short introduction to cell and tissue-like P systems: As a first step, a formal specification of the system is presented, followed by an examination of some of its past performance in terms of both computational completeness and computational efficiency (Lajoie et al. 2020).

8.2.3 DNA Computing

In nature, DNA serves as a kind of memory. All of an organism's genetic information is contained in its DNA. In the DNA computing paradigm, synthetic (or natural)

DNA molecules are used as data store material. Several real-world DNA properties and functionalities, such as polymerase chain reaction, gel electrophoresis, enzymatic processes, and enormous parallelism, are used in DNA computing (Chandler 2020). To replicate, sort, and divide/concatenate the information in DNA molecules, molecular biology procedures, including polymerase chain reaction (PCR), gel electrophoresis, and enzyme reactions, are utilized as computational operators. DNA-based computing provides several benefits over traditional solid-state semiconductor-based computing. While DNA does one action at a time, computers concurrently conduct billions of operations (Emanuelson et al. 2021).

When compared to the speed of modern digital computers, where each process is performed in parallel, the difference is striking. Since information is encoded at the molecular level, DNA has a large storage capacity. It is critical to understand how to encode numerical data in DNA strands to increase DNA computing's capacity to tackle numerical optimization issues (Chandler 2020).

8.2.4 Gene Expression Programming

Gene expression programming (GEP) is a full-fledged genotype/phenotyping system using linear chromosomes with defined lengths to encode computer programs. GEP is a full-fledged genotype/phenotyping system. It is possible to operate essential genetic operators like mutation, transposition, and recombination in an unrestricted and productive manner since each gene's expression results in a valid program at all times (Gil and Ulitsky 2020). GEP's genotype/phenotype system is the first artificial/phenotype system with a sounding translation mechanism, even if it is fundamental. Genotype and phenotype are intertwined in GEP, and this interaction has led to a dramatic improvement in performance (Schoenfelder and Fraser 2019).

The exact form of tree representation is shared by genetic programming (GP) and GEP, making it feasible to quickly go back and retrace the steps made by GP and explore the new frontiers that opened up, bypassing the phenotypic threshold (Ruhr et al. 2021). The distinctions between genetic algorithms and genetic programming are briefly reviewed in this course better to understand the evolutionary benefits of gene expression programming. An extensive explanation of the algorithm's key actors is given, concentrating on their interconnections and how the chromosomes' basic but innovative structure permits fast and unfettered search space exploration (Frenk and Houseley 2018).

8.2.5 Multi-Expression Programming

In multi-expression programming (MEP), chromosomes are represented in a linear fashion, a variation on GP. Gene strings encode complicated computer programs for MEP persons. Compiled C or Pascal code is represented similarly by MEP people when encoding expressions in their programs. Multiple solutions to a problem may be stored on a single chromosome as a unique feature of MEP. When it comes to fitness assignments, the most acceptable option is often selected (Ramakrishnan et al. 2020).

Problems involving symbolic regression or classification need the use of MEP since it uses a single solution to be stored in a chromosome (as does any other issue for which the training set is known before solving the problem) (such as GP, CGP, GEP, or GE) (Travaglini et al. 2020). A single parsing of the chromosome may evaluate the expressions stored in an MEP individual. In the case of crossover and mutation, the offspring is always MEP individuals with the correct syntax (computer programs). As a result, fixing freshly acquired persons does not need further procedures (Wei and Zhong 2020).

8.2.6 Drug Development

Problems involving symbolic regression or classification need the use of MEP since it uses a single solution to be stored in a chromosome (as does any other issue for which the training set is known before solving the problem) (such as GP, CGP, GEP, or GE) (Raza and Alam 2016). A single parsing of the chromosome may evaluate the expressions stored in an MEP individual. In the case of crossover and mutation, the offspring is always MEP individuals with the correct syntax (computer programs). As a result, fixing freshly acquired persons does not need further procedures (Alam et al. 2022).

Companies like turbine and recursion pharmaceuticals are using AI and deep genomics to create new drug candidates and unique therapeutic solutions in record time and speed up the time to market while saving money and human lives as a side benefit. In silico drug testing is another potential healthcare technique (Fási et al. 2019). A medical product, gadget, or intervention may be subjected to these simulations as part of the development process or a regulatory examination. Even though existing technology and biological knowledge do not allow for totally replicated clinical trials, organs-on-a-chip, which are currently in use, has made tremendous progress in this sector. Multiple research initiatives use the "most thorough, mathematical model of human physiology yet produced," referred to as humor. The Virtual Physiological Human (VPH) Institute has also developed virtual models for studying heart disease and osteoporosis (Gupta et al. 2021).

9 Conclusion

This chapter clearly shows that nature-inspired computing is a vast and growing area. An overview of recent achievements in this fascinating field of study, with an emphasis on physics and biology-based techniques and algorithms, was offered in this chapter article for the reader's convenience. Additionally, a parallel development has occurred in the field of computational intelligence (Subramanian et al.

2020; Gupta et al. 2021) over the past twenty years that demonstrated how a multi-paradigm approach and integration of the three computational intelligence computing paradigms could lead to more effective solutions to complex and intractable pattern recognition and learning problems. Cross-fertilization between NIIC and computational intelligence may be expected based on current trends.

It is critical to underline that cognitive components and human thinking are fundamentally significant natural systems. As a result, symbolic artificial intelligence is included in the scope of natural computing. However, the method's main drawback is its sensitivity to parameter selection since there is a chance that no convergence may occur. Another drawback of the PSO technique is that it might prematurely converge to a local minimum. Particle dispersion, on the other hand, is a possibility. It is advised that future studies include adjusting weights and selecting the optimal settings for various ANN-based procedures using the other techniques mentioned in this chapter.

References

Adler J (1966) Chemotaxis in bacteria: motile Escherichia coli migrate in bands that are influenced by oxygen and organic nutrients. Science 153(3737):708–716

Agbehadji IE et al (2020) Review of big data analytics, artificial intelligence and nature-inspired computing models towards accurate detection of COVID-19 pandemic cases and contact tracing. Int J Environ Res Public Health 17(15):5330

Ahmad S et al (2021) Mobile technology solution for COVID-19: surveillance and prevention. Computational intelligence methods in COVID-19: surveillance, prevention, prediction and diagnosis. Springer, pp 79–108

Akram M, Raza A (2018) Towards the development of robot immune system: a combined approach involving innate immune cells and T-lymphocytes. Biosystems 172:52–67

Alam M et al (2022) Potential therapeutic implications of caffeic acid in cancer signaling: past, present, and future. Front Pharmacol 13:845871. https://doi.org/10.3389/fphar

Aličković E, Subasi A (2017) Breast cancer diagnosis using GA feature selection and Rotation Forest. Neural Comput Appl 28(4):753–763

Beerenwinkel N, Greenman CD, Lagergren J (2016) Computational cancer biology: an evolutionary perspective. PLoS Comput Biol 12(2):e1004717

Bejnordi BE et al (2017) Diagnostic assessment of deep learning algorithms for detection of lymph node metastases in women with breast cancer. JAMA 318(22):2199–2210

Best M, Sol N (2017) In't Veld, SGJG; Vancura, A.; Muller, M.; Niemeijer, A. LN, Fejes, A. V, Tjon Kon Fat, L.-A., Huis In't Veld, AE, Leurs, C., et al, p 238–252

Bishop JM (1989) Stochastic searching networks. In: 1989 First IEE international conference on artificial neural networks (Conf Publ No 313). IET

Bishop JM (2007) Stochastic diffusion search. Scholarpedia 2(8):3101

Blum C (2005) Ant colony optimisation: introduction and recent trends. Phys Life Rev 2(4):353–373

Blum C, Vallès MY, Blesa MJ (2008) An ant colony optimisation algorithm for DNA sequencing by hybridisation. Comput Oper Res 35(11):3620–3635

Chandler DL (2020) DNA-like materials could open new computing frontiers. IEEE Pulse 11(3):12–15

Chen H et al (2020) An enhanced bacterial foraging optimisation and its application for training kernel extreme learning machine. Appl Soft Comput 86:105884

Coleto-Alcudia V, Vega-Rodríguez MA (2021) A metaheuristic multi-objective optimisation method for dynamical network biomarker identification as pre-disease stage signal. Appl Soft Comput 109:107544
De MK, Slawomir NJ, Mark B (2006) Stochastic diffusion search: partial function evaluation in swarm intelligence dynamic optimization. In: Stigmergic optimisation. Springer, pp 185–207
de Jong TV, Moshkin YM, Guryev V (2019) Gene expression variability: the other dimension in transcriptome analysis. Physiol Genomics 51(5):145–158
De Meyer K, Bishop J, Nasuto S (2000) Attention through self-synchronisation in the spiking neuron stochastic diffusion network. Consc Cogn 9(2):81–81
Derangula A, Edara SR (2021) Identification of optimised features using nature-inspired meta-herustics based optimisations in breast cancer detection. In: Materials Today: Proceedings
Dix A et al (2016) use of systems biology to decipher host–pathogen interaction networks and predict biomarkers. Clin Microbiol Infect 22(7):600–606
Dorigo M (1992) Optimisation, learning and natural algorithms. Ph. D. Thesis, Politecnico di Milano
Emanuelson C, Bardhan A, Deiters A (2021) DNA computing: NOT logic gates see the light. ACS Synth Biol 10(7):1682–1689
Esteva A et al (2017) Dermatologist-level classification of skin cancer with deep neural networks. Nature 542(7639):115–118
Fási L et al (2019) Antioxidant-inspired drug discovery: antitumor metabolite is formed in situ from a hydroxycinnamic acid derivative upon free-radical scavenging. J Med Chem 62(3):1657–1668
Feller W (1949) On the theory of stochastic processes, with particular reference to applications. In: Proceedings of the [First] Berkeley symposium on mathematical statistics and probability. University of California Press
Fischer R (2015) Java 8: It's a whole new Java. Java closures and lambda. Springer, pp 1–10
Frenk S, Houseley J (2018) Gene expression hallmarks of cellular ageing. Biogerontology 19(6):547–566
Gambardella LM, Dorigo M (2000) An ant colony system hybridised with a new local search for the sequential ordering problem. INFORMS J Comput 12(3):237–255
Gawehn E, Hiss JA, Schneider G (2016) Deep learning in drug discovery. Mol Inf 35(1):3–14
Gil N, Ulitsky I (2020) Regulation of gene expression by cis-acting long non-coding RNAs. Nat Rev Genet 21(2):102–117
Grenander L, Heppner C (1990) A stochastic nonlinear model for coordinated bird flocks, The ubiquity of chaos, E. Krasner
Gupta R et al (2021) Artificial intelligence to deep learning: machine intelligence approach for drug discovery. Mol Diversity 25(3):1315–1360
Hall M et al (2009) The WEKA data mining software: an update. ACM SIGKDD Explor Newsl 11(1):10–18
Hartenfeller M, Proschak E, Schüller A, Schneider G (2008) Concept of combinatorial de novo design of drug-like molecules by particle swarm optimization. Chem Biol Drug Des 72(1)
Haynes WA, Higdon R, Stanberry L, Collins D, Kolker E (2013) Differential expression analysis for pathways. PLoS Comput Biol 9(3):e1002967
Hermeking H (2003) Serial analysis of gene expression and cancer. Curr Opin Oncol 15(1):44–49
Hippisley-Cox J et al (2008) Predicting cardiovascular risk in England and Wales: prospective derivation and validation of QRISK2. BMJ 336(7659):1475–1482
Holmes G, Nevill-Manning CG (1995) Feature selection via the discovery of simple classification rules
Hosseini HS (2007) Problem solving by intelligent water drops. In: 2007 IEEE congress on evolutionary computation. IEEE
Ibarra OH (2009) Keynote 3: computing with membranes: an overview. In: 2009 International conference on parallel and distributed computing, applications and technologies. IEEE
Iorio F et al (2016) A landscape of pharmacogenomic interactions in cancer. Cell 166(3):740–754
Jones DT (1999) Protein secondary structure prediction based on position-specific scoring matrices. J Mol Biol 292(2):195–202

Keller L, Gordon E (2009) The lives of ants. OUP Oxford

Kennedy J, Eberhart R (1995) Particle swarm optimisation. In: Proceedings of ICNN'95-international conference on neural networks. IEEE

Khazaee A (2013) Heart beat classification using particle swarm optimisation. Int J Intell Syst Appl 5(6):25

Korb O, Stützle T, Exner TE (2007) An ant colony optimisation approach to flexible protein–ligand docking. Swarm Intell 1(2):115–134

Kumar L, Greiner R (2019) Gene expression based survival prediction for cancer patients—a topic modeling approach. PLoS ONE 14(11):e0224446

Lajoie MJ et al (2020) Designed protein logic to target cells with precise combinations of surface antigens. Science 369(6511):1637–1643

Latha NR et al (2020) Gene expression signatures: a tool for analysis of breast cancer prognosis and therapy. Crit Rev Oncol Hematol 151:102964

Leclercq M et al (2019) Large-scale automatic feature selection for biomarker discovery in high-dimensional OMICs data. Front Genet 10:452

Liu R et al (2014) Early diagnosis of complex diseases by molecular biomarkers, network biomarkers, and dynamical network biomarkers. Med Res Rev 34(3):455–478

Majid A et al (2014) prediction of human breast and colon cancers from imbalanced data using nearest neighbor and support vector machines. Comput Methods Programs Biomed 113(3):792–808

Manning T, Sleator RD, Walsh P (2014) Biologically inspired intelligent decision making: a commentary on the use of artificial neural networks in bioinformatics. Bioengineered 5(2):80–95

Marcus S (2002) Bridging P systems and genomics: a preliminary approach. In: Workshop on membrane computing. Springer

McMillan EA et al (2018) Chemistry-first approach for nomination of personalised treatment in lung cancer. Cell 173(4):864–878. e29

Miho E et al (2018) Computational strategies for dissecting the high-dimensional complexity of adaptive immune repertoires. Front Immunol 9:224

Moskowitz HR, Saguy IS, Straus T (2009) An integrated approach to new food product development. CRC Press

Motta S, Pappalardo F (2013) Mathematical modeling of biological systems. Brief Bioinform 14(4):411–422

Muhammad Y et al (2020) Early and accurate detection and diagnosis of heart disease using intelligent computational model. Sci Rep 10(1):1–17

Odell JJ et al (2002) Modeling agents and their environment. In: International workshop on agent-oriented software engineering. Springer

Pagadala NS, Syed K, Tuszynski J (2017) Software for molecular docking: a review. Biophys Rev 9(2):91–102

Pal SK, Meher SK (2013) Natural computing: a problem solving paradigm with granular information processing. Appl Soft Comput 13(9):3944–3955

Păun G (2000) Computing with membranes. J Comput Syst Sci 61(1):108–143

Păun G, Pérez-Jiménez MJ (2006) Membrane computing: brief introduction, recent results and applications. Biosystems 85(1):11–22

Pérez-Jiménez MJ (2014) The P versus NP problem from the membrane computing view. Eur Rev 22(1):18–33

Qazi S et al (2020) A coadunation of biological and mathematical perspectives on the pandemic COVID-19: a review

Qazi S, Sheikh K, Raza K (2021) In silico approach to understand the epigenetic mechanism of SARS-CoV-2 and its impact on the environment. Virusdisease 32(2):286–297

Rabanal Basalo PM, Rodríguez Laguna I, Rubio Díez F (2017) Applications of river formation dynamics

Ragoza M et al (2017) Protein–ligand scoring with convolutional neural networks. J Chem Inf Model 57(4):942–957

Rahmatizadeh SS-H, Torkaman H (2009) The ant-bee routing algorithm: a new agent based nature-inspired routing algorithm. J Appl Sci 9(5):983–987
Ramadan RM, Abdel-Kader RF (2009) Face recognition using particle swarm optimisation-based selected features. Int J Signal Process Image Process Pattern Recogn 2(2):51–65
Ramakrishnan K et al (2020) Transcriptional control of gene expression in Pichia pastoris by manipulation of terminators. Appl Microbiol Biotechnol 104(18):7841–7851
Ramlal A et al (2021) From molecules to patients: the clinical applications of biological databases and electronic health records. Translational bioinformatics in healthcare and medicine. Elsevier, pp 107–125
Raza K, Alam M (2016) Recurrent neural network based hybrid model for reconstructing gene regulatory network. Comput Biol Chem 64:322–334
Reynolds CW (1987) Flocks, herds and schools: a distributed behavioral model. In: Proceedings of the 14th annual conference on Computer graphics and interactive techniques
Ruhr I et al (2021) Developmental programming of DNA methylation and gene expression patterns is associated with extreme cardiovascular tolerance to anoxia in the common snapping turtle. Epigenetics Chromatin 14(1):1–42
Schoenfelder S, Fraser P (2019) Long-range enhancer–promoter contacts in gene expression control. Nat Rev Genet 20(8):437–455
Schott-Verdugo S, Gohlke H (2019) PACKMOL-memgen: a simple-to-use, generalised workflow for membrane-protein–lipid-bilayer system building. J Chem Inf Model 59(6):2522–2528
Schranz M (2020) Swarm robotic behaviors and current applications. Front Robot AI 7
Schulman J et al (2014) Motion planning with sequential convex optimisation and convex collision checking. Int J Robot Res 33(9):1251–1270
Schultz KT, Grieder F (1987) Structure and function of the immune system. Toxicol Pathol 15(3):262–264
Shanthi D, Sahoo G, Saravanan N (2009) Designing an artificial neural network model for the prediction of thrombo-embolic stroke. Int J Biometric Bioinform (IJBB) 3(1):10–18
Sharma S, Singh G, Singh D (2019) Role and performance of different traditional classification and nature-inspired computing techniques in major research areas. EAI Endorsed Trans Scalable Inf Syst 6(21)
Sheikh K, Raza K (2021) Viroinformatics and viral diseases: a new era of interdisciplinary science for a thorough apprehension of virology. Translational bioinformatics applications in healthcare. CRC Press, pp 109–132
Sheikh K, Ramlal A, Khan FN (2022) Computational resources for oncology research: a comprehensive analysis. Comput Intell Oncol 65–92
Siddique N, Adeli H (2015) Nature inspired computing: an overview and some future directions. Cogn Comput 7(6):706–714
Smith KP, Kang AD, Kirby JE (2018) Automated interpretation of blood culture gram stains by use of a deep convolutional neural network. J Clin Microbiol 56(3):e01521-e1617
Subramanian M et al (2020) Precision medicine in the era of artificial intelligence: implications in chronic disease management. J Transl Med 18(1):1–12
Tay D, Poh CL, Kitney RI (2015) A novel neural-inspired learning algorithm with application to clinical risk prediction. J Biomed Inform 54:305–314
Thimbleby H (2013) Technology and the future of healthcare. J Public Health Res 2(3)
Travaglini KJ et al (2020) A molecular cell atlas of the human lung from single-cell RNA sequencing. Nature 587(7835):619–625
Tzanetos A, Fister I Jr, Dounias G (2020) A comprehensive database of Nature-inspired algorithms. Data Brief 31:105792
Vamathevan J et al (2019) Applications of machine learning in drug discovery and development. Nat Rev Drug Discovery 18(6):463–477
Vu TM (2021) Software review: Pony GE2. Genet Program Evolvable Mach 22(3):383–385
Wang Z et al (2021) A comparative study of common nature-inspired algorithms for continuous function optimisation. Entropy 23(7):874

Wei T, Zhong J (2020) A preliminary study of knowledge transfer in multi-classification using gene expression programming. Front Neurosci 1396
Weng SF et al (2017) Can machine-learning improve cardiovascular risk prediction using routine clinical data? PLoS ONE 12(4):e0174944
Yadav MK et al (2021) Predictive modeling and therapeutic repurposing of natural compounds against the receptor-binding domain of SARS-CoV-2. J Biomol Struct Dyn 1–13
Yang X-S (2009) Firefly algorithms for multimodal optimisation. In: International symposium on stochastic algorithms. Springer
Yang X-S, Karamanoglu M (2020) Nature-inspired computation and swarm intelligence: a state-of-the-art overview. Nat Inspired Comput Swarm Intell 3–18
Yu K-H, Beam AL, Kohane IS (2018) Artificial intelligence in healthcare. Nat Biomed Eng 2(10):719–731
Yu Z et al (2021) Elucidating transcriptomic profiles from single-cell RNA sequencing data using nature-inspired compressed sensing. Briefings Bioinform 22(5):bbab125
Yuan Y, Pei J, Lai L (2020) LigBuilder V3: a multi-target de novo drug design approach. Front Chem 8:142
Zhang J, Zhang Y, Ma Z (2019) In silico prediction of human secretory proteins in plasma based on discrete firefly optimisation and application to Cancer biomarkers identification. Front Genet 10:542

Nature-Inspired Computing Techniques in Drug Design, Development, and Therapeutics

Sarra Akermi, Abira Dey, Nicholas Franciss Lee, Ruoya Lee, Nathalie Larzat, Jean Bernard Idoipe, Ritushree Biswas, Jasbir Kaur Simak, Suparna Dey, Subrata Sinha, Surabhi Johari, Chandramohan Jana, Anshul Nigam, Sunil Jayant, Ahmet Kati, and Ashwani Sharma

1 Introduction

Drug designing is a complex process that is very expensive and time-consuming. It involves several rounds of screening such as lead identification, which involves finding of potential active drug candidates from millions of new compounds. These compounds need to pass assessment of their physiochemical properties. Several

Sarra Akermi and Abira Dey—contributed equally.

S. Akermi · A. Dey · R. Biswas
Indian Science and Technology Foundation (ISTF), C-1/31, Yamuna Vihar, New Delhi 110053, India
e-mail: sarra.akermi@isto-india.org

A. Dey
e-mail: abira.dey@isto-india.org

R. Biswas
e-mail: ritushree.biswas@isto-india.org

N. F. Lee
Technical University of Munich, Munich, Germany
e-mail: n.lee@tum.de

R. Lee · N. Larzat · J. B. Idoipe · A. Sharma (✉)
Insight Biosolutions (IBS), Biopole Rennes 35000, Rennes, France
e-mail: ashwani.sharma@insight-biosolutions.com

R. Lee
e-mail: ruoya.li@insight-biosolutions.com

N. Larzat
e-mail: nathalie.larzat@insight-biosolutions.com

J. B. Idoipe
e-mail: jean.idoipe@insight-biosolutions.com

K. Raza (ed.), *Nature-Inspired Intelligent Computing Techniques in Bioinformatics*, Studies in Computational Intelligence 1066,
https://doi.org/10.1007/978-981-19-6379-7_14

computational methods such as molecular docking, QSAR, and molecular dynamics have been exploited to perform high throughput screening of the compounds. After the lead identification step, it is very important to under safety of the drugs using pre-screening method. Several methods have been reported to predict toxicity of the drugs; however, the development of the safer drug is a very time-consuming task. Recently, FDA and EMA guidelines recommended to use alternative methods to test toxicity of drugs instead of using animal testing under the concept of 3R (reduced, replacement, and refinement). Therefore, it became important to find some non-animal testing methods to predict wide range of drugs toxicities such as carcinogenicity, genotoxicity, mutagenicity, oral toxicity, development toxicity, and reproductive toxicity. Battery of tests is required for complete safety assessment of drugs which consume 4–5 years; however, thanks to the development of in silico methods, which instantly produces the several toxicities end points in short interval of time.

Among all in silico methods, nature-inspired computing (NIC) such as swarm intelligence is new advanced tools implemented to optimize the drug designing process, explore drug development, and integrate *in-vitro/in-vivo* methods to develop efficient drug response, and prediction models. These methods help in reducing the cost of drug development by 30%, increase drug screening capacities for larger-scale compounds databases and improve the accuracy, sensitivity, and specificity of

J. K. Simak · A. Nigam
Amity University Mumbai, Mumbai–Pune Expressway, Bhatan Post, Somathne, Panvel, Mumbai, Maharashtra 410206, India

S. Dey
St. Xavier's College (Autonomous), Kolkata, India

S. Sinha
Centre for Biotechnology and Bioinformatics, Dibrugarh University, Assam, India
e-mail: subratasinha@dibru.ac.in

S. Johari
Institute of Management Studies (IMSUC), Ghaziabad, Uttar Pradesh, India

C. Jana
Department of Chemistry, Michael Madhusudan Memorial College, Durgapur, West Bengal 713216, India

A. Nigam
FBE Technologies Pvt. Ltd., 128/64 Y Block, Kidwai Nagar, Kanpur 208011, India

S. Jayant
Annotation Analytics Pvt. Ltd., 36, Ward no-14, Biswa, Amrpura, Gurgaon 122001, India
e-mail: sunil.jayant@annotationanalytics.com

A. Kati
Experimental Medicine Research and Application Center, University of Health Sciences Turkey, Uskudar, Istanbul, Turkey
e-mail: ahmet.kati@sbu.edu.tr

Department of Biotechnology, Institution of Health Sciences, University of Health Sciences Turkey, Uskudar, Istanbul, Turkey

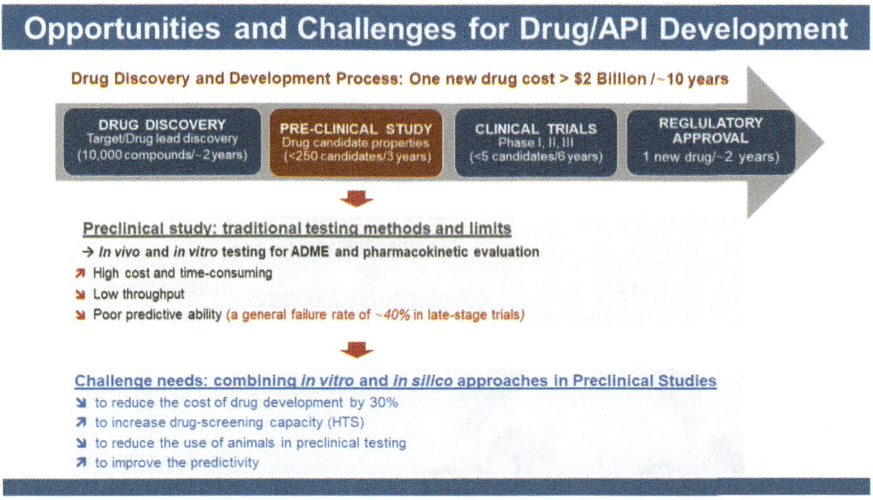

Fig. 1 Drug designing and development cycle and roles of in silico methods

the prediction (Fig. 1) (Yang and Deb 2009; Wang et al. 2019; Sikdar et al. 2015; Corne et al. 2001; Yang et al. 2013; Deb et al. 2002; Bader and Zitzler 2011; Moore et al. 2000; Moustakidis et al. 2012; Rojas 2009). Several tools have been expanded recently, to summarize the use of NIC tools in drug designing, drug development, and therapeutics, which we are describing in the book chapter.

2 Nature-Inspired Intelligent Computing Tools

Multi-objective ensemble cuckoo search-based algorithm

A cuckoo search algorithm, established in 2009 by Suash Deb and Xin-She Yang, is an optimization algorithm. It was included in the area which is concerned with the growth and implementation of advanced analytical techniques to boost up decision making which is known as operation research (Yang and Deb 2009). The eggs laying behavior of the cuckoo birds had been the inspiration for proposing this algorithm. A cuckoo bird chooses a random nest to lay its egg; therefore, the egg is considered to be a solution. So, the solution with more desirable fitness value enters the next generation. Now, a probability exists that the host bird is able to identify the cuckoo's egg and may throw the cuckoo's egg out of its nest or it may set up a new nest for itself. Thus, this algorithm searches the whole decision space and lists the fitness value of all the solutions for finding the optimal solutions (Fig. 2).

Y. Wang et al. proposed a decomposition-based multi-objective ensemble cuckoo search algorithm (MOECSA) in order to identify different types of cancers (Wang

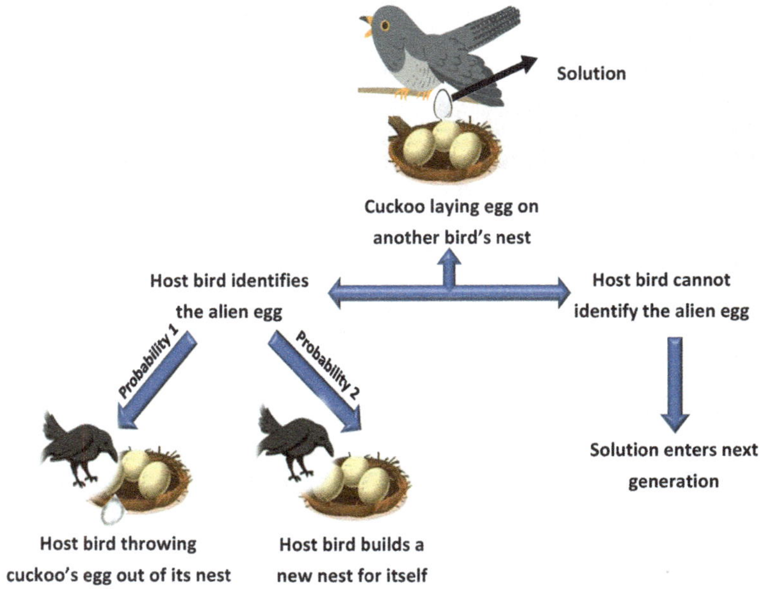

Fig. 2 Schematic representation of the concept behind designing the cuckoo search algorithm

et al. 2019). This algorithm was studied for the optimization of four objective functions which include two entropy-based measures which are redundancy and relevance concurrently, accuracy, and the number of features. The Tchebycheff perspective was utilized for decomposing four objectives with the multi-objective classification problem. In order to form scalar classification subproblems.

The main contribution of MOECSA was as follows:

1. Two advanced search methods were suggested, motivated by the differential evolutionary algorithm, in order to trade off the exploration and exploitation which was based on the present individual and its surrounding.
2. A productive ensemble method was established that can assist the algorithm in the extraction of native complexity details from the expression data of the cancer gene. In this ensemble method, the two advanced search methods were modified in order to search highly qualified possible solution.
3. A novel binary encoding method was suggested for the selection of gene subsets in order to calculate the fitness of possible solutions.

To judge the productivity of MOECSA in comparison with the other multi-objective algorithms, already existing six futuristic algorithms which were multi-objective differential evolution (MODE) (Sikdar et al. 2015), region-based Pareto envelope-based selection algorithm (PESA-II) (Corne et al. 2001), grid-based evolutionary algorithm (GrEA) (Yang et al. 2013), non-dominated sorting genetic algorithm II (NSGA-II) (Deb et al. 2002), hypervolume-based algorithm (HypE) (Bader and Zitzler 2011), and multi-objective particle swarm optimization (MOPSO) (Moore et al. 2000) were used on expression datasets of thirty-five standard cancer gene. It was observed that MOECSA produced the best solutions while the worst solutions were produced by NSGA-II.

To judge the effectiveness of MOECSA in the classification of expression data of cancer genes into different subtypes, seven classification algorithms were utilized such as support vector machine (SVM) (Moustakidis et al. 2012), Adaboosting (Rojas 2009), K-nearest neighbors algorithm (KNN) (Denoeux 1995), artificial neural network (ANN) (Xue et al. 2016), extreme learning machine (ELM) (Huang et al. 2006), and generalized learning vector quantization classification (GRLVQ) (Gonzalez et al. 1995) were used for comparison. Thirty-five expression datasets of cancer genes were used for the test procedure. Each algorithm was allowed to run thirty times on each dataset independently. It had been observed that MOECSA produced the best results for most of the datasets in comparison with the other classification algorithms.

3 MicroRNA-Disease Association Prediction Method-Tuned with Surrogate Model Assisted EA

MicroRNA-disease association prediction method (MiRAI) is based on distributional semantics for predicting connections between miRNAs and diseases whose parameters are modified using an evolutionary approach. Vectors from the constructed vector space model were processed using latent semantic analysis (LSA). With the help of singular value decomposition (SVD), the original matrix was broken into matrices based on the row and column elements as vectors of arbitrary size and equal size, respectively. The SVD also helped in reducing the dimension of the original matrix while maintaining a similar structure. Different matrices represented miRNA as rows and various sources of data, such as target genes, associated disorders, miRNA family, genomic location, and scientific papers as columns. The product of ith and jth elements of all the matrices was equal to 1 if the miRNA at position i targets the corresponding column characteristic at position j or else 0. In terms of vector similarities, the MiRAI technique benefited in uncovering information between miRNAs and diseases. If a miRNA relates to a term, it is also related to all the subsumed terms, which might aid in inferring diseases. A miRNA's closeness to other miRNAs is also important as it necessitates the consideration of a weighting method that will aid in the comprehension of the coexpression pattern. The dimension of the reduced space was the final parameter that affected the MiRAI. The 12 binary parameters that regulated the MiRAI method containing real values and an integer were converted in the binary space.

3.1 MiRAI Tuned with Surrogate Model

The area under the ROC curve (AUC) was used to determine the method's accuracy. The AUC was maximized by utilizing the Gaussian process (GP) as the surrogate model that assisted the evolutionary algorithm (EA) to help identify promising areas of the search space while minimizing computational costs significantly (Pallez et al. 2017; Pasquier and Gardès 2016). The Gaussian process (Buche et al. 2005) opted as the surrogate model because it could estimate any function, could anticipate the mean and standard deviation of any new individual's fitness score, and could feature a modest number of hyperparameters that can be specified by the user or by an optimizer unlike artificial neural network (ANN). The EA could be hybridized with a surrogate model through evolution control, which regulated the number of people as a real fitness function and others as a model. The other method of hybridization was to optimize the model using EA. The resulting optima were then re-evaluated using the true fitness function, and the surrogate model was updated accordingly.

One disadvantage of the model is the computational cost of learning, forecasting the standard deviation, and mean of a new individual's fitness value. After getting the value of EA as optimum in the surrogate model, the evaluation of the individual of the population is determined and analyzed using the real fitness function, and late the surrogate model is further updated as compared to the previous real evaluations. The algorithm used for tuning MiRAI framework using the Gaussian process is referred to as the parallel GP model-assisted binary differential evolution (PGPABDE). On the basis of the number of evaluations that considered the expensive fitness function, the PGPABDE model was compared with parallel binary differential evolution (PBDE) which lacked a surrogate model. Both algorithms were run 9 times and used over 18 nodes (288 cores) over the course of 11 days. The best individual EA population utilizing the PGPABDE algorithm gave a score of 0.895 after 1000 real evaluations, whereas the PBDE algorithm required over 2000 real evaluations. The overall best MiRAI configuration using the PGPABDE algorithm received a score of 0.90 after 1750 real evaluations, whereas the best configuration using the PBDE algorithm received a score of 8.89 after 2000 real evaluations.

Different steps are performed to divide the dataset such as the fivefold cross-validation step that divides a specific disease dataset into five equal-sized subsets. Later, the validation process with repeats of 5 times was performed to predict disease-miRNA associations. The higher the ranking of miRNAs related to certain diseases, the better the performance. The PGPABDE-tuned MiRAI was tested on a set of 83 human diseases from the HMDD that were related to almost 20 miRNAs and on a set of 15 diseases that were connected to the largest number of miRNAs. The average AUC value found was 0.897 and 0.880 for the collection of 83 and 15 diseases, respectively. The R-precision was 0.625 and 0.691 when considering a set of 83 diseases and 15 diseases, respectively. In comparison with different methods of MiRAI. The performance was found to be slightly better than all the other compared methods. More significant results were obtained when compared with the results of all 83 diseases. Even though MiRAI is optimized with EA, it is still outperformed by some of the new approaches, such as NCPMDA (AUC score 0.9173), PBMDA (AUC score 0.9172), and ensemble-based method (AUC score 0.9226).

Regardless of this, identification of putative false associations as possible in about 86 miRNA and disease associations by characterizing highly distant vectors in HMDD. Each association's connected articles were checked manually of which 54 of them were identified prior. Among checking the 126 newly found associations between miRNA and diseases, most of them implied that mir-188 and mir-765 are linked to many diseases. The two miRNAs have been connected to a variety of major biological processes and pathways, including cell cycle, phosphorylation cascades, and epigenetics. Mir-188 is reported to be a tumor suppressor miRNA, particularly in the case of human lung cancer, and is also known to play role in various diseases such as Alzheimer's disease, azoospermia, breast cancer, and prostate cancer (Pallez et al. 2017; Pasquier and Gardès 2016). Whereas, mir-765 is reported to be an oncogene, particularly in the case of human lung cancer. It is reported to alter the MAPK pathway involved in cell cycle regulation. After getting the output along with the analysis report, a connection link is identified between mir-188 and mir-765 with various diseases, including neural disorders, cardiovascular diseases, and autoimmune diseases (Xie et al. 2016).

3.2 Applications and Future Prospective

MiRNA dysregulation has been observed in a variety of diseases, including cancer and neurological disorders. The use of MiRAI on human miRNA-disease database (HMDD) data revealed potential novel connections between miRNAs and diseases. MiRAI has the potential to identify different areas of the search space while minimizing computational costs significantly by using evolutionary computation. Of 126 putative associations, 72 of them imply mir-188 and mir-765 with several diseases. More laboratory research is required to confirm these theories. Several miRNAs have been discovered as tumor suppressors, thereby playing important role in various diseases. In contrast to existing cancer therapies, which focus on one or two oncogenes, miRNAs are known to focus on several genes. MiRAI increases the probability of discovering novel connections by assisting scientists in focusing on high-potential targets. Increased research on miRNAs will assuredly lead to the development of microRNA-based biotechnology that will pave way for advancements in human health applications (Pallez et al. 2017; Pasquier and Gardès 2016; Buche et al. 2005; Zhao et al. 2018; Xie et al. 2016).

4 Nature-Inspired Computing Models Toward Accurate Detection of COVID-19

Computational approaches could be applied to provide deep insights along with prospects of different models for COVID-19 contact tracing. These methods are included artificial intelligence, big data, and NIC methods. The aim of the development of new models for accurate case detection and contact tracing is to try and keep a rough, real-time view of the rate of new infections, and hot spots in high-risk areas of a given area, which would prove valuable for the allocation of health care and appropriate response aids to that area. This problem is especially important for countries that have yet to implement an effective data infrastructure which is the case for poorer countries, which may suffer from a lack of self-test kit availability (Agbehadji et al. 2020).

The use of big data analytics platforms to allow for the rapid analysis of unprocessed data from a variety of data sources into a data architecture system that is readily available and understandable is an important conduit to provide a foundation to store and process the data for contact tracing. The current big data leaders implement machine learning models whose purpose is monitoring and prediction of events which can encapsulate information from medical imagery to patient data records (Mohamed et al. 2019). For contact tracing, big data platforms can process the data extracted from different sources to trace an infected person's social historical contact to provide a tracking model. The advantage of big data platforms manifests itself in the 5Vs which are: volume, veracity, value, variety, and velocity (Longbottom and Bamforth 2013). Volume refers to the large quantity of data that can be processed. Veracity is essentially the accuracy of the results. Value is the value of the information which is gained. Variety is the different kinds of data that can be inputted/outputted. Velocity is how quickly the data can be processed, updated, and created. All these data can help in contact tracing of COVID-19. The problem is that human social contact is stochastic in nature and is far too varied for any reliable data collector to provide a fully accurate estimation of contact data. Such an example of a data platform is Apache Spark which processes all the data in large volumes in a short amount of time in streams. An exemplar implementation of big data platforms for contact tracing was attempted by Qin et al. (2020), who used keyword occurrences on social media pertaining to COVID symptoms such as 'cough' and 'fever' in a method called social media search indexes (SMSI). However, such attempts to use social media posts for big data platforms are the unreliability and inaccuracy of the information, leading to misleading output data (Keeling et al. 2020).

Closely intertwined with big data methods for contact tracing are artificial intelligence methods. This also includes deep learning and machine learning models which are prevalent in many intelligence systems around the world (Salah et al. 2019). Machine learning encompasses the meaningful identification and application of patterns in data which depends a lot on the representative features of the presented data, whereas deep learning is machine learning algorithms that search several more layers of the data to find more complex and inherent patterns in the data and use backpropagation to optimize the model through parameter learning, with applications such as predicting the effects of drug interaction and non-coding mutations in DNA on gene expression (Leung et al. 2014). Rahmatizadeh et al. (2020), ELGhamrawy and Hassanien (2020) have proposed AI models for the purpose of identifying COVID-caused pneumonia from CT scans and X-ray images with accuracies such as 99.56% on their test set and all classes at 91.4%. Algorithms such as convolutional neural network (CNN), recurrent neural networks (RNN), and long short-term memory (LSTM) are very accurate, especially when trained with a large dataset, which then can be used for image classification (such as medical imagery) and even predict the epidemic spread of COVID-19. Even though the deep learning models that are under use for COVID purposes are extensions of current ML models, they can be applied for feature extraction without human input which may enable a deeper understanding of more complex forms of input data. The key comparison between ML and DL models for image classification and identification ranges from 85 to 98.3% with DL approaches and 79.3–95% with ML approaches. Although from these percentages one would be inclined to use deep learning algorithms, the implementation of data for use in DL models can be quite particular and not worth the slight increase in accuracy in comparison with well-established machine learning models. The limitation of AI-based methods reduces from over-fitting and under-fitting of data when training the model. The advantages of deep learning in the context of contact tracing lie in its recent advances that make the analysis of multi-dimensional data better. Current AI usages for contact tracing include applications that are in use in Singapore, South Korea, and Europe which combine Bluetooth contact, as well as information from public space surveillance monitors, to create a community-derived application that allows a level of privacy in local detection of COVID cases.

NIC is a wide field that computing with other scientific domains such as biochemistry, physics, and engineering. NIC models are optimized and are inspired by the behaviors found in nature and take from an organism's built-in capability to find the easiest and most effective solution to a problem. The most common NIC branches include evolutionary and swarm intelligence-based methods. Examples of algorithms include the ant colony algorithm (ACO), artificial bee colony (ABC), and BAT algorithms. Similar algorithms have been implemented for the application of diagnosing and predicting COVID-19 from X-ray image segmentation features. Interestingly, NIC methods are under review to provide performance optimizations to current existing AI-based algorithms for the detection accuracy of pandemic cases. Although fascinating in concept, NIC has much fewer use cases in comparison with the previously mentioned approaches and their framework makes it so that they are usually used as an addition to other methods rather than being a method of itself.

Being 2 years deep into the COVID-19 pandemic has shown the innovation and possibilities that the three outlined methods have to offer. The most hopeful programs that can be applied to contact tracing involve a combination of two or more. The contact tracing big data analytics model uses the features of NIC and AI-based models to select and filter relevant features that can be used to predict contact spread, with the help of fine-tuning random parameters and optimization that comes with NIC approaches and the power of artificial intelligence such as the whale optimization algorithm. The flock by leader algorithm can be applied to proximity detection to trace contacts based on the hierarchical representative nature that leaders can be traced to their followers, in the way that an infected person is followed by those who have been in close contact. Such intuitive approaches to solving the contact tracing problem, if proven to be fully effective and implementable in poorer environments, can bear fruit in helping to curb the spread of pandemics.

5 Role of PSO Techniques in Medical Disease Diagnosis

Data mining and processing, also known as knowledge discovery in data (KDD), is conceptualized as the process of extracting different patterns, values, and other useful information from massive datasets in a database or available resources. Increased data volumes prompted the development of data warehousing, which in turn spurred the adoption of data mining techniques.

When it comes to diagnosing diseases, large patient datasets find the application of computational intelligence that employs state-of-the-art algorithms in detecting a particular ailment. Such data processing and prediction algorithms include genetic algorithm (GA), bat algorithm (BA), and particle swarm optimization (PSO), in combination with data mining tools like decision tree and artificial neural networks (ANNs). Concerning the classifications of medical datasets having several variables that affect disease, a pattern is recognized to detect or diagnose whether a person is actually a patient or not. Modeling and optimization techniques like linear programming and ANNs are a few classification models that help both doctors and patients to become understand the biomarkers and attributes that have been considered for the classification purpose. One downside of such classification models is, however, the lack of explanation of the decision methods and the inability to analyze the important attributes. In order to solve this problem, scientists have used the PSO algorithm paired with various other approaches, such as a random forest with PSO and k-nearest neighbor with PSO. PSO has been found to provide better results than GA in terms of each criterion's fitness function value. PSO and its improved approaches, namely C4.5, Naïve Bayes, SVM, and NN, are currently employed to detect heart disease, hepatitis, Parkinson's, thyroid, breast cancer, and lung cancer. This algorithm is useful in deciding support systems for diagnosing the disease, assisting doctors to determine the right treatment, and caring for their patients at the right time (Pervaiz et al. 2021; Suamanda and Wibisono 2020). Moreover, it has also helped in determining disease prevention methods in the short and long terms.

5.1 The Technique—Particle Swarm Optimization (PSO)

The PSO is a bio-inspired swarm-based intelligent stochastic search technique inspired by the natural behavior of a swarm of bees or a flock of birds looking for food. Kennedy and Eberhart invented this iterative technique in 1995, and it is used to find the suitable and most preferable solution in the solution space (Tam 2021). PSO is the best fit when it comes to determining the maximum or minimum of a function specified on a multi-dimensional vector space. To better understand this, please follow Fig. 3 where, we start with several random points or particles on the plane and enable them to hunt for the smallest point in random directions, analogous to a flock of birds looking for food. Every particle should seek in each direction around the lowest position it has ever found (personal best), as well as the lowest point found by the entire swarm of particles (global best). This is an iterative process, and after a few cycles, the minimum point of the function is regarded as the swarm of particles' lowest place ever discovered. The PSO algorithm is summarized in the flowchart shown in Fig. 3 (Wang et al. 2017).

Fig. 3 Flowchart of the particle swarm optimization (PSO) algorithm

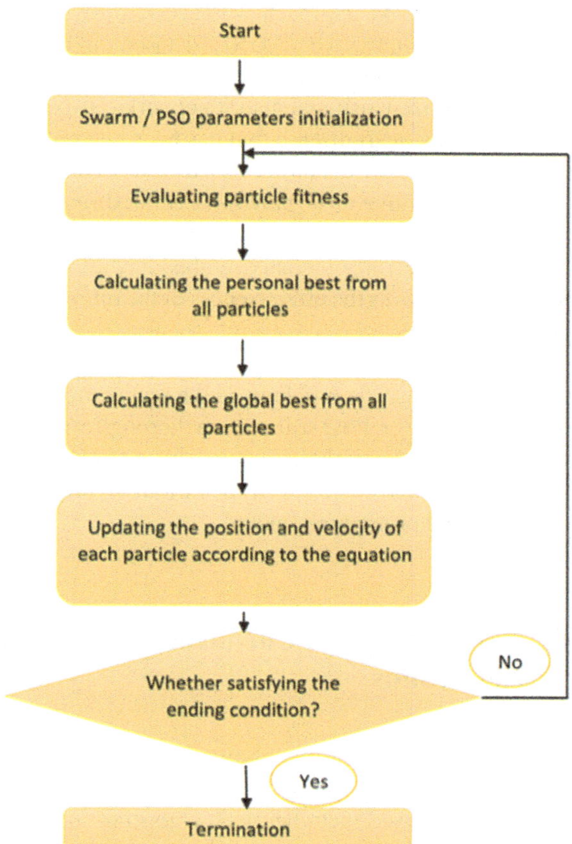

5.2 PSO in Disease Identification

A novel PSO and artificial neural network (PSO-ANN)-based model for early-stage diagnosis of dengue illness is created, which uses PSO to optimize the bias and weight factors of the ANN method. The findings of the proposed model are compared to other standard approaches such as decision tree (DT), PSO, Naive Bayes (NB), and ANN based on characteristics such as sensitivity, error rate, accuracy, specificity, and area under the curve (AUC) parameters, proving PSO-ANN to be superior to others (Tam 2021).

In the instance of Alzheimer's disease (AD), a unique method based on magnetic resonance imaging (MRI) image pre-processing, principal component analysis, feature extraction, and the SVM model has been developed and named as switching delayed particle swarm optimization–support vector machine (SDPSO-SVM). The new SDPSO-SVM approach was successfully tested on the Alzheimer's disease Neuroimaging Initiative (ADNI) dataset, and it achieved greater classification accuracy on 6 standard cases (Tam 2021; Wang et al. 2017).

The optimized artificial neural network (OANN) is an enhanced ANN approach for diagnosing heart disease that has been suggested and tested on medical datasets relating to the disease. A filter-based ANN is utilized here for binary disease classification (as positive or negative) based on disease features. When the performance plot, ROC values, confusion matrix, and regression of the OANN and the traditional approaches are compared, embedding the suggested PSO with ANN for feature reduction improves the effectiveness of the introduced model (Tam 2021; Wang et al. 2017).

As a result, we can see that PSO has been widely used to address nonlinear difficult optimization problems in the real world, including those involving medical datasets. However, PSO still needs a thorough evaluation to improve its performance. Researchers in the healthcare domain can apply upgraded forms of neural networks with PSO by-and-by, as well as various hybridized versions of PSO for disease diagnosis and prompt treatment.

6 Nature-Inspired Chemical Reaction Optimization Algorithms

A meta-heuristic algorithm is an advanced approach for finding good solutions to a complex problem. It can be classified as follows (Fig. 4). Lam and Li proposed a population-based meta-heuristic algorithm for chemical reaction optimization. It describes what happens to molecules during a chemical reaction and mimics the interaction between molecules in a chemical reaction to reach a low stable energy state. Certain criteria are required to be fulfilled for designing a meta-heuristic algorithm. These are (a) exploration of the search scope and (b) exploitation of the best

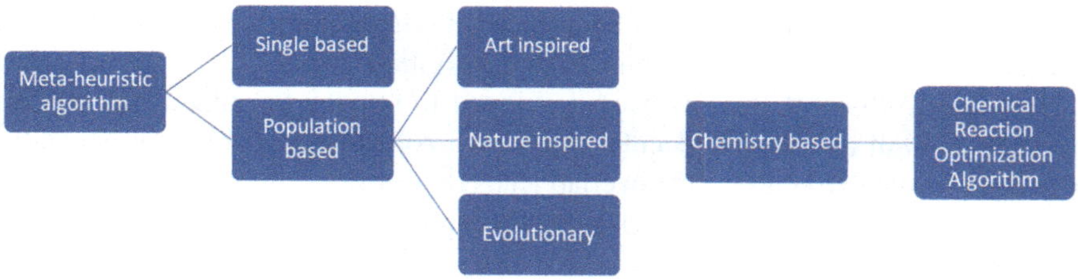

Fig. 4 Flow of meta-heuristic algorithm

solution found (Siddique and Adeli 2017; Brückner and Ckner 2004; Yu et al. 2011; Bolourchi et al. 2015; Reyes et al. 2014).

The elementary reactions serve as an operator in chemical reaction operation (CRO). Chemical reactions are classified into four types:

(a) *Uni-molecular reactions*

  (i) On-wall ineffective collision
  (ii) Decomposition

(b) *Multi-molecular reactions*

  (i) Inter-molecular ineffective collision
  (ii) Synthesis

6.1 Variants of CRO

The use of different CRO algorithms in various fields of application is seen in recent days. Many modifications have also been introduced which came to be known as variants of CRO, viz. real-coded CRO, orthogonal CRO, and opposition-based CRO.

CRO put emphasis on solving discrete optimization problems. Lam et al. put forward an algorithm for continuous problems, which is known as real-coded CRO. In a standard CRO solution, an optimization problem is expressed by a molecular structure where each individual is an integer, whereas in real-coded CRO each individual can be represented as a floating-point number. For making it a continuous problem continuous search ability is included in the neighboring search operator.

When an optimization problem includes a huge number of decision variables, it becomes difficult for standard CRO, as it screens the search space in a random manner. The orthogonal CRO algorithm finds a way better amalgamation of different factors within a small number of experiments. The efficacy of orthogonal CRO has been tested on 23 uni-modal, high, and low-dimensional multi-modal functions.

Tizhoosh suggested the concept of opposition-based CRO. The idea of opposition-based CRO is that the opposite set of numbers has a greater possibility of satisfying a problem than a set of randomly generated numbers. Computational potency is found to improve in an opposition-based approach. The efficiency of opposition-based CRO has been established on short-term hydrothermal scheduling problems. It enhances the speed of convergence in the optimization algorithms.

Adaptive collision CRO has been developed to remove the functional overlap between on-wall and inter-molecular ineffective collision operators that generally is present in standard CRO. This overlap leads to needless computation times. The efficacy of such CRO has been experimentally confirmed on 16 benchmark functions (Lam and Li 2010; Lin and Ku 2013; Lam et al. 2012; Bhattacharjee et al. 2014a, b; Tizhoosh 2005; Montgomery 2008; Ho et al. 2004; Jung et al. 2016; Yu et al. 2015).

There are a few other algorithms that are inspired by chemistry. On the basis of artificial chemical processes, an artificial chemical process algorithm has been developed. From the principles of uni-molecular and bimolecular chemical reactions, artificial chemical reaction optimization algorithm has been developed. Depending on the laws of Brownian motion gases Brownian motion optimization has evolved. Still, certain areas like parameter selection, convergence speed, and searching behavior are some principal issues that need more research.

7 Cuckoo Search Method for Drug Design and Discovery in Chemoinformatics Using Hybrid Harris Hawks' Optimization

The ability to anticipate and study the structures of molecules is critical in cheminformatics. When you explore the databases, however, you will see that there is a lot of information in the datasets and that not all of it is relevant to drug design and discovery. Therefore, to predict and analyze the structure of the molecules, the traditional methods like protein bank database and quantum mechanical calculations are not sufficient. To overcome this, a hybrid method is introduced, called **CHHO–CS**. The method CHHO-CS is a hybrid meta-heuristic algorithm where Harris hawk's optimizer (HHO) (Heidari et al. 2019) combines with two operators: one of them is the cuckoo search (CS) (Yang and Deb 2009) and another one is chaotic maps.

The chaotic maps are used to avoid premature convergence as well as slipping into a local optimum. This operator basically updates the control energy parameters. On the other hand, CS's job is to manage the HHO algorithm's main position vectors to balance the exploration and exploitation phases.

The primary idea of the CHHO-CS strategy is to enhance the HHO searching process that helps to find near-optimal solutions. This approach is examined and backtested with diverse datasets to verify the prevalence of the proposed CHHO–CS approach.

The impact of chaotic maps

Incorporating chaos theory into the random search process of MAs boosts the random search effect significantly. Because based on the randomness of chaotic local search, MAs can avoid slipping into local optimums and premature convergence.

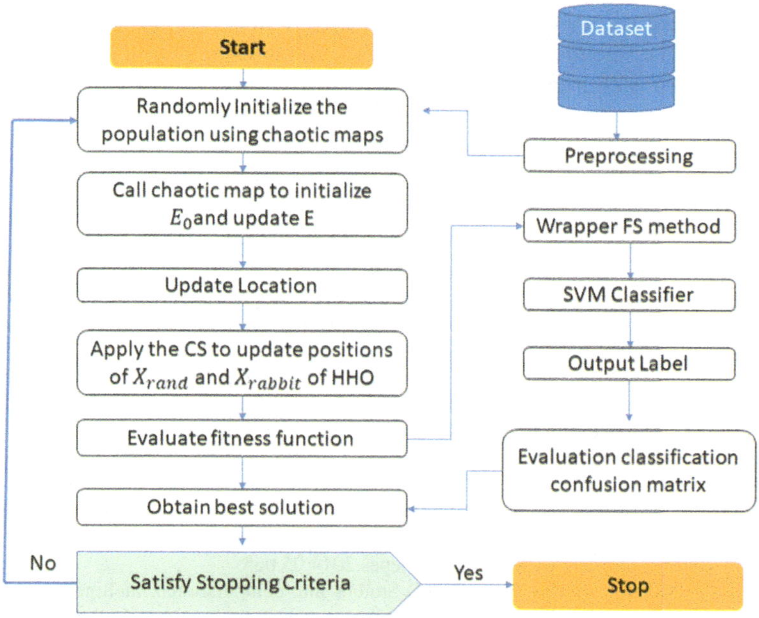

Fig. 5 Flowchart of the proposed CHHO–CS method

The impact of CS

The local vectors in the basic HHO are responsible for balancing exploitation and exploration. Position vectors with larger values speed global exploration, whereas smaller values expedite exploitation. As a result, a suitable choice of local vectors should be determined to achieve a consistent equilibrium between local exploitation and global exploration. As a result, we use the merits CS approach to control the HHO position vectors in the CHHO–CS algorithm. CS tries to discover a better solution after each iteration T. A simple flowchart of the CHHO–CS algorithm is given below (Fig. 5).

This method has a wide range of applications; for example, you will be able to predict the number of iterations and the fitness function, crystal protein structure analysis, analysis of protein database, and more. This method always gives better results as compared to the competitor algorithms.

8 Conclusion

This book chapter provides a deep insight into using nature-inspired intelligent computing tools for drug designing, drug development, and application in therapeutics. These methods escalate the number of druggable compounds prediction within a short time and reduce cost and time. We believe that our book chapter provides knowledge to the scientist working in the pharma industry to implement NIC-based technology for faster drug design and development.

References

Agbehadji IE, Awuzie BO, Ngowi AB, Millham RC (2020) Review of big data analytics, artificial intelligence and nature-inspired computing models towards accurate detection of COVID-19 pandemic cases and contact tracing. Int J Environ Res Public Health 17(15):5330. https://doi.org/10.3390/ijerph17155330

Bader J, Zitzler E (2011) HypE: an algorithm for fast hypervolume-based many-objective optimization. Evol Comput 19(1):45–76. https://doi.org/10.1162/evco_a_00009

Bhattacharjee K, Bhattacharya A, Dey SH (2014a) Real coded chemical reaction based optimization for short-term hydrothermal scheduling. Appl Soft Comput 24:962–976. https://doi.org/10.1016/j.asoc.2014.08.048

Bhattacharjee K, Bhattacharya A, Dey SH (2014b) Oppositional real coded chemical reaction based optimization to solve short-term hydrothermal scheduling problems. Int J Electr Power Energy Syst 63:145–157. https://doi.org/10.1016/j.ijepes.2014.05.065

Bolourchi AA, Masri S, Aldraihem O (2015) Studies into computational intelligence and evolutionary approaches for model-free identification of hysteretic systems. Comput Aided Civil Infrastruct Eng 30(5):330–346. https://doi.org/10.1111/mice.12126

Brückner R, Ckner RB (2004) Reaktionsmechanismen: Organische Reaktionen, Stereochemie, Moderne Synthesemethoden (in German), 3rd edn. Spektrum Akademischer Verlag

Buche D, Schraudolph NN, Koumoutsakos P (2005) Accelerating evolutionary algorithms with Gaussian process fitness function models. IEEE Trans Syst Man Cybern Part C (Appl Rev) 35(2):183–194. https://doi.org/10.1109/TSMCC.2004.841917

Corne D, Jerram N, Knowles J, Oates M (2001) PESA-II: region-based selection in evolutionary multiobjective optimization. In: Proceedings of 6th international conference parallel problem solving from nature PPSN-VI.

Deb K, Pratap A, Agarwal S, Meyarivan T (2002) A fast and elitist multiobjective genetic algorithm: NSGA-II. IEEE Trans Evol Comput 6(2):182–197. https://doi.org/10.1109/4235.996017

Denoeux T (1995) A k-nearest neighbor classification rule based on Dempster-Shafer theory. IEEE Trans Syst Man Cybern 25(5):804–813. https://doi.org/10.1109/21.376493

Elghamrawy SM, Hassanien AE (2020) Diagnosis and prediction model for COVID19 patients response to treatment basedon convolutional neural networks and whale optimization algorithm using CT images. MedRxiv, 1–23. https://doi.org/10.1101/2020.04.16.20063990

Gonzalez AI, Grana M, D'Anjou A (1995) An analysis of the GLVQ algorithm. IEEE Trans Neural Netw 6(4):1012–1016. https://doi.org/10.1109/72.392266

Heidari AA, Mirjalili S, Faris H, Aljarah I, Mafarja M, Chen H (2019) Harris hawks optimization: algorithm and applications. Future Gener Comput Syst 97:849–872. https://doi.org/10.1016/j.future.2019.02.028

Ho S-J, Ho S-Y, Shu L-S (2004) OSA: orthogonal simulated annealing algorithm and its application to designing mixed $rm H_2 /rm H_infty$ optimal controllers. IEEE Trans Syst Man Cybern Part A Syst Hum 34(5):588–600. https://doi.org/10.1109/TSMCA.2004.832834

Huang G-B, Zhu Q-Y, Siew C-K (2006) Extreme learning machine: theory and applications. Neurocomputing 70(1–3):489–501. https://doi.org/10.1016/j.neucom.2005.12.126

Jung J, Jayakrishnan R, Park JY (2016) Dynamic shared-taxi dispatch algorithm with hybrid simulated annealing. Comput Aided Civil Infrastruct Eng 31(4):275–291. https://doi.org/10.1111/mice.12157

Keeling MJ, Hollingsworth TD, Read JM (2020) Efficacy of contact tracing for the containment of the 2019 novel coronavirus (COVID-19). J Epidemiol Community Health 74(10):861–866. https://doi.org/10.1136/jech-2020-214051

Lam AYS, Li VOK (2010) Chemical-reaction-inspired meta-heuristic for optimization. IEEE Trans Evol Comput 14(3):381–399. https://doi.org/10.1109/TEVC.2009.2033580

Lam AYS, Li VOK, Yu JJQ (2012) Real-coded chemical reaction optimization. IEEE Trans Evol Comput 16(3):339–353. https://doi.org/10.1109/TEVC.2011.2161091

Leung MK, Xiong HY, Lee LJ, Frey BJ (2014) Deep learning of the tissue-regulated splicing code. Bioinformatics (oxford, England) 30(12):i121–i129. https://doi.org/10.1093/bioinformatics/btu277

Lin DY, Ku YH (2013) Using genetic algorithms to optimize stopping patterns for passenger rail transportation. Comput Aided Civil Infrastruct Eng 29(4):264–278. https://doi.org/10.1111/mice.12020

Longbottom C, Bamforth R (2013) Optimising the data warehouse: dealing with large volumes of mixed data to givebetter business insights. Quocirca: Maidenhead

Mohamed AH, Najafabadi MK, Yap BW, Kamaru-Zaman EA, Maskat R (2019) The state of the art and taxonomy of big data analytics: view from new big data framework. Artif Intell Rev 53:989–1037

Montgomery DC (2008) Design and analysis of experiments. Wiley

Moore J, Chapman R, Dozier G (2000) Multiobjective particle swarm optimization. In: Proceedings of the 38th annual on southeast regional conference, Clemson, South Carolina. ACM Press the 38th annual, pp 56–57

Moustakidis S, Mallinis G, Koutsias N, Theocharis JB, Petridis V (2012) SVM-based fuzzy decision trees for classification of high spatial resolution remote sensing images. IEEE Trans Geosci Remote Sens 50(1):149–169. https://doi.org/10.1109/TGRS.2011.2159726

Pallez D, Gardès J, Pasquier C (2017) Prediction of miRNA-disease associations using an evolutionary tuned latent semantic analysis. Sci Rep 7(1):10548. https://doi.org/10.1038/s41598-017-10065-y

Pasquier C, Gardès J (2016) Prediction of miRNA-disease associations with a vector space model. Sci Rep 6(1):27036. https://doi.org/10.1038/srep27036

Pervaiz S, Ul-Qayyum Z, Bangyal WH, Gao L, Ahmad J (2021) A systematic literature review on particle swarm optimization techniques for medical diseases detection. Comput Math Methods Med 2021:5990999. https://doi.org/10.1155/2021/5990999

Qin L, Sun Q, Wang Y, Wu KF, Chen M, Shia BC, Wu SY (2020) Prediction of number of cases of 2019 novel coronavirus (COVID-19) using social media search index. Int J Environ Res Public Health 17(7):2365. https://doi.org/10.3390/ijerph17072365

Rahmatizadeh S, Valizadeh-Haghi S, Dabbagh A (2020) The role of artificial intelligence in management of critical COVID-19 patients. J Cell Mol Anesth 5:16–22. https://doi.org/10.22037/jcma.v5i1.29752

Reyes O, Morell C, Ventura S (2014) Evolutionary feature weighting to improve the performance of multi-label lazy algorithms. Integr Comput Aided Eng 21(4):339–354. https://doi.org/10.3233/ICA-140468

Rojas R (2009) AdaBoost and the super bowl of classifiers a tutorial introduction to adaptive boosting

Salah K, Rehman HU, Nizamuddin N, Al-Faquha A (2019) Blockchain for AI: review and open researchchallenges. IEEE Access 7:10127–10149. https://doi.org/10.1109/ACCESS.2018.2890507

Siddique N, Adeli H (2017) Nature-inspired chemical reaction optimisation algorithms. Cogn Comput 9(4):411–422. https://doi.org/10.1007/s12559-017-9485-1

Sikdar UK, Ekbal A, Saha S (2015) MODE: multiobjective differential evolution for feature selection and classifier ensemble. Soft Comput 19(12):3529–3549. https://doi.org/10.1007/s00500-014-1565-5

Suamanda IN, Wibisono IS (2020) Particle swarm optimization for improved accuracy of disease diagnosis. J Appl Intell Syst 5(2):57–68. https://doi.org/10.33633/JAIS.V5I2.4242

Tam A (2021) A gentle introduction to particle swarm optimization. Machine Learning Mastery. https://machinelearningmastery.com/a-gentle-introduction-to-particle-swarm-optimization/

Tizhoosh HR (2005) Opposition-based learning: a new scheme for machine intelligence. In: International conference on computational intelligence for modelling, control and automation and international conference on intelligent agents, web technologies and internet commerce (CIMCA-IAWTIC'06), pp 695–701. https://doi.org/10.1109/CIMCA.2005.1631345

Wang D, Tan D, Liu L (2017) Particle swarm optimization algorithm: an overview. Soft Comput 22(2):387–408. https://doi.org/10.1007/s00500-016-2474-6

Wang Y, Liu B, Ma Z, Wong KC, Li X (2019) Nature-inspired multiobjective cancer subtype diagnosis. IEEE J Trans Eng Health Med 7:4300112. https://doi.org/10.1109/JTEHM.2019.2891746

Xie B-H, He X, Hua R-X, Zhang B, Tan G-S, Xiong S-Q, Liu L-S, Chen W, Yang J-Y, Wang X-N, Li H-P (2016) Mir-765 promotes cell proliferation by downregulating INPP4B expression in human hepatocellular carcinoma. Cancer Biomark 16(3):405–413. https://doi.org/10.3233/CBM-160579

Xue B, Zhang M, Browne WN, & Yao X (2016) A survey on evolutionary computation approaches to feature selection. IEEE Trans Evol Comput 20(4):606–626. https://doi.org/10.1109/TEVC.2015.2504420

Yang X-S, Deb S (2009) Cuckoo Search via Lévy flights. In: 2009 World Congress on Nature & Biologically Inspired Computing (NaBIC), pp 210–214. https://doi.org/10.1109/NABIC.2009.5393690

Yang S, Li M, Liu X, Zheng J (2013) A grid-based evolutionary algorithm for many-objective optimization. IEEE Trans Evol Comput 17(5):721–736. https://doi.org/10.1109/TEVC.2012.2227145

Yu JJQ, Lam A, Li V (2011) Evolutionary artificial neural network based on chemical reaction optimization. In: 2011 IEEE congress of evolutionary computation, CEC 2011, pp 2083–2090.https://doi.org/10.1109/CEC.2011.5949872

Yu JJQ, Lam AYS, Li VOK (2015) Adaptive chemical reaction optimization for global optimisation. In: 2015 IEEE congress on evolutionary computation, pp 3192–3199

Zhao L, Ni X, Zhao L, Zhang Y, Jin D, Yin W, Wang D, Zhang W (2018) MiroRNA-188 acts as tumor suppressor in non-small-cell lung cancer by targeting MAP3K3. Mol Pharm 15(4):1682–1689. https://doi.org/10.1021/acs.molpharmaceut.8b00071

Illustrious Implications of Nature-Inspired Computing Methods in Therapeutics and Computer-Aided Drug Design

Shaban Ahmad, Kayenat Sheikh, Nagmi Bano, Misbahuddin M. Rafeeq, Mohammed Razeeth Shait Mohammed, Manoj Kumar Yadav, and Khalid Raza

1 Introduction

Living organisms showcase extraordinarily state-of-the-art learning, choice-making, and processing abilities that allow them to live to tell the tale and proliferate. Nature has continually served as an idea for numerous medical and technological developments. In computer technological know-how, synchronisation, parallelisation, distributiveness, versatility, heartiness, flexibility, sensibility, overt repetitiveness,

S. Ahmad · K. Sheikh · N. Bano · K. Raza (✉)
Department of Computer Science, Jamia Millia Islamia, New Delhi 110025, India
e-mail: kraza@jmi.ac.in

S. Ahmad
e-mail: Shaban184343@st.jmi.ac.in

K. Sheikh
e-mail: Kayenat1903945@st.jmi.ac.in

N. Bano
e-mail: Nagmi2001337@st.jmi.ac.in

M. M. Rafeeq
Department of Pharmacology, Faculty of Medicine, King Abdulaziz University, Jeddah 21589, Kingdom of Saudi Arabia
e-mail: marafeeq@kau.edu.sa

M. R. S. Mohammed
Moores Cancer Center, University of California, San Diego, 3855 Health Sciences Dr, La Jolla, CA 92037, USA
e-mail: Mshaitmohammed@ucsd.edu

M. K. Yadav
Department of Biomedical Engineering, SRM University, Sonepat, Haryana, India
e-mail: Manoj.yadav@srmuniversity.ac.in

K. Raza (ed.), *Nature-Inspired Intelligent Computing Techniques in Bioinformatics*, Studies in Computational Intelligence 1066,
https://doi.org/10.1007/978-981-19-6379-7_15

collaboration, and coordination are the traits for software development. Nature has a similar, offbeat, decentralised, and aggregate way of behaving. The nature-inspired techniques are exceptional in shape for figuring conditions that feature those qualities. Disciplined medical and engineering investigations must be undertaken to efficiently switch those algorithms, strategies, and infrastructures into rising computing environments. These nature-stimulated strategies expand numerous algorithms to remedy seek and optimisation problems.

Nature-inspired computing (NIC) is an entirely new region that endeavours to increment new computational strategies through gazing at how happening phenomena behave to remedy complicated issues in numerous natural circumstances. This has delivered momentous studies that have made new branches, such as neural networks, swarm intelligence, evolutionary computation, and synthetic immunology. NIC techniques are used withinside physics, biology, engineering, economics, or event management. The natural system is a wealthy reasserts of ideas for inventing new sensible structures. Swarm intelligence is one of the clinical fields that might be primarily based on natural swarms, ant colonies, bee colonies, organisation behaviour of bacterium, and rivers (organisation of water drops). Evolutionary computation, neural networks, time adaptive self-organising maps, DNA computing electromagnetism-like optimisation, and sensible water drops are several the techniques which might be to a few quantities primarily based totally on or take an idea from naturally taking place phenomena.

The most recent example of NIC technique application in therapeutics was the COVID-19 pandemic. The health systems were kept on red alert as the virus grew exponentially globally. No drugs or vaccines were available to control the spread. Computer-aided drug design (CADD) was employed to identify drugs to treat COVID-19. According to a recent study, machine learning techniques such as complex neural networks (CNN) are much faster than typical molecular docking methods for drug discovery. Virtual screening of more than 100,000 compounds was carried out due to the computational intelligence methods. It is a valuable tool for fast-track drug discovery, developing and treating new diseases (Onawole et al. 2020). Let us take a deeper look into the NIC techniques' historical growth and wide applications.

2 Historical Background of NIC Techniques

2.1 Nature-Inspired Techniques in Ancient Therapeutics to Drug Designing

The present biomedical arena has a modern strategy to discover and develop the drug. CADD is a trending field that facilitates optimising, designing, and discovering new and successful curative drugs. Drug designing is a very complex and vast process. The drug is identified based on an excellent biological profile and further optimised, processed, and so on (Aarthy et al. 2017). However, drug design using computational methods did not develop over time. It was a long, gradual, and step-by-step process that took more than a few decades. Numerous NIC standards have arisen recently. They can be assembled into three expansive classes: physical-based calculations (PBA), chemistry-based calculations (CBA), and biology-based calculations (BBA). Let us discuss the developmental history of NIC techniques.

2.1.1 Physical-Based Calculations (PBA)

This look at nature has its beginning withinside the vestige. At some point in the Hellenic age, the three mathematicians who have become famous above the maximum in their predecessors and successors are Euclid, Archimedes and Apollonius Perga (Himabindu and Jyothi 2017). The twelfth-century Persian mathematician, astronomer, and geographer, Al-Khwarizmi delivered decimal point range gadgets and immediate orderly answers to straight and quadratic equations and was considered the unique inventor of algebra. The expression set of rules starts from his Latin name, Algoritmi. In Europe, major essential natural science, the mechanics of terrestrial and heavenly bodies become, evolved withinside the fifteenth century. The fundamental functions of many essential mathematical strategies have evolved within the sixteenth and seventeenth centuries, including logarithms using John Napier (1550–1617). Physics-inspired algorithms appoint fundamental ideas of physical science, for instance, Newton's laws, laws of motion, and Coulomb's law of electrical charge (Siddique and Adeli 2015).

The formula of Coulomb's law is

$$|Fe| = k|q_1 q_2 / r_2|$$

and the formula of laws of motion is

$$F = m.a$$

They are essentially founded on deterministic actual thoughts. These calculations might be sorted widely as follows:

- Inspired by Newton's law of motion.
- Inspired by Newton's gravitational force, gravitational search algorithm, central force optimisation, space gravitation optimisation, and gravitational interaction optimisation.
- Inspired by celestial mechanics, Big Bang crunch search, black hole search, galaxy-based search algorithm, artificial optimisation, and integrated radiation search.
- Inspired by electromagnetic.
- Inspired with acoustics.
- Inspired by thermodynamics.

Newton's law of gravitation: Newton published the law of universal gravity in 1687 (Newton 1990). As per the law of universal gravity, we are all attracted to bodies in the universe. The enchantment power between bodies is directly proportional to their mass and inversely proportional to the square of the distance between them. Current quantifiable technology found expression in unique methods over almost three centuries until the work by Einstein in the early hundred years. Newton's principle has so far had sufficient accuracy for maximum applications (Newton 1990).

Law of motion and Coulomb's law: Newton published his three laws in his book in 1687 (Newton 1990). The three physical laws laid the foundation for classical mechanics, which describes the relationship between a body and forces acting on it. And its movement in response to these forces. Laws include the law of inertia, the law of force, and the law of action and reaction. The first law of inertia states that a body at rest or moving in a straight line at a constant velocity is at rest or moving at a steady speed moving in the same direction. The second law of force states that the acceleration is directly proportional to force F and inversely proportional to the mass m of the body. The third law, otherwise called the law of action and reaction, expresses that each activity has an equivalent inverse response.

2.1.2 Chemistry-Based Calculations (CBA)

There are many miracles of science. The incredible strength of the jewel, the most costly stone, and the most challenging material come from carbon's tetragonal space structure. Plant leaves are green because of chlorophyll, a vital part of photosynthesis, which allows plants to utilise energy from daylight to change over water and carbon dioxide into sugars (Zekri and Obreza 2003; Mikkilä-Erdmann 2001). Flames, which contain CO_2, water vapour, O, and N, provide heat and light. These are basic examples of substance reactions that occur in the natural world. There can be no chemical compounds at the planet's birth at a temperature as high as the sun's temperature. As they chilled off, the transaction of the actual types of development changed. Continuous, he arrived where compound fondness started to develop. Elements that were previously chemically indifferent were successively chemically differentiated. They have acquired chemical properties and combined. The elements pass from gaseous matter to a liquid and then to a solid state with the new conditions created.

2.1.3 Biology-Based Calculations (BBA)

Biology was revolutionised during the European Renaissance, and numerous discoveries in the early current period (Himabindu and Jyothi 2017). Cells are more effectively perceived in plants found by Grew and Malpighi, and the cell hypothesis started to foster in the seventeenth hundred years. Antonie Philips van Leeuwenhoek (1632–1723) became known for his commitment to improving the magnifying lens (Pijper 1939). He could see something a lot more modest than a cell. He was known for finding the unicellular organism. The paper by Robert Brown was the

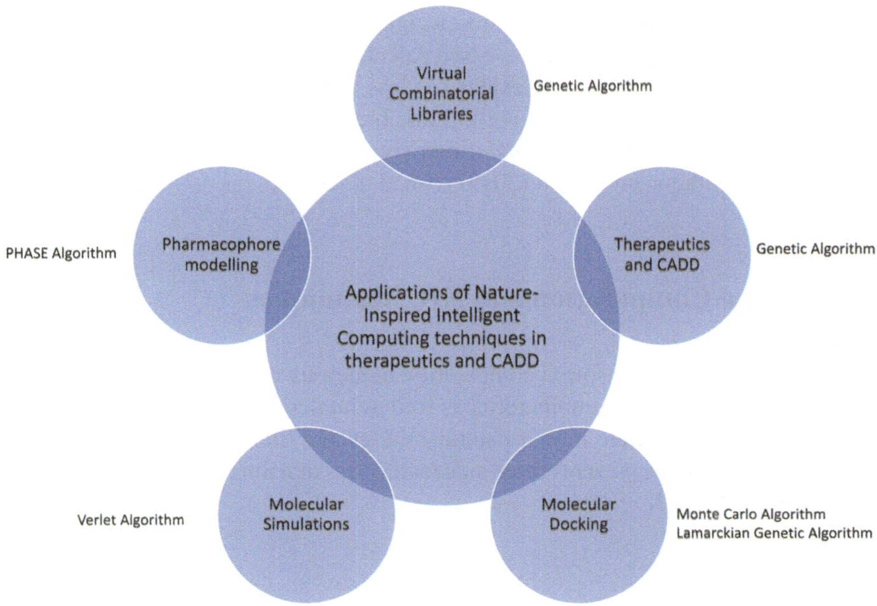

Fig. 1 Applications of nature-inspired intelligent computing techniques in therapeutics and CADD

principal portrayal of the cell nucleus (Pijper 1939). There are many nature-inspired techniques used in computational biology, especially in cancer and diabetic-related therapeutics. In Fig. 1, we have scored the global effect on different countries and its risk to symptoms obtained from Gautam et al. (2019).

MacArthur and Wilson created numerical models of biogeography that depict how species move from one island to the next, how new species emerge and how they become terminated (Siddique and Adeli 2015; Losos et al. 2010). Since the 1960s, biogeography has arisen as a significant field of exploration concentrating on the geographic dissemination of natural species. Given the idea of biogeography, Simon proposed biogeography-based optimisation (BBO). Farmer et al. proposed counterfeit resistant frameworks (AIS) models during the 1980s, which numerically resolved the communication between antibodies in the light of the organic, safe framework standards. In 1968, Lindenmayer initiated the formalism to re-enact the advancement of multicellular creatures, first known as Lindenmayer systems and later called L-system, which pulled in the light of a legitimate concern for hypothetical PC researchers. Aono, Kunii, and Smith utilised L. frameworks to make practical pictures of trees and plants. Other bio-propelled search and enhancement calculations are accounted for in the writing that has not drawn much consideration in the examination of the local area, like B. the climatic cloud model, dolphin echolocation, Japanese tree frog calls, Egyptian vulture, blossom fertilisation calculation, huge salmon run, intrusive weed enhancement, paddy field calculation, cockroach pervasion calculation, and arbitrary frog hopping calculation.

Science has been a motivation for most weighty developments since the nineteenth 100 years; the worldview of science-based calculations is enormous and different in numerous ways. It shapes a more extensive class of search and enhancement calculations. Gatherings: evolutionary algorithms, bio-inspired algorithms, and swarm intelligence-based algorithms. Transformative calculations have been shrouded in a new book by Siddique and Adeli (2015) (Fan et al. 2020).

3 Advance Computation and NIC Techniques

Classification of nature-inspired computation numerous nature-inspired algorithms has been created to solve combinatorial as well as numerical optimisation problems. These algorithms simulate how a natural biological system works (Keerthi et al. 2015). Building on this presents three main categories of nature-inspired approaches, also shown in Table 1.

3.1 Swarm Intelligence

The collective behaviour of birds, bacteria, and insects such as ants, termites, and bees demonstrates problem-solving abilities. The corresponding behaviour is a consequence of self-organisation and indirect communication between insects. It also inspires the field of artificial intelligence due to the possibility of simulating this behaviour and exploiting it to solve real-world applications (Nedjah and Macedo Mourelle 2006). This area of data processing is referred to as swarm intelligence. All algorithms possible in this area are implemented to find the best-optimised solution.

3.2 Biological Cells

Life is without delay associated with cells; the whole thing alive includes cells or must do rapidly along with cells. A cell is a minute unit each, and everyone is taken into consideration as alive. It may be minimal and problematic in its shape and functioning, has an intricate inner pastime and a first-rate interplay with the neighbouring cells and lifestyles in an equation form. Many cells reside alone; however, in well-known, the cells are prepared in tissue, organ system, organisms, and a group of organisms. Many think of a particular agency, beginning with the direct communication and cooperation among neighbouring cells, and finishing together with the interplay with the surroundings at numerous levels. With the inner shape and agency of the cell, many of these advocates quite a few ideas, interesting as a maths factor of sight and doubtlessly beneficial from a measurable factor of sight. The processing of the cells may be explored using membrane computing.

Table 1 Nature-inspired computing techniques used in different biological and allied sciences

Nature-Inspired Computing Techniques			
Based on human	Genetic algorithm	Based on plants	Invasive weed optimisation
	Evolutionary algorithm		Artificial algae algorithm
	Evolutionary programming		Flower pollination algorithm
	Evolution strategies		Runner-root algorithm
	Biogeography-based optimisation	Based on insect	Artificial bee colony
Based on animal	Monkey algorithm		Glow-worm swarm optimisation
	Jaguar algorithm		Ant colony optimisation
	Grey wolf optimiser		Firefly algorithm
	Lion optimisation algorithm		Ant lion optimisation
	Artificial buffalo optimisation		Dragonfly algorithm
	Elephant herding optimisation	Based on birds	Particle swarm optimisation
Based on natural phenomena	Key cutting algorithm		Bat algorithm
	Fireworks algorithm		Migrating birds optimisation algorithm
	Mine blast algorithm		Green herons optimisation algorithm
	Hydrological cycle algorithm		Laying chicken algorithm
	Car tracking optimisation algorithm		Crow search algorithm

3.3 *Biological Neural Network*

One of the maximum inspiring natural intelligence is the human thoughts themselves. There are many theories of the way minds work. This is a massive query, so making a system that possesses thoughts will never be viable. If we remember the general shape of the human mind and the factors we discover are nerve cells or neurons (Boers 1992). Neurons technique electrical signals constitute all mind activities linked up into complex networks. This simultaneous cooperative behaviour of many straightforward processing gadgets is the basis of the widespread sophistication and computational electricity. The simulation of this behaviour may be performed through an artificial neural network (Boers 1992; Gupta 2013).

There are various NIC techniques that humans inspire; genetic algorithms, evolutionary algorithms, biogeography-based optimisation, and inspired by animals, monkeys, and jaguar algorithms and a few are based on lion, buffalo, and elephant herding optimisation (Gautam et al. 2019). Plant-inspired techniques are invasive weed, artificial algae, and flower pollination algorithms. The birds'-based NIC techniques are bat algorithms and laying chicken and crow search algorithms, while insect-based NIC techniques are bee colony, glow-worm swarm optimisation, and anti-lion optimisation (Gautam et al. 2019).

4 NIC in Therapeutics and Computer-Aided Drug Designing

Computer-aided drug designing (CAAD) assumes a focal part in the modern discovery of therapeutically basic little atoms. Refers to calculation methods that can help speed up lead identification and optimisation processes. CADD is tools and resources for collecting, managing, analysing, and modelling connections in their broadest sense (Yadav 2021; Llorach-Pares et al. 2017; Song et al. 2009; Seth et al. 2011). All in all, these tools and resources could help improve the efficiency of new drug development and reduce costly late-stage clinical trial failures. An outline of computational and therapeutic techniques has been conveyed to facilitate the medication plan and revelation process.

4.1 Virtual Connectional Libraries

Virtual connectional libraries (VCL) offer the opportunity to improve the design by optimising the diversity or similarity of a library with a target. These proceeds can help single out molecules along with the desired composition by systematically examining the properties space of compounds. Sub-atomic variety, inclusion, representativeness, physicochemical, and pharmacokinetic properties are frequently applied to guarantee significant testing in the item region with negligible particles (Seth et al. 2011). Lately, much accentuation has been put on creating libraries that permit agreement choice of appropriate particles by improving different properties. These compounds assist with concentrating on organic systems and as rules for streamlining drug properties (Tong 2017).

4.2 NIC Techniques for Therapeutics and Computer-Aided Drug Designing

Drug discovery is a technique for figuring out current molecules and positive healing activities. CADD techniques have become extra famous and have therapeutically performed chief function over the past three decades with inside improvement. CADD strategies cover numerous components of the drug discovery pipeline, starting from the choice of the molecules to the optimisation of lead compounds (Ahmad et al. 2021; Pathak et al. 2021). Like, virtual profiling techniques can expect the organic material in addition to mechanisms of action of positive molecules as molecular modelling strategies, inclusive of docking as well as molecular dynamics, can be expected protein–ligand interactions in the phrases of binding affinity, permitting difference among compounds, virtual screening methods is capable of locating similar analogue molecules for a given compound and construct compound collection from entering molecules, optimisation strategies used for layout new molecule, enhancing a current compound for absorption, distribution, metabolism, excretion, and toxicity (ADMET) prediction methods (Yadav 2021). A common type of those methods is primarily based on the enter molecule's character. There are preferred forms of CADD types: structure-based drug design (SBDD) and ligand-based drug design (LBDD) (Llorach-Pares et al. 2017).

4.3 Molecular Docking and Dynamics Simulations

Docking is one of the best combinations of the algorithms combinations, and there is no such replacement available for now that can screen and provide from lacs of compounds to just a few best ones that can be used for further studies. The docking is ultimately a nature-inspired technique like algorithmic design with genetic algorithm-based sampling (both satisfied and continuous sampling). Some molecular docking software available is glide with three sampling algorithms (HTVS, SP, and XP), AutoDock, FlipDock, and GOLD (Halgren et al. 2004; Repasky et al. 2007; Verdonk et al. 2003; Huey et al. 2012).

Further, docking-generated complexes are used to validate the next level of molecular dynamics (MD) simulation algorithms. Molecular dynamics allows scientists to read for macromolecules such as proteins, nucleic acids, and biological membranes. These steps in the MD simulation are

(i) Preparation of the instance model of the protein–ligand system.
(ii) Estimation of the molecular forces reserved on each atom.
(iii) Position of an atom according to forces based on Newton's law of motion.
(iv) Increase the simulation time by one or two.

Various tools and open source codes are available for the MD simulation like Desmond (considered the most standard and optimised one), Gromacs, Amber NAMD, FATSLiM, and many more (Fiorin et al. 2013; Buchoux 2017; Christen et al. 2005; Bergdorf et al. 2021).

4.4 Pharmacophore Modelling

This technique determined the tiny molecular properties, such as the biological activity of a compound of interest (Najjar et al. 2019). The International Union of Pure and Applied Chemistry accordingly interpret a pharmacophore as "the set of steric and electronic properties required to ensure optimal supramolecular interactions with a given biological structure and to initiate (or block) its biological response" (Zhang 2011). List to develop a pharmacophore model involves-

(a) A database can be used to identify the best active sites that can bind to the target.
(b) The essential atoms and their interactions are defined for a 2D pharmacophore model.
(c) Defined using the IUPAC nomenclature for the 3D pharmacophore model.
(d) The basic properties required in the binders are known through the alignment of ligands.
(e) Construction of pharmacophore model.
(f) Top ranking of pharmacophore models and selection of the good models.
(g) Validation of pharmacophore modelling.

Further, the model used to be tested using various docking-related programmes against the drug library.

5 Advantages of Implementing the NIC Techniques

Several NIC ideas and metaheuristic-based techniques, such as particle swarm optimisation, ant colony optimisation, and harmony search, are applied to many medical field problems. AI and ML have been globally used for data classification and predicting diseases and pandemics. Most commonly used in detecting myocardial infarctions, coronary artery diseases, and heart-related problems (Salehahmadi and Manafi 2014). NIC techniques such as swarm intelligence are found in the human body and the environment. Artificial immune systems have overcome challenges such as self-healing of robots, anomaly detection, intrusions, and optimisation. AI-based techniques have imparted efficient analytical abilities, phenotype detection, prognosis, treatment, and antigen detection (Timmis et al. 2010).

Some notable dramatic advances in therapeutics come from developing physics-based computational systems to predict various endpoints accurately. Artificial intelligence and deep learning method improvements also contribute to drug discovery. The enhanced computational power of cloud computing and GPUs have significantly contributed to the exploration of drug profiles. In addition, cryogenic electronic microscopy (cryo-EM) is a rapid advancement in structural biology. The protein-structure prediction via computational methods allows access to high-resolution drug-receptor complex 3D structures, thus discovering novel small therapeutic molecules (Frye et al. 2021) (Fig. 2).

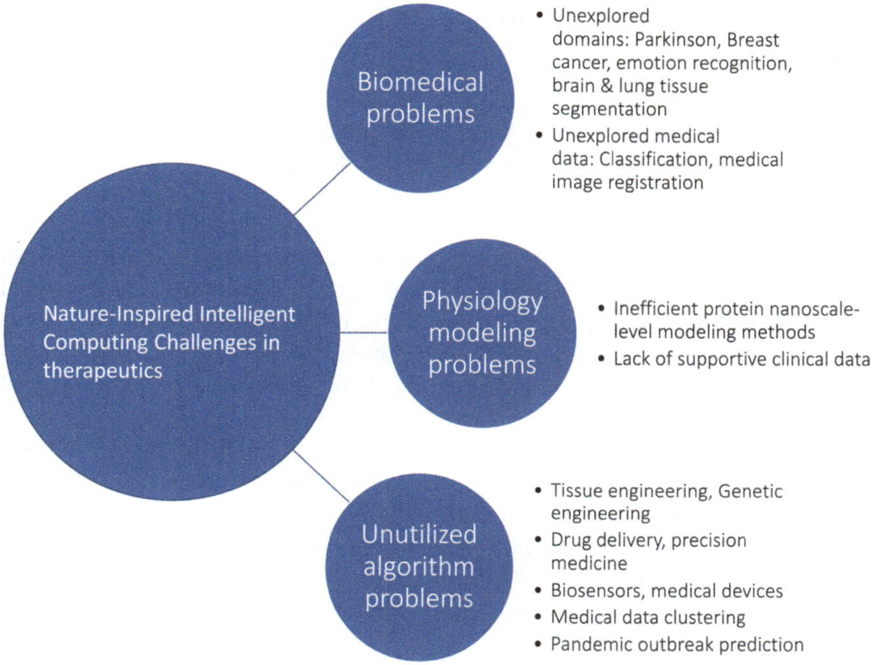

Fig. 2 Advantage of nature-inspired intelligent computing techniques in therapeutics and health care

An efficient application of the NIC technique, the firefly algorithm, has a promising future in the healthcare industry. Telemedicine in health care premises allows access to and analysis of patient datasets and cures them from a distance. This methodology involves exchanging medical data through an electric communication system in different locations. Thus, the health status of patients is constantly updated. Some vital advantages of telemedicine in health therapeutics include sessions with specialists, medical imaging, video assessments, health diagnosis, and more (Nayak et al. 2020). The BCA is an exciting algorithm with tremendous flexibility. It facilitates the medical decision-making process with ease. Complex interactions among biological systems are translated into comprehensible and understandable formats—a robust and global optimisation algorithm with undiscovered potential and a promising future (Salehahmadi and Manafi 2014). The significant advantages of NIC techniques are seen in biomedical signal processing, cancer disease detection and diagnosis, brain-related disorders, medical image processing, face and emotion recognition, ECH and EEG detection and analysis, heart disease detection, and telemedicine (Nayak et al. 2020).

6 Challenges of NIC Techniques in Biomedical Computation

Undoubtedly, the NIC techniques are a challenging area of research and need the significant attention of research groups to solve biomedical problems more systematically using robust setups. Problems with proper emotion recognition, brain tissue segmentation, lung segmentation, biomedical data classification, medical image registration, Parkinson's disease diagnosis, breast cancer analysis, and more remain hindrances (Nayak et al. 2020). In further research, many issues need to be addressed in intelligent computing. Firefly algorithms have not been fully utilised to resolve medical issues such as tissue engineering, genetic engineering, drug delivery, biosensors, medical devices, precision medicine, pandemic outbreak prediction, medical data clustering, analysis, and more (Nayak et al. 2020) (Fig. 3). There is a need to imply powerful NIC algorithms to cope with similar application problems. NIC area can be explored more to resolve biomedical problems (Emambocus et al. 2021).

Designing computational models of physiological functions in disease and health is a great challenge. The failure of prediction does not necessarily result in model refinement and experimental testing. However, the model predictions and outcomes sometimes suggest that more biological processes should be considered in the model

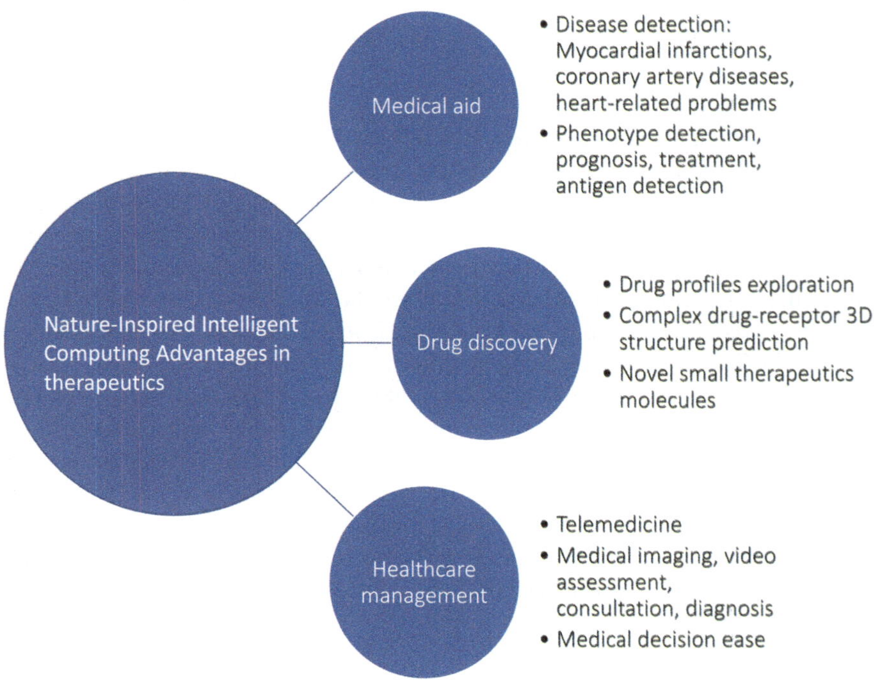

Fig. 3 Limitations and biomedical challenges of nature-inspired intelligent computing techniques in therapeutics

or that some components need better characterisation (Overgaard and Mogensen 2011). New experimental approaches would be needed, which is quite daunting work. Experimental results do not derive refined models. It becomes complex and challenging to incorporate multiple biological phenomena into the models. Another significant limitation is developing modelling methods for protein function at the nanoscale level and further simplifying these models and inclusion in a multi-scale modelling framework (Overgaard and Mogensen 2011). Genome sequencing is a powerful way to revolutionise the healthcare system. It takes us towards the concept of precision medicine.

Nonetheless, the interpretation of genetic variations is a significant challenge that limits the usage of DNA sequencing as a diagnostic tool. The recent algorithms applied for genetic variation prediction are not enough to be employed without additional supportive clinical data. However, more biomedical data allows machine learning algorithms, specifically NIC-based techniques, to predict genetic variations with optimum accuracy (McInnes et al. 2021).

7 Future Perspective and Scope of NIC Techniques in Therapeutics and Drug Design

Various chronic diseases such as cancer and diabetes have been considered in the computational techniques for therapeutic approaches, and most algorithms are nature inspired. The optimised algorithms in chronic disease therapeutics are ACO, artificial bee colony (ABC), glow-worm swarm optimisation (GWSO), firefly algorithm, and antlion optimisation. Various techniques are used for diabetes (type I and II) and cancers like lung, breast, and ovarian cancer therapeutics (Gautam et al. 2019). The machine learning-based modelling or predictive modelling incorporates nature-inspired algorithms like genetic algorithms that are now also seen in various areas of physics like the overall sampling and quantum mechanics-based calculation and accelerator facility at GSI/FAIR (Appel et al. 2019; Baell et al. 2013). The current sciences are again focusing on finding novel drug candidate that is supposed to be obtained from natural sources, and for that the screening machines used are nature-inspired by sampling (Mishra and Tiwari 2011). The drugs' natural sources can be plant extracts, biological macromolecules, or small molecule substances. The semi-synthesised compound's concepts are widely spreading in the market that can be automated by using predictive modelling concepts as these concepts are learning from the source of the previous natural compound's formation and orientation. The prediction is about adding a few atoms that can widely change the dynamics of the drug, and this can lead to developing a complete database for the fragments that should not get out of the ADMET filtering criteria (Najjar et al. 2019; Mishra and Tiwari 2011). The simulative algorithms are getting intelligent and advanced after incorporating a multidimensional view of physics in biological science, like the concept of quantum mechanics, which was initially developed in physics and has now been utilised by advanced computational chemists to design and validate drugs after analysing the electronic movement in the orbitals (Sachan et al. 2014). These techniques, like quantum-mechanics-based density function theory, HOMO–LUMO, and time-dependent density-functional theory, revolutionise computational therapeutics (Sachan et al. 2014; Choudhary et al. 2014).

8 Conclusion

While summarising the NIC techniques in computational therapeutics and drug designing, various sampling algorithms were inspired by the nature that has been used in molecular docking, molecular dynamics simulation, and other model generation for pharmacophore modelling. The Verlet velocity algorithms used widely in simulative parameters or the genetic algorithms used in the sampling reduce the computational consumption burden. It was more accessible to opt-in the computations than to design a whole new concept. In addition to widely used algorithms, there are a few sets of algorithms used for specific purposes, and we have classified them in Table 1 to make them easier to understand and list correctly. The last few decades have been experimentally successful in designing and implementing different NIC techniques to diagnose various human disorders. This chapter presented various methods and approaches in therapeutics and drug designing.

References

Aarthy M et al (2017) Advantages of structure-based drug design approaches in neurological disorders. Curr Neuropharmacol 15(8):1136–1155

Ahmad S et al (2021) Molecular dynamics simulation and docking studies reveal NF-κB as a promising therapeutic drug target for COVID-19

Appel S et al (2019) Application of nature-inspired optimization algorithms and machine learning for heavy-ion synchrotrons. Int J Mod Phys A 34(36):1942019

Baell J et al (2013) Ask the experts: past, present and future of the rule of five. Future Med Chem 5(7):745–752

Bergdorf M et al (2021) Desmond/GPU Performance as of April 2021. DE Shaw Research, Tech Rep DESRES/TR–2021–01

Boers EJ, Kuiper H (1992) Biological metaphors and the design of modular artificial neural networks

Buchoux S (2017) FATSLiM: a fast and robust software to analyze MD simulations of membranes. Bioinformatics 33(1):133–134

Choudhary N et al (2014) Quantum chemical calculations of conformation, vibrational spectroscopic, electronic, NBO and thermodynamic properties of 2, 2-dichloro-N-(2, 3-dichlorophenyl) acetamide and 2, 2-dichloro-N-(2, 3-dichlorophenyl) acetamide. Comput Theor Chem 1032:27–41

Christen M et al (2005) The GROMOS software for biomolecular simulation: GROMOS05. J Comput Chem 26(16):1719–1751

Emambocus BAS et al (2021) Dragonfly algorithm and its hybrids: a survey on performance, objectives and applications. Sensors 21(22):7542

Fan X et al (2020) Review and classification of bio-inspired algorithms and their applications. J Bionic Eng 17(3):611–631
Fiorin G, Klein ML, Hénin J (2013) Using collective variables to drive molecular dynamics simulations. Mol Phys 111(22–23):3345–3362
Frye L et al (2021) From computer-aided drug discovery to computer-driven drug discovery. Drug Discov Today Technol 39:111–117
Gautam R, Kaur P, Sharma M (2019) A comprehensive review on nature inspired computing algorithms for the diagnosis of chronic disorders in human beings. Prog Artif Intell 8(4):401–424
Gupta N (2013) Artificial neural network. Netw Complex Syst 3(1):24–28
Halgren TA et al (2004) Glide: a new approach for rapid, accurate docking and scoring. 2. Enrichment factors in database screening. J Med Chem 47(7):1750–1759
Himabindu K, Jyothi S (2017) Nature inspired computation techniques and its applications in soft computing: survey. Int J Res Appl Sci Eng Technol 5(7):1906–1916
Huey R, Morris GM, Forli S (2012) Using AutoDock 4 and AutoDock vina with AutoDockTools: a tutorial. Scripps Res Inst Mol Graph Lab 10550:92037
Keerthi S, Ashwini K, Vijaykumar M (2015) Survey paper on swarm intelligence. Int J Comput Appl 115(5)
Llorach-Pares L et al (2017) Computer-aided drug design applied to marine drug discovery: Meridianins as Alzheimer's disease therapeutic agents. Mar Drugs 15(12):366
Losos JB, Ricklefs RE, MacArthur RH (2010) The theory of island biogeography revisited. Princeton University Press, Princeton
McInnes G et al (2021) Opportunities and challenges for the computational interpretation of rare variation in clinically important genes. Am J Hum Genet 108(4):535–548
Mikkilä-Erdmann M (2001) Improving conceptual change concerning photosynthesis through text design. Learn Instr 11(3):241–257
Mishra BB, Tiwari VK (2011) Natural products: an evolving role in future drug discovery. Eur J Med Chem 46(10):4769–4807
Najjar A et al (2019) Fragment-based drug design of nature-inspired compounds. Phys Sci Rev 4(9)
Nayak J et al (2020) Firefly algorithm in biomedical and health care: advances, issues and challenges. SN Comput Sci 1(6):1–36
Nedjah N, de Macedo Mourelle L (2006) Swarm intelligent systems, vol 26. Springer
Newton I (1990) Mathematical principles of natural philosophy. Encyclopaedia Britannica
Onawole AT et al (2020) COVID-19: CADD to the rescue. Virus Res 285:198022
Overgaard M, Mogensen J (2011) A framework for the study of multiple realizations: the importance of levels of analysis. Front Psychol 2:79
Pathak N et al (2021) Phytochemical analysis and antifungal activity of weed extracts against Rhizoctonia root rot in Buckwheat (Fagopyrum tataricum)
Pijper A (1939) The microscope in biology. S Afr J Sci 36(12):58–72
Repasky MP, Shelley M, Friesner RA (2007) Flexible ligand docking with Glide. Curr Protoc Bioinform 18(1):8.12.1–8.12.36
Sachan AK et al (2014) Molecular structure, vibrational and electronic properties of 4-Phenyl-3H-1, 3-thiazol-2-ol using density functional theory and comparison of drug efficacy of keto and enol forms by QSAR analysis. Spectrochim Acta Part A Mol Biomol Spectrosc 132:568–581
Salehahmadi Z, Manafi A (2014) How can bee colony algorithm serve medicine? World J Plast Surg 3(2):87
Seth D et al (2011) Nature-inspired novel drug design paradigm using nanosilver: efficacy on multi-drug-resistant clinical isolates of tuberculosis. Curr Microbiol 62(3):715–726
Siddique N, Adeli H (2015) Nature inspired computing: an overview and some future directions. Cogn Comput 7(6):706–714
Song CM, Lim SJ, Tong JC (2009) Recent advances in computer-aided drug design. Brief Bioinform 10(5):579–591
Timmis J, Andrews P, Hart E (2010) On artificial immune systems and swarm intelligence. Swarm Intell 4(4):247–273

Tong JC (2017) Applications of computer-aided drug design. Drug design: principles and applications. Springer, pp 1–7
Verdonk ML et al (2003) Improved protein–ligand docking using GOLD. Proteins Struct Funct Bioinform 52(4):609–623
Yadav MK et al (2021) Predictive modeling and therapeutic repurposing of natural compounds against the receptor-binding domain of SARS-CoV-2. J Biomol Struct Dyn 1–13
Zekri M, Obreza TA (2003) Plant nutrients for citrus trees. University of Florida Cooperative Extension Service, Institute of Food and…
Zhang S (2011) Computer-aided drug discovery and development. Drug Des Discov 23–38

Nature-Based Computing Bioinformatics Approaches in Drug Discovery Against Promising Molecular Targets Carbonic Anhydrases and Serine/Threonine Kinases for Cancer Treatment

Mudasir Nabi Peerzada, Masood Ahmad Rizvi, Ajeesh Kumar KK, Ankita Sahu, and Saurabh Verma

1 Introduction

In drug discovery, nature-inspired computing (NIC) techniques such as particle swarm optimization, evolutionary algorithms, and ant colony optimization are treated as significant optimization methods. Nature-inspired algorithms are useful for bioisosteric replacement or scaffold hopping of complex natural products to develop synthetically more easily accessible molecules with the help of medicinal chemistry. The computer-aided drug design strategy (CADD) uses nature analog algorithms such as evolutionary algorithms (genetic algorithm and evolution strategy) ant colony optimization (ACO), and particle swarm optimization (PSO) to prepare novel bioactive molecules (Schwefel 2002). These computational techniques are widely used for the design of the different libraries of compounds, combinatorial building-block assembly de novo design, and several optimization tasks in molecular modeling and drug design (Schüller and Schneider 2008). For drug design, besides optimization of binding affinity of the molecule, features such as aqueous solubility, lipophilicity, or pharmacokinetic properties are important (Hutter 2009). These properties can be assessed in population-based algorithms for the optimization of fitness functions (Gillet 2004). Depending on the particular implementation, the medicinal chemist has an option to evaluate different features of the fitness function depending upon the requirement.

M. N. Peerzada · A. K. KK
Tumor Biology Department, National Institute of Pathology, Indian Council of Medical Research, New Delhi 110029, India

M. A. Rizvi
Department of Chemistry, University of Kashmir, Srinagar 190006, India

A. Sahu · S. Verma (✉)
Tumor Biology, ICMR-National Institute of Pathology, New Delhi, India
e-mail: saurabhverma.nip@gov.in; svarmasv1@rediffmail.com

K. Raza (ed.), *Nature-Inspired Intelligent Computing Techniques in Bioinformatics*,
Studies in Computational Intelligence 1066,
https://doi.org/10.1007/978-981-19-6379-7_16

Adaptive parameters can address large chemical search spaces through population search agents like digital ants, and swarm particles and converge on chemically significant optima. The term "adaptation" here describes the ability of a population to reply to changes within the environment. The phenomenon of a population of simple individuals producing complex behavior and developing additional abilities that cannot be readily explained by simply adding the individuals' abilities together is called "emergence". Both adaptation and emergence properties are sought for drug design strategy because drug design involves compound synthesis and testing till the desired molecule is obtained. The natural computing paradigms are used in virtual screening and drug design. Evidently, the combination of techniques leads to improved accuracy and applicability, therefore, nature-inspired adaptive techniques also contribute to producing improved therapeutic molecules and play a significant role in drug design (Hiss et al. 2010).

Overwhelming advances in informatics and computational biology have amplified efficiency at various stages of the drug discovery pipeline including the anticancer drug discovery. Multi-omics data such as transcriptomics, genomics, metabolomics, and proteomics data in addition to microarray and next-generation sequencing data were produced with the help of bioinformatics and are useful in developing new cancer agents (Argelaguet et al. 2018; Bakker et al. 2018). Multi-omics data provides an essential tool for computer-aided drug discovery, particularly in drug repositioning (Huang et al. 2018). For cancer treatment, combination therapies are used to improve the outcome of treatment. Bioinformatics approaches are useful in calculating the possible combinations iteratively and identifying a proficient multi-component therapy. Particularly, on the basis of biological theories and methodologies, numerous studies have systematically tested pair-wise drug combinations and predicted efficacious drug combinations for cancer therapy (Du et al. 2019; Sheng et al. 2017).

Computational methods are significant tools to accelerate the effective drug discovery process. CADD is one of the promising technologies for the design and development of effective drug candidates. Recently, CADD has been extended to the development of anticancer therapies, and it has proved to be useful in cancer therapy. Evolutionary algorithms, particle swarm optimization, and ant colony optimization are important optimization methods of the CADD method of molecular modeling. CADD approaches such as structure-based drug design (SBDD) and ligand-based drug design (LBDD) are two common approaches that are widely used in drug discovery. Commercially available CADD software including Discovery Studio, Schrödinger, MOE, and OpenEye cover most of the properties required for both SBDD and LBDD methods (Usha et al. 2018).

Virtual database screening (VS) methods are commonly used to screen large in silico compound databases to find out the promising binders of a particular target. For in silico screening, the commonly used software's are DOCK, AutoDock, AutoDock Vina, program Pharmer, and Schrödinger. The in silico database of drug-like compounds is one of the important requirements for CADD ligand identification. The database of compounds for VS is ZINC which has about 90 million compounds. In silico or molecular docking involves posing a compound in the putative binding site of the target optimized by a scoring function along with a conformational sampling method. Various available docking programs fluctuate based on the scoring function used to portray the interaction between ligand and target protein. The conformational sampling method is employed to ascertain the binding poses of the ligand with protein (Agarwal et al. 2016).

MD simulations can be used to study target-ligand interactions at an atomic level and to generate various favorable conformations of target or ligand taking into account both SBDD and LBDD studies. In combination with other studies, MD simulations are used to find potential low-molecular weight binders to the target protein. Commonly used MD simulation codes include AMBER, CHARMM, NAMD, OpenMM, and GROMACS. These programs run on a variety of computer architectures including running in parallel on multicore central processing units and graphics processing units (Durrant and McCammon 2011).

Since computational methods play a significant role in the drug discovery process, the applications of these methods in anticancer drug discoveries have shown great advantages. The development of cMet/ALK inhibitor crizotinib is a successful example of using structure-based design techniques (Cui et al. 2011). It was approved by FDA in 2011 (Cui et al. 2013). Axitinib was approved by the FDA in 2012 as a new therapy for advanced renal cell carcinoma (Meadows and Hurwitz 2012). It was developed with the help of a structure-based drug design strategy and it inhibits the VEGF kinase domain in the DFG-out conformation (Kania 2009). The use of the computational approach also leads to the development of Luminespib (NVP-AUY922) as one of the strong HSP90 inhibitors currently undergoing phase two clinical trials and it exhibited positive results for patients with ALK rearrangements (Felip et al. 2018). Luminespib also exhibited potent anti-tumor activity in lung adenocarcinomas targeting EGFR (Piotrowska et al. 2018). Some of the recently FDA-approved anticancer drugs (Cui et al. 2020) that were designed with the aid of computational methods are enlisted in Fig. 1.

2 Computer-Aided Drug Design for Tumor-Associated Human Carbonic Anhydrases

Human carbonic anhydrases are zinc-based metalloenzymes involved in catalyzing the reversible hydration of CO_2 to bicarbonate ion and proton ($CO_2 + H_2O \leftrightarrows HCO_3^- + H^+$). Carbonic anhydrase isoforms IX and XII have been described as tumor-associated proteins due to their limited presence in normal tissues and high over-expression in a large number of solid tumors, where these isoforms are involved with survival and metastasis (Aggarwal et al. 2013). CA IX and CA XII are expressed in several cancers, particularly in breast cancer and both play important roles in tumor progression, acidification, and metastasis, suggesting that both make important therapeutic cancer targets (Mboge et al. 2018). Several inhibitors have been developed and some of them showed promising inhibition of these tumor-associated isoforms. The inhibition of carbonic anhydrase is a well-understood field of anticancer drug

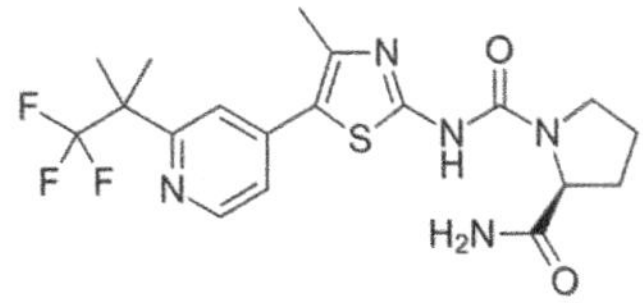

Alpelisib
(PI3K inhibitor)

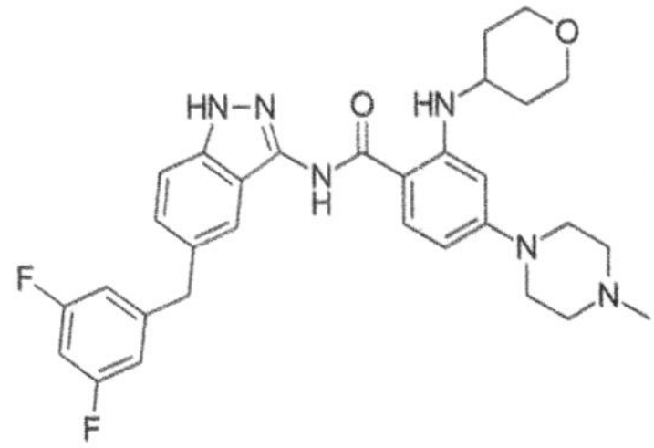

Entrectinib
(Tyrosine kinase inhibitor)

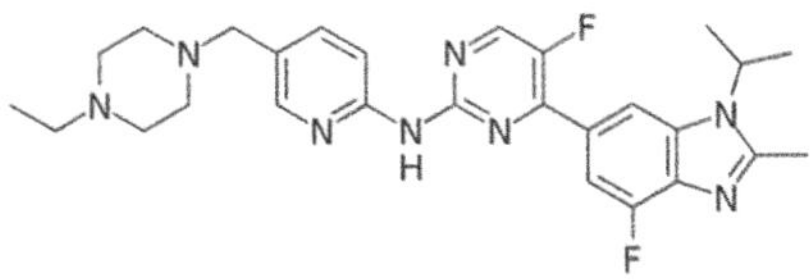

Abemaciclib
(CDK inhibitor)

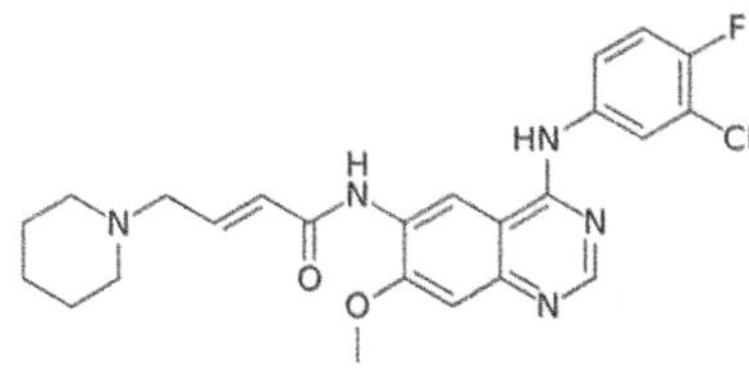

Dacomitinib
(EGFR kinase inhibitor)

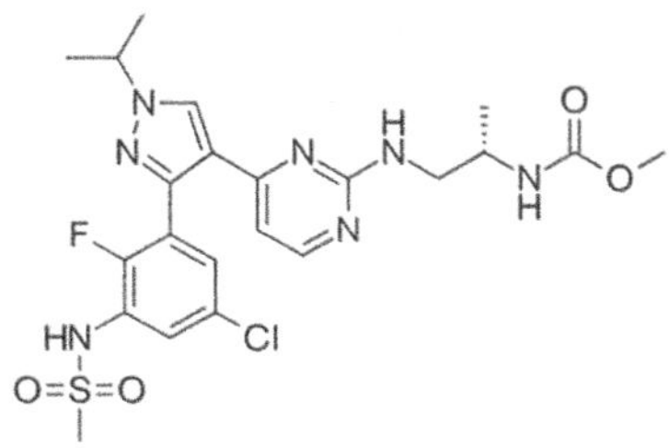

Encorafenib
(BRAF kinase inhibitor)

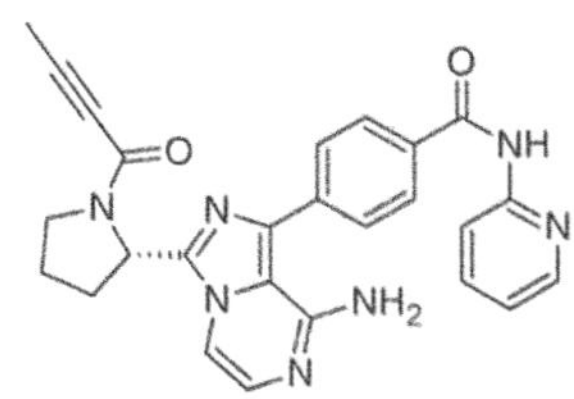

Acalabrutinib
(Tyrosine kinase inhibitor)

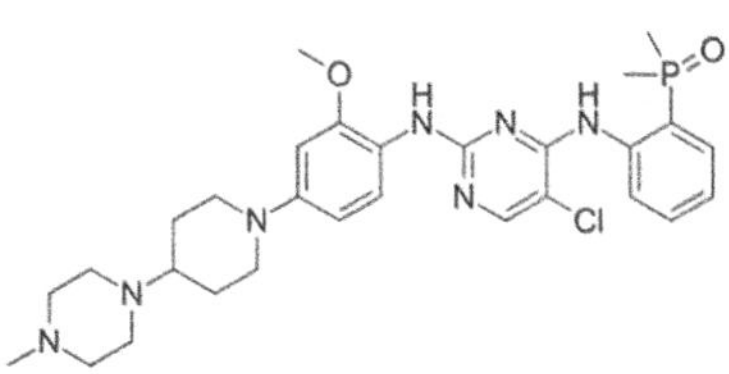

Brigatinib
(ALK, EGFR inhibitor)

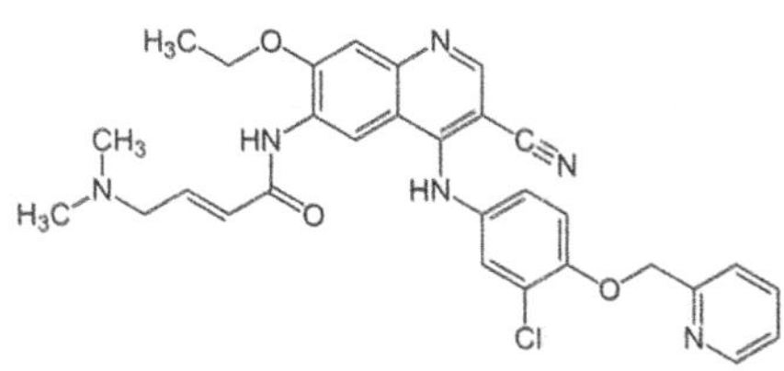

Neratinib
(EGFR, HER2 inhibitor)

Fig. 1 Chemical structures of some of the anticancer drugs developed through computational methods

discovery with most classes of inhibitors binding to the metal center. Carbonic anhydrase IX and XII inhibitors generally bind to the zinc ion from the enzyme active site in a deprotonated state. Computer-aided drug design is widely used in developing potent CA inhibitors (Alterio et al. 2012; Kumar et al. 2021).

Using molecular docking as a tool, a series of carbonic anhydrase IX were designed and screened in vitro. Among several analogs of this series, compounds 1–3 showed significant binding interactions with CAIX. Molecular docking showed that compounds 1, 2, and 3 bind effectively to the active site of CAIX having ΔG values of -10.7 kcal/mol, -11.5 kcal/mol, and -10.1 kcal/mol, respectively. It was found that compound 1 binds with CAIX effectively and is stabilized by eight H-bond interactions occurring with residues Trp5, Pro11, Arg60, Gln67, Gly98, and Arg246. Residues Gln28, Thr199, His224, Leu148, and Glu150 are involved in hydrophobic interaction, whereas the His224 is stabilized by phenyl moiety of compound 1 via π–π interaction. Most active compound 2 formed four H-bonds with Arg60, Thr69, Thr200, and Pro11 residues of the active site. The residues Gln67, Thr125, Phe128, Leu135, and Pro138 are involved in hydrophobic interaction, whereas Trp209 and His94 residues are stabilized by indole and phenyl scaffolds of compound 2, respectively, through π–π interaction. Compound 3 forms six H-bond interactions with Gln67, His94, His96, and Thr199, and show hydrophobic interactions with Leu70, Leu81, Tyr88, Arg89, and Gln92 residues. The indole ring of compound 3 is stabilized by the π–π interaction of Tyr88 (Fig. 2). All these compounds were least active against CAII off-target. Molecular docking results suggested that interactions with the active site of CAIX are more favorable for 1, 2, and 3 as compared to other compounds of the series. Compound 2 showed effective CAIX inhibition and the observed IC50 was 0.13 μM and inhibited the CAII with 9.10 μM (Peerzada et al. 2018).

A potent series of benzene-sulfonamides was designed and validated by computational studies. It was found that the compound SLC-0111 is a selective inhibitor hCA IX over the cytosolic off-target form hCA II. SLC-0111 is a ureido analog of benzene-sulfonamide in Phase II phase second clinical trials for the treatment of hypoxic, advanced-stage solid tumors (McDonald et al. 2020). The selectivity inhibition of CAIX by SLC-0111 renders it an ideal lead candidate for designing more potent hCA IX inhibitors. SLC011 was docked with hCA II and hCA IX and corresponding Gibbs energies of binding were calculated. SLC-0111 shows three common interactions with CAIX and CAII isoform and the interactions include: (1) nitrogen atom of the sulfonamide group coordinates to the Zn(II) ion, (2) formation of hydrogen bond ((Thr199)NH...O=S(SLC-0111)), and (3) π–π stacking interaction of the benzene ring attached to the sulfonamide group with His94. Molecular docking showed that ligand SLC-0111 makes a stronger hydrogen bond with Thr199 in the hCA IX active site, and the sulfonyl oxygen group of SLC-0111 makes a hydrogen bond also with HN(His94).

The 2D ligand interaction diagram shows that SLC-0111 makes hydrophobic interactions with the residues Val131, Leu135, and Pro202 in the hCA IX binding pocket. In addition, the tail moiety of this ligand occupies the hydrophobic pocket

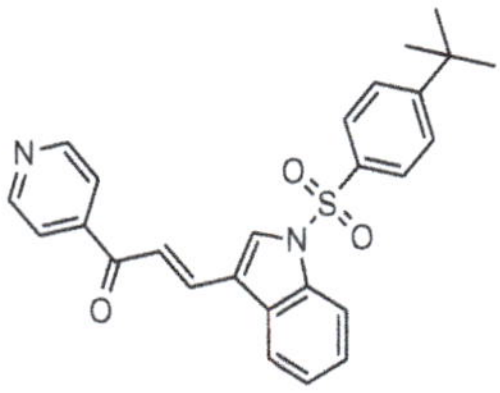

(1)

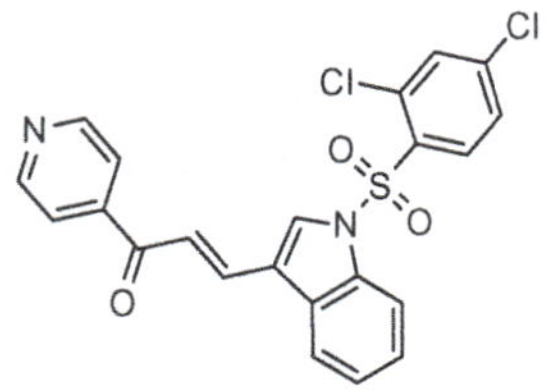

(2)

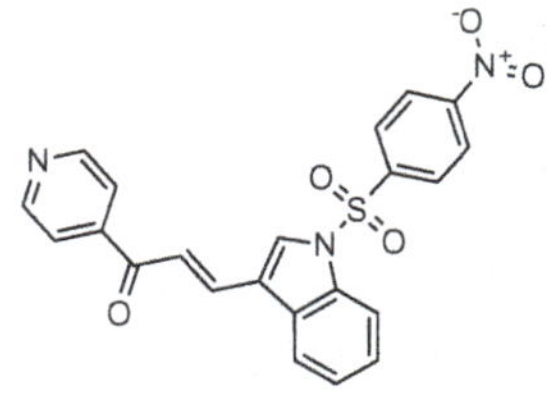

(3)

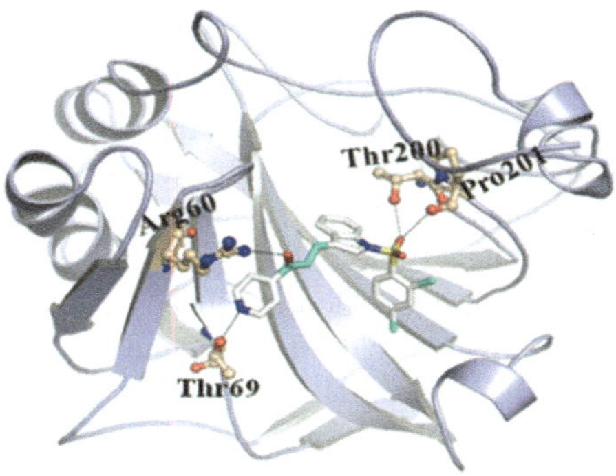

(a)

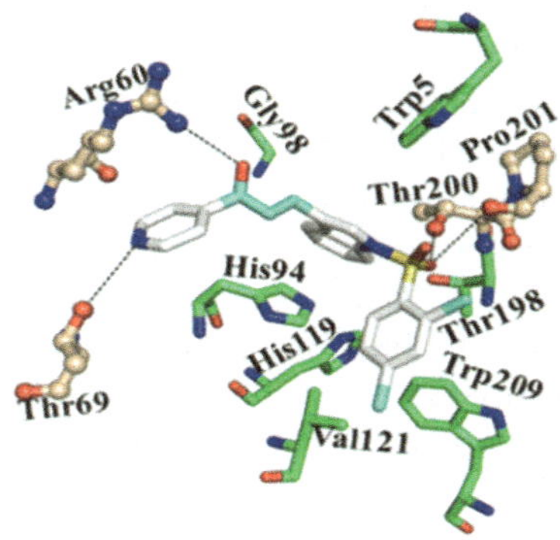

(b)

Fig. 2 Chemical structures of compounds 1–3 are shown. **a** Cartoon view of compound 2 docked with CAIX. **b** Active site residue interactions with compound 2 are presented. Residues are shown with a ball and stick whereas compound 2 is shown with the stick model. Hydrogen bonds are shown as broken black lines

formed by the residues Val131, Leu135, and Pro202 and shows van der Waals interactions. The ligand is found to chelate Zn(II) in a bidentate way in the hCA IX—SLC-0111 complex. Zn^{2+} ion is also complexed by imidazole nitrogen of His119 (Fig. 3). These studies were helpful in designing further analogs of SLC-0111 (McDonald et al. 2020).

Another series of compounds were designed and molecular docking proved to be helpful in the optimization of analogs and finding the binding interactions of designed compounds with hCA IX. Ligand 5 was docked in the active site of hCA IX, and it was found that the negatively charged nitrogen atom of the sulfonamide group chelates the zinc divalent positive ion. The sulfonamide group forms H-bond with T200. Benzimidazole interacts with the V121, L199, and T201 side chains. The phenyl ring and 3-methoxycarbonyl moiety interact with lipophilic side-chain residues L91 and V130. The 4-hydroxy group of the ligand is pointing outside of the binding site and forming the H-bonding interactions with the solvent water molecules. The ethyl linker has greater flexibility and allows the phenyl ring to interact with L91 and V130 (Fig. 4) (Milite et al. 2019).

Coumarin-based compounds were designed and molecular docking was used as a tool to determine the binding affinity of compounds toward CAIX. Molecular

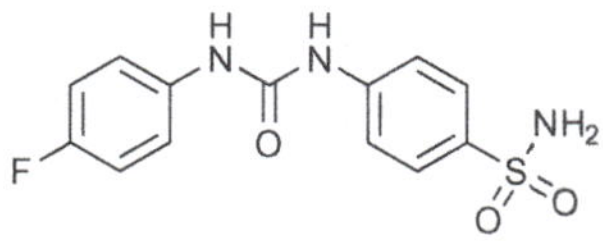

SLC 0111 (4)

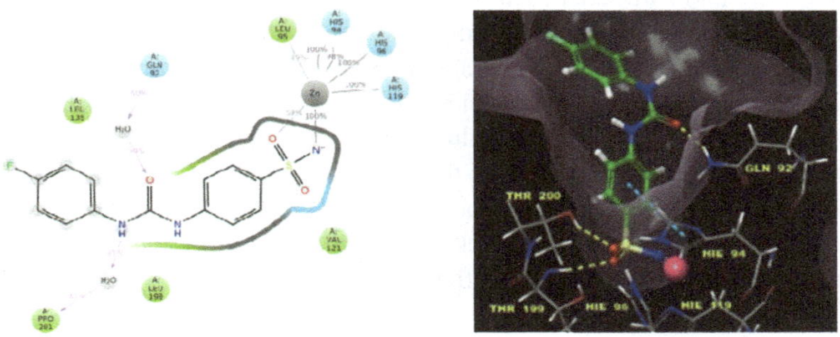

(a) **(b)**

Fig. 3 Chemical structure and binding modes of SLC-0111 (compound 4) are shown. **a** Ligand–protein interaction diagram for (left) hCA IX. **b** Hydrogen bonding of compound 4 with hCA IX is shown in yellow dashed lines, and π–π stacking interactions are shown in cyan dashed lines

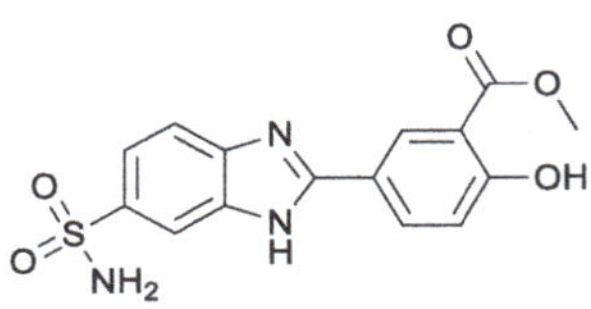

(5)

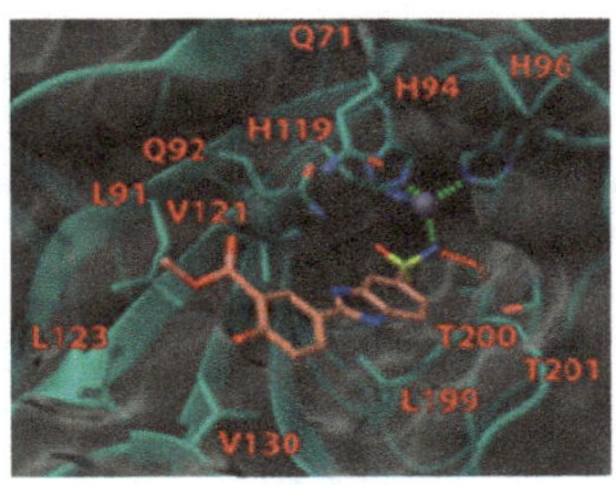

Fig. 4 Chemical structure of compound 5 and interactions with CAIX are shown. The protein is shown as cyan ribbons and sticks while the ligand is represented as salmon sticks. H-bonds are depicted as red dashed lines, while coordination bonds as green dashed lines

docking predicted that hydrolyzed coumarin products have the lowest mean binding energies as compared to their mimic closed forms, and the 2-hydroxycinnamic acid moiety of hydrolyzed form shows additional interactions with the CA IX. Docking analysis showed that the open form of coumarin derivative 6b exhibits additional interactions with the side residues and is linked by five H-bonds to residues Gln203, Arg196, Gln224, and The205. The side-chain residues that are in close contact with the ligand are Leu223, Vall262, and Arg194. Further, additional interaction between residue Asn198 and a 2-hydroxycinnamic acid moiety of the open form of the ligand is observed and such interaction is absent in the corresponding closed form of the coumarin ligand (Figs. 5 and 6a). It was found that rhenium containing coumarin compound of the designed series of compounds exhibits additional polar and hydrophobic interactions between the fac-Re$(CO)_3$ unit and residues such as Pro334, His200, Pro335, and Leu331 (Fig. 5b) (Aimene et al. 2021).

Molecular docking was performed with the purpose of identifying a reliable binding mode with hCA XII and hCA IX for the novel series of saccharin-derived ligands. Compound 7 was docked into the structures of the two enzyme isoforms hCA XII and hCA IX. It was found that compound 7 interacted with hCA XII by coordinating the zinc ion through one sulfonamide oxygen atom belonging to the saccharin scaffold, while the other oxygen of the sulfonamide motif forms a hydrogen bond with the hydroxyl group of T227 residue. A hydrogen bond is also formed between the carbonyl oxygen of the ligand with the Q117 side chain of CA XII. The saccharin phenyl ring occupies a hydrophobic pocket having N92, H121, W32, H119, and H94, residues interacting through π–π stacking with the imidazole ring of H119 residue and showed lipophilic interactions with the other residues.

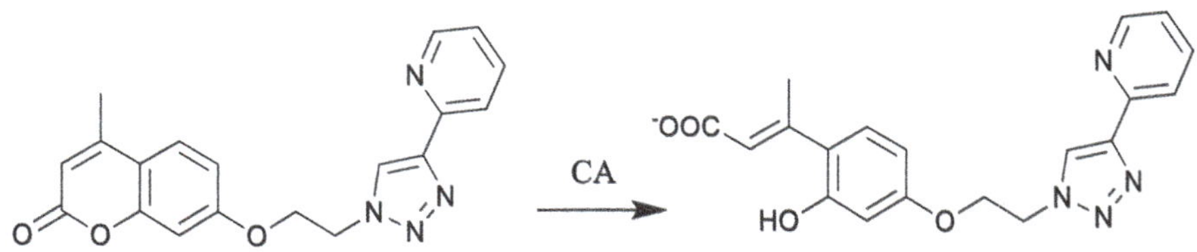

(6a) **(6b)**

Fig. 5 Coumarin compound 6a and corresponding hydrolysis product 6b produced on catalysis by the CA enzyme are shown

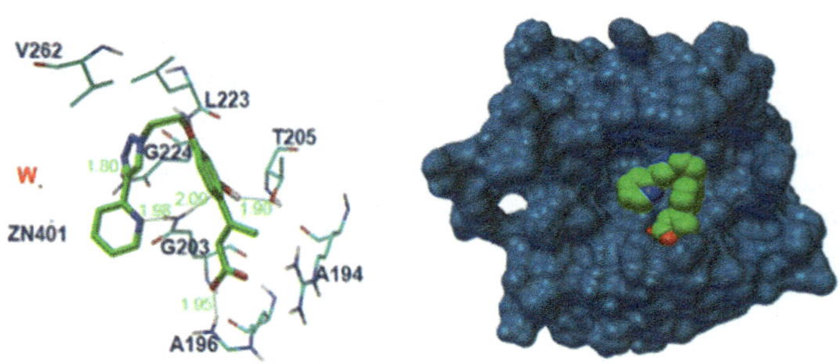

(a) **(b)**

Fig. 6 **a** Alignments of the docked organic ligand (compound 6b) within the hCA IX binding site, compared with the native 2-hydroxycinnamic acid ligand: H-bonds are shown in green color. W in red color represents the water molecule. Carbon atoms are shown in green color; oxygen atoms are represented in red; sulfur atoms in yellow; nitrogen atoms in blue; chlorine and rhenium atoms are shown in green color. **b** Surface view of hCA IX actives site and ligand. Carbon atoms are shown in green color; sulfur atoms in yellow; nitrogen atoms in blue; oxygen atoms in red; chlorine and rhenium atoms are represented in green color

The phenylisoxazole group of ligand forms hydrophobic interactions with V147, S158, S161, L167, L225, and P229. The binding interactions of ligand for CA IX were almost similar to that of CA XII. Compound 7 showed the same orientation in the hCA IX binding cavity. The phenylisoxazole moiety of compound 6 occupied the hydrophobic portion of the catalytic site, the saccharin scaffold occupies a portion near the zinc ion, and the phenyl ring is located in the pocket having W141, N198, H200, Q203, H226, and H228 residues. Compound 7 formed a hydrogen bond with Q224 through the carbonyl oxygen and interacts with the zinc metal ion through the oxygen atom of the sulfonamide group. Ligand 7 formed two hydrogen bonds with the nitrogen atom and the hydroxyl group of T333 residue (Fig. 7) (D'Ascenzio et al. 2020).

In addition, several sulfonamides and benzoic acid-containing compounds were designed for the inhibition of carbonic anhydrase activity and it was found that sulfonamides show a better inhibition profile (Eldeeb et al. 2021). Similarly, 4-(3-(2-arylidenehydrazine-1-carbonyl)-5-(thiophen-2-yl)-1*H*-pyrazole-1-yl) benzene-sulfonamides-based compounds were designed and the binding pattern with hCA I, hCA II, and hCA IX was predicted through computational methods to synthesize the potent anticancer novel compounds (Yamali et al. 2021). Another series of benzene-sulfonamides containing 3-methylthiazolo[3,2-*a*]benzimidazole having ureido, enaminone, hydrazone, or hydrazide linkers were designed and binding was predicted using molecular docking studies for II, IX, and XII hCA isoforms. Some of the compounds of this series showed inhibition of hCA IX in nanomolar concentration during invitro evaluation (Alkhaldi et al. 2020). Several other benzene-sulfonamides were designed recently through a computer-aided drug

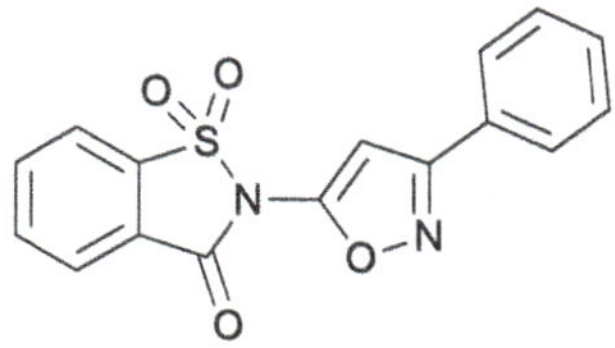

(7)

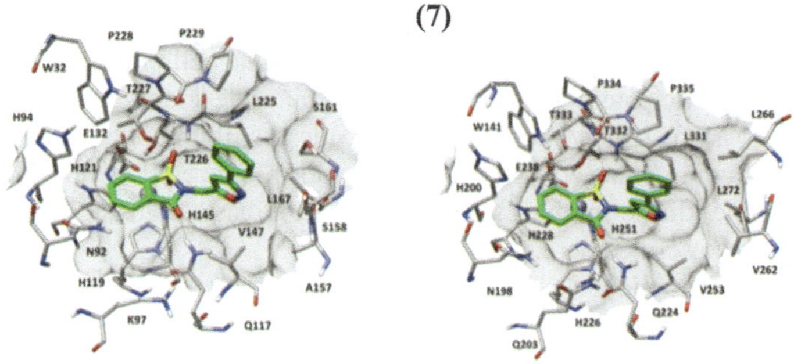

Fig. 7 Chemical structure of compound 7 is shown. Interaction of compound 7 (green) in complex with hCA XII and hCA IX is presented. The protein surface surrounding the ligand is shown in gray color

design strategy that showed significant inhibition of various carbonic anhydarses. Molecular docking studies indicated that sulfonamide group of the ligands forms the efficient binding with the zinc metal atom of the CAs (Milite et al. 2019; Khushal et al. 2022).

3 Computer-Aided Drug Design for Tumor-Associated Serine/Threonine Protein Kinase

Protein kinases are key regulators of cell function and catalyze the covalent addition of phosphate to substrate proteins. The deregulated kinase activities lead to the development and progression of several types of cancers and, therefore, are of great clinical interest in drug discovery. Bioinformatics including genetic, computational technologies, and mathematical and genomic-based strategies have become the most promising option for effective drug discovery against kinases. There is still a long way to go by combining molecular biology, genetics, bioinformatics, and biochemistry for modern drug discovery including kinase inhibitors (Chen and Chen 2006). Bioinformatics has been employed to investigate the mechanisms of protein kinase regulation along with functional and structural features. Bioinformatics has been helpful in building human protein kinase gene family and repository, to identify kinases on data analysis that are important in drug discovery. Several drugs have been developed as kinase inhibitors with significant selectivity, potency, and pharmacokinetic properties indicating the role of bioinformatics in drug development against kinases (Chen et al. 2015). Kinases are involved in the phosphorylation process, and abnormal protein phosphorylation has been proved to be a primary cause of some cancers. Therefore, there has been an increasing growth of interest in developing kinase inhibitors for cancer treatment. Protein kinases have now become the second most important group of drug targets, and serine-threonine kinases are exploited in drug discovery for tumor management (Cohen 2002).

Serine-threonine kinases cause phosphorylation of serine or threonine residues of the protein substrate and interact with backbone residues near the end of the activation segment. Some serine/threonine kinases regulate cell proliferation, survival, and death and participate in many processes of tumor development. To identify the inhibitors of various serine/threonine kinases, the use of computational techniques resulted in a finding of several potential inhibitors (Gagic et al. 2020). Some of the serine-threonine kinase inhibitors developed on the basis of using bioinformatics approaches are described here.

A series of nitroimidazole-derived compounds having aryl hydroxylamine scaffold were designed and screened through molecular docking to investigate the binding affinity compounds with MARK4 serine-threonine kinase. It was observed that all the compounds showed good affinity toward MARK 4; however, it was predicted that compound 8 binds the active site of protein effectively. This compound showed the binding energy of −7.5 kcal/mol on interaction with MARK 4 protein. The hydroxyl group of compound 8 forms a hydrogen bond with Asn193, and the oxygen atom of the nitro group also makes a hydrogen bond with Asp196. Imidazole ring is interacting through a pi-sigma bond with Val70; however, the phenyl ring is also forming the same bonding with Val70, Ly85, and Ala195 residues. The nitrogen atom of –NOH is held with Glu182 and Asp196 by attractive charges whereas the phenyl ring is interacting with Met132 through pi-sulfur interactions (Fig. 8). The compounds showed better inhibition of MARK4 in vitro, and the results were in agreement with the predictions of molecular docking (Peerzada et al. 2020b).

In another attempt to develop the potent seine threonine MARK4 potential inhibitors, molecular docking was used to determine the affinity of designed compounds. In silico approach proved to be helpful in the optimization of various analogs to develop promising inhibitors. The unsubstituted compound 9 containing the phenyl ring, hydroxylamine core, and morpholine fragment was docked with MARK4. Compound 9 demonstrated the binding energy of −6.9 kcal/mol on interaction with MARK 4 protein. It was found that compound 9 exhibited favorable interactions with the various residues of MARK4. The morpholine ring formed π-alkyl interactions with residues Val70 and Ala195, whereas the phenyl ring interacts with Val116, Ala83, Met111, Ala195, Lys85, and Val70 through π-alkyl interaction. The free nitrogen atom of the free –NOH group undergoes polar interaction with Asp196, whereas a hydrogen bond is formed between Lys85 residue and the hydrogen atom of the –NOH group (Fig. 9). On the basis of in silico data, various derivatives of compounds were prepared and examined in vitro against MARK4 and compounds showed significant inhibition of MARK4 protein (Peerzada et al. 2020a).

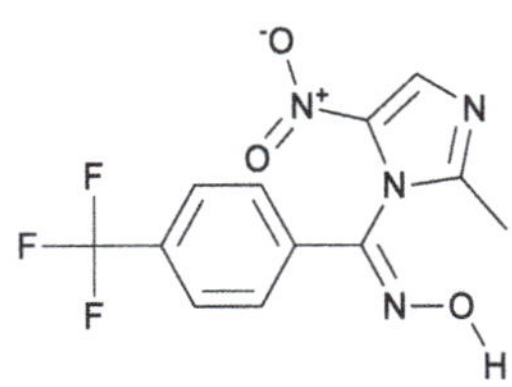

(8)

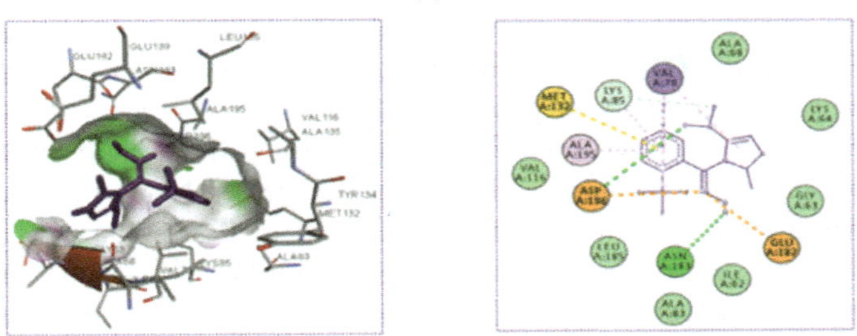

(a) **(b)**

Fig. 8 Structure of compound 8 and interactions with serine/threonine kinase MARK4 is shown. **a** A focused view of the MARK4 binding pocket with compound 8 depicts the residues undergoing hydrogen bonding. **b** 2D representation of various residues undergoing diverse interactions

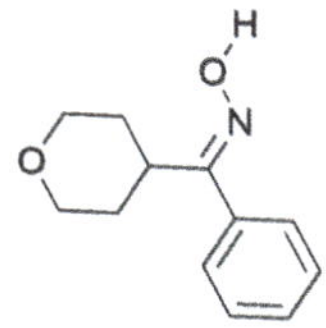

(9)

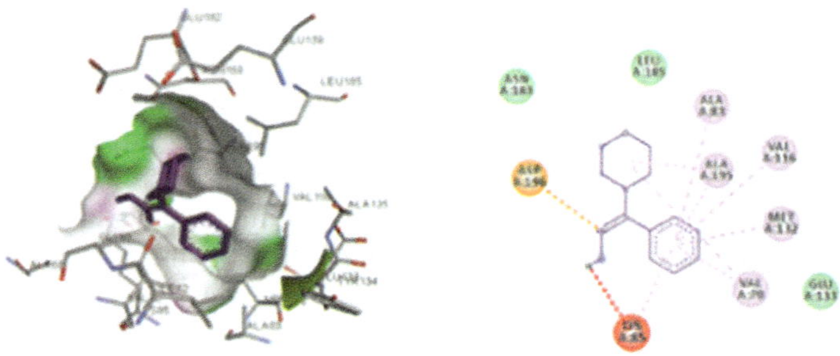

(a) **(b)**

Fig. 9 Chemical structure of compound 9 and interactions are shown. **a** View of the catalytic pocket of MARK4 serine-threonine kinase with compound 9. **b** 2D representation of the docking models of compound 9. Dotted lines in different colors reflected various types of interaction

Using CADD, purine-based compounds were designed and docked with serine-threonine Aurora kinase to develop the novel inhibitors. It was observed that compound 10 formed the hydrogen bond with various residues of the active site of Aurora kinase. The ligand-Aurora kinase complex showed that compound 10 interacts with Val107, Leu223, Lys101, Ala120, Leu154, Pro174, Leu170, Gly176, Phe172, Glu171, Gly100, Leu99, Arg97, Ala173, Glu181, Lys180, Arg175, and Lys184 residues effectively. The ligand forms four hydrogen bonds and two π–π stacking with Aurora kinase (Fig. 10) (Khalifa 2021).

IM-1 is a serine/threonine kinase comprising two domains that are connected by a hinge region. The active site residues in the groove are formed by these domains. Pyrazolopyrimidine derivatives were designed and screened in silico for predicting the interactions with PIM-1 serine/threonine kinase. Computational studies showed that compound 11 has significant physical features such as LogP, molecular weight, and topological polar surface area. Compound 11 showed the MOE docking score of −19.54 kcal/ mol with human PIM-1 kinase. The –NH group attached to the phenyl ring showed π–π interactions with Phe49 of PIM-1 kinase. Compound 11 showed hydrophobic interactions with Ile185 and Leu174 residues. Molecular docking suggested target compound 11 and analogs represent potential anticancer candidates (Fig. 11) (Philoppes et al. 2020).

CDK2 and GSK-3b are the serine/threonine protein kinases that are treated as important targets in breast cancer drug development. Therefore, various compounds of oxindole-benzofuran hybrids were designed and synthesized. The designed

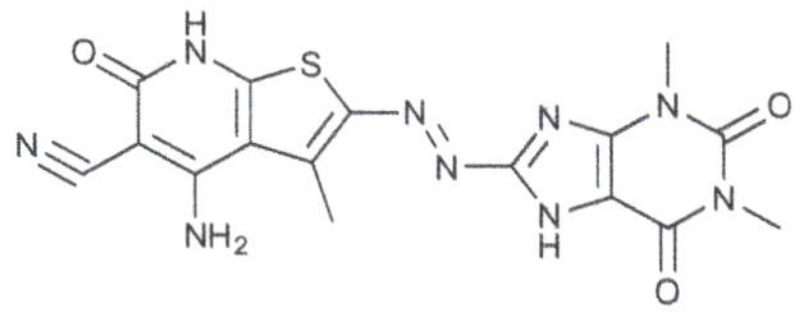

(10)

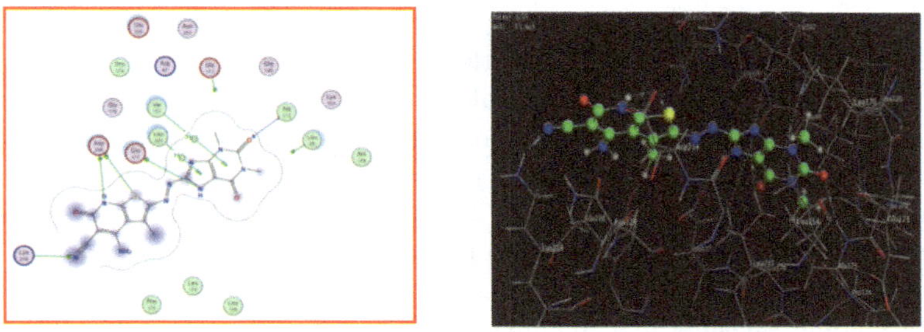

(a) (b)

Fig. 10 Chemical structure of compound 10 and interactions with Aurora kinase are shown. **a** Representation of 2D interaction of compound 10 with Aurora kinase; **b** 3D representation of compound 10 with Aurora serine-threonine kinase showing various interactions

Fig. 11 Chemical structure of compound 11 and interactions with PIM-1 serine-threonine kinase

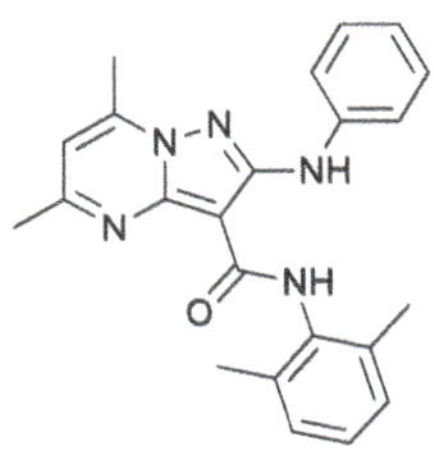

(11)

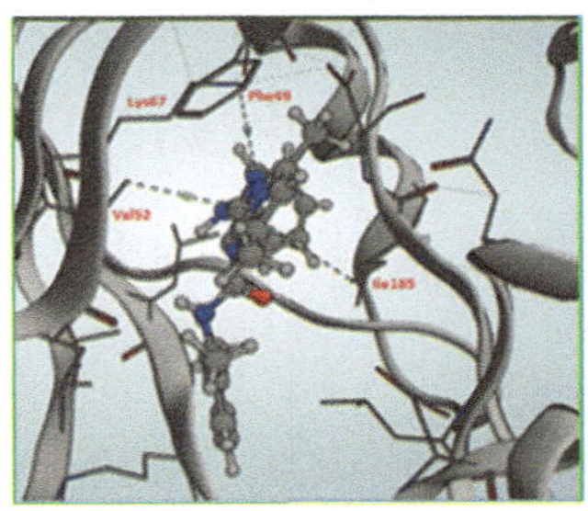

compounds were evaluated by using molecular docking to find out the interactions with CDK2 and GSK-3b kinases. It was found that compound 12, the *N1*-unsubstituted oxindole hybrid showed similar interactions with CDK2 and GSK-3b kinases. The oxindole ring occupies the hinge region and forms a hydrogen bond with Glu81 and Leu83 residues in CDK2 and with Asp133 and Val135 residues in GSK-3b.

The phenyl ring attached oxindole nucleus interacts through hydrophobic interaction with Ala31, Val18, Val64, Leu134, Ala144, and Phe80 residues in CDK2 and with Cys199, Val70, Val110, Leu188, Leu132, and Ala83 residues in GSK-3b. The –NH group of hydrazone linkers interacts through hydrogen bonding with Leu83 and Val135 both in CDK2 and GSK-3b. The benzofuran moiety forms hydrophobic interactions with Ile10, Leu298, and Phe82 residues in CDK2 and Pro136, Ile62, and

Tyr134 residues in GSK-3b. The –NH of hydrazone interacts via hydrogen bonding with residues Val135 and Leu83 in GSK-3b and CDK2, respectively. The benzofuran moiety orients outward toward the bulk solvent and is involved in a hydrophobic interaction with amino acids available at the gate of the hinge region; Phe82, Leu298, and Ile10 in CDK2 and Tyr134, Pro136, and Ile62 in GSK-3b kinase (Figs. 12 and 13) (Eldehna et al. 2021).

Using a computational approach, pyrazole containing new compounds was designed and screened in silico to determine the binding affinity of CDK2 serine-threonine kinase. Docking results revealed that compound 13 formed residues on one Pi–Pi bond, and two hydrogen bonds with Phe80, Asp86, and Lys89 residues. Pi–Pi bond was formed between phenyl ring which directly attached to the nitrogen atom of pyrazole and Phe80 residue of CDK2. One hydrogen bond was between the –NH group and Asp86, whereas another hydrogen bond was formed between sulfur of thiourea linker and Lys89 (Fig. 14).

This binding revealed that thiourea-containing compounds bind tightly in the active site of CDK2. Compound 13 showed IC_{50} of 25 nM in vitro against CDK2 and validated the predicted inhibition through the docking tool. The surface model showed that compound 13 occupies the ATP-binding pocket of CDK2 kinase (Sun et al. 2013).

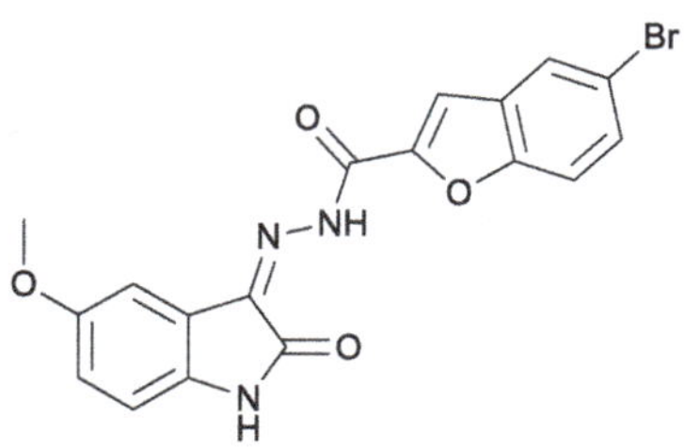

(12)

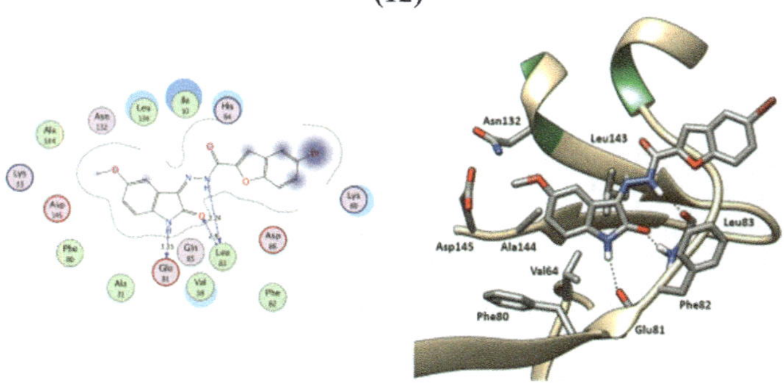

(a) **(b)**

Fig. 12 Chemical structure of compound 12 and interactions are shown. **a** 2 D diagram and **b** 3D representation of compound 12 showing various interactions with the CDK2 serine-threonine kinase

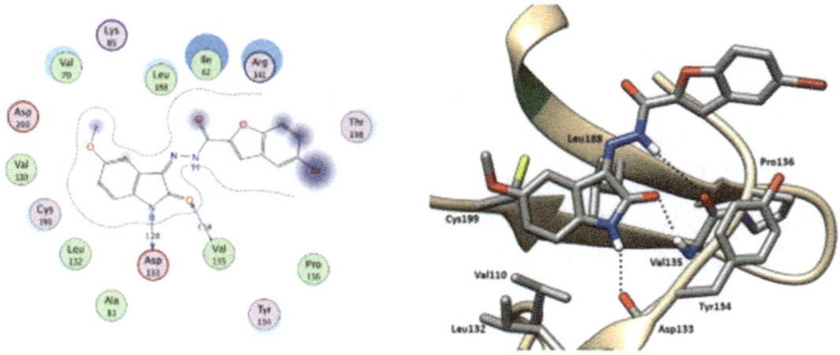

(a) **(b)**

Fig. 13 **a** 2D diagram and **b** 3D representation of compound 12 showing interaction with the GSK-3b serine-threonine kinase active site

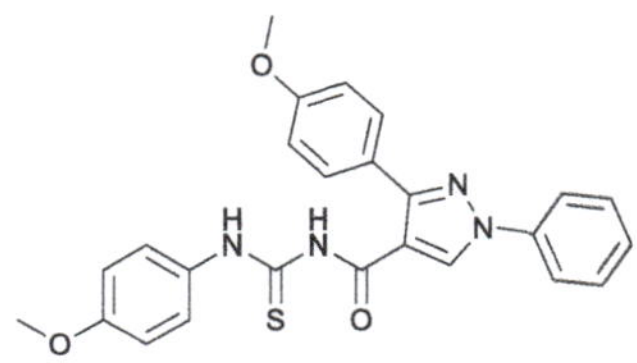

(13)

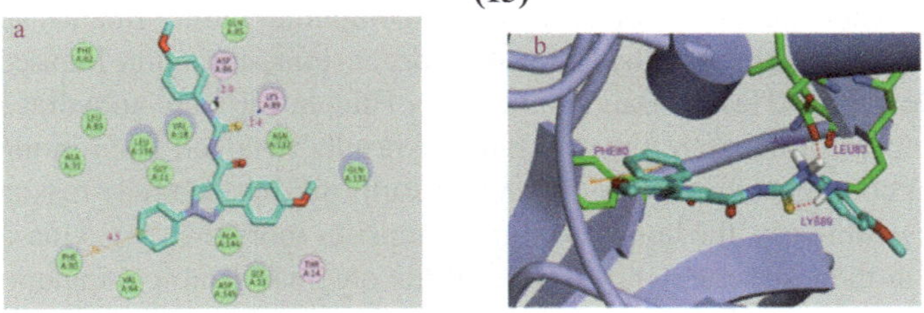

Fig. 14 Chemical structure of compound 13 and interactions with CDK2 are shown. **a** 2D molecular docking model of compound 13 with CDK2. **b** Model of compound 13 bound to CDK2 serine-threonine kinase

Through a computational approach, Galuteolin and Linarin were identified as the potential inhibitors of Akt1 and Akt2 serine/threonine kinases. Galuteolin and Linarin exhibited the respective binding affinities of −12.3 and −11.5 kcal/mol against Akt1 whereas for Akt2 the binding affinities were −11.4 and −11.1 kcal/mol (Sharif Siam et al. 2021). Similarly, 3- substituted indolin-2-one derivatives were identified as potent serine-threonine kinase inhibitors for cancer therapy through computational methods (Xu et al. 2022). A series of 3,5-disubstituted-2-aminopyrazines were synthesized as serine/threonine protein kinases Aurora kinases inhibitors. Compound 4-((3-amino-6-(3,5-dimethylisoxazol-4-yl)pyrazin-2-yl)oxy)-*N*-(3-chlorophenyl) benzamide inhibited the Aurora A and B kinases with the IC_{50} values of 90 and 152 nM, respectively. Bioinformatics revealed the significant binding affinity and drug-likeness of this compound (Bo et al. 2020). Bioinformatics has been helpful in identifying the nilotinib, vemurafenib, and idelalisib as serine/threonine protein Pim-1 kinase for consideration as potential inhibitors (Arrouchi et al. 2019). Several potential serine/threonine protein kinase inhibitors were identified through computational techniques recently and some of them have shown nanomolar efficacy in vitro (Parate et al. 2021; Swamy et al. 2021; Zhong et al. 2022).

4 Limitation of Nature-Based Computing Bioinformatics Approaches in Drug Discovery

Although the nature-based computing bioinformatics approaches proved to be useful in the drug repurposing, design, and development of several anticancer agents, there are significant limitations associated with these methods. In general, the binding pattern and the affinities shown by the compounds through various computational methods do not correlate with results either at in vitro or in vivo stage of screening due to remarkable differences in biological systems and computer-based algorithms. One of the significant challenges for nature-based computing bioinformatics approaches in discovering the carbonic anhydrase inhibitors or serine-threonine kinase inhibitors is target flexibility (Hassan Baig et al. 2016). Human carbonic anhydrases exist in 15 different isoforms, therefore, the binding results obtained from bioinformatics approaches contradict the in vitro or in vivo results due to selectivity reasons (Pinard et al. 2015). Similarly, the human kinome comprises more than 500 kinases and the similarity in their structures makes the results obtained from computational techniques to correlate with the in vitro or in vivo results challenging (Duong-Ly and Peterson 2013). This is because it is unlikely to simulate the complete biological system on a computer system using state-of-the-art techniques (Hassan Baig et al. 2016).

5 Conclusion

Nature-driven computing-based bioinformatics approaches including SBDD and LBDD have proven useful in finding the potent inhibitors of tumor-associated human carbonic anhydrases and serine-threonine kinases. CADD is widely used at present in the discovery of effective new molecules and in increasing the efficacy of exiting anticancer compounds. Medicinal chemists use the CADD as a tool in optimizing the various substituents, and privileged scaffolds for designing novel anti-tumor drugs. Bioinformatics has been useful in the discovery of SLC-0111 and other analogs benzene-sulfonamides as carbonic anhydrase inhibitors for cancer treatment. Similarly, several serine-threonine kinase inhibitors containing pyrazole, pyrazolopyrimidine, purine, hydroxylamines, and quinazoline were developed for cancer treatment with the aid of computational techniques. In nutshell, bioinformatics serves as a useful tool in early-stage drug design and development of potent anticancer molecules targeting carbonic anhydrases and serine-threonine kinases. The advancement in the algorithms derived from nature-inspired strategy could be helpful in designing drug candidates with improved efficacy for tumor-associated carbonic anhydarses and serine-threonine human kinases in the future.

Acknowledgement M. N. Peerzada is thankful to the Indian Council of Medical Research (ICMR) New Delhi, Department of Health Research, Government of India for a postdoctoral fellowship (No. 3/1/3PDF(24)/2021-HRD-6).

References

Agarwal R, Singh A, Sen S (2016) Role of molecular docking in computer-aided drug design and development, pp 1–28

Aggarwal M, Boone CD, Kondeti B, McKenna R (2013) Structural annotation of human carbonic anhydrases. J Enzyme Inhib Med Chem 28:267–277. https://doi.org/10.3109/14756366.2012.737323

Aimene Y, Eychenne R, Rodriguez F et al (2021) Synthesis, crystal structure, inhibitory activity and molecular docking of coumarins/sulfonamides containing triazolyl pyridine moiety as potent selective carbonic anhydrase IX and XII inhibitors. Curr Comput-Aided Drug Des 11:1076. https://doi.org/10.3390/cryst11091076

Alkhaldi AAM, Al-Sanea MM, Nocentini A et al (2020) 3-Methylthiazolo[3,2-a]benzimidazole-benzenesulfonamide conjugates as novel carbonic anhydrase inhibitors endowed with anticancer activity: design, synthesis, biological and molecular modeling studies. Eur J Med Chem 207:112745. https://doi.org/10.1016/j.ejmech.2020.112745

Alterio V, Di Fiore A, D'Ambrosio K et al (2012) Multiple binding modes of inhibitors to carbonic anhydrases: how to design specific drugs targeting 15 different isoforms? Chem Rev 112:4421–4468. https://doi.org/10.1021/cr200176r

Argelaguet R, Velten B, Arnol D, et al (2018) Multi-Omics Factor Analysis—a framework for unsupervised integration of multi-omics data sets. Mol Syst Biol 14. https://doi.org/10.15252/msb.20178124

Arrouchi H, Lakhlili W, Ibrahimi A (2019) Re-positioning of known drugs for Pim-1 kinase target using molecular docking analysis. Bioinformation 15:116–120. https://doi.org/10.6026/97320630015116

Bakker OB, Aguirre-Gamboa R, Sanna S et al (2018) Integration of multi-omics data and deep phenotyping enables prediction of cytokine responses. Nat Immunol 19:776–786. https://doi.org/10.1038/s41590-018-0121-3

Bo Y-X, Xiang R, Xu Y et al (2020) Synthesis, biological evaluation and molecular modeling study of 2-amino-3,5-disubstituted-pyrazines as Aurora kinases inhibitors. Bioorg Med Chem 28:115351. https://doi.org/10.1016/j.bmc.2020.115351

Chen Q, Chen Y-PP (2006) Mining frequent patterns for AMP-activated protein kinase regulation on skeletal muscle. BMC Bioinform 7:394. https://doi.org/10.1186/1471-2105-7-394

Chen Q, Luo H, Zhang C, Chen Y-PP (2015) Bioinformatics in protein kinases regulatory network and drug discovery. Math Biosci 262:147–156. https://doi.org/10.1016/j.mbs.2015.01.010

Cohen P (2002) Protein kinases—the major drug targets of the twenty-first century? Nat Rev Drug Discov 1:309–315. https://doi.org/10.1038/nrd773

Cui JJ, Tran-Dubé M, Shen H et al (2011) Structure based drug design of Crizotinib (PF-02341066), a potent and selective dual inhibitor of mesenchymal-epithelial transition factor (c-MET) kinase and anaplastic lymphoma kinase (ALK). J Med Chem 54:6342–6363. https://doi.org/10.1021/jm2007613

Cui JJ, McTigue M, Kania R, Edwards M (2013) Case history, pp 421–434

Cui W, Aouidate A, Wang S et al (2020) Discovering anti-cancer drugs via computational methods. Front Pharmacol 11. https://doi.org/10.3389/fphar.2020.00733

D'Ascenzio M, Secci D, Carradori S et al (2020) 1,3-Dipolar Cycloaddition, HPLC Enantioseparation, and docking studies of Saccharin/Isoxazole and Saccharin/Isoxazoline derivatives as selective carbonic anhydrase IX and XII inhibitors. J Med Chem 63:2470–2488. https://doi.org/10.1021/acs.jmedchem.9b01434

Du D, Chang C-H, Wang Y et al (2019) Response envelope analysis for quantitative evaluation of drug combinations. Bioinformatics 35:3761–3770. https://doi.org/10.1093/bioinformatics/btz091

Duong-Ly KC, Peterson JR (2013) The human kinome and kinase inhibition. Curr Protoc Pharmacol 60. https://doi.org/10.1002/0471141755.ph0209s60

Durrant JD, McCammon JA (2011) Molecular dynamics simulations and drug discovery. BMC Biol 9:71. https://doi.org/10.1186/1741-7007-9-71

Eldeeb AH, Abo-Ashour MF, Angeli A et al (2021) Novel benzenesulfonamides aryl and arylsulfone conjugates adopting tail/dual tail approaches: synthesis, carbonic anhydrase inhibitory activity and molecular modeling studies. Eur J Med Chem 221:113486. https://doi.org/10.1016/j.ejmech.2021.113486

Eldehna WM, Al-Rashood ST, Al-Warhi T et al (2021) Novel oxindole/benzofuran hybrids as potential dual CDK2/GSK-3β inhibitors targeting breast cancer: design, synthesis, biological evaluation, and in silico studies. J Enzyme Inhib Med Chem 36:271–286. https://doi.org/10.1080/14756366.2020.1862101

Felip E, Barlesi F, Besse B et al (2018) Phase 2 study of the HSP-90 inhibitor AUY922 in previously treated and molecularly defined patients with advanced non-small cell lung cancer. J Thorac Oncol 13:576–584. https://doi.org/10.1016/j.jtho.2017.11.131

Gagic Z, Ruzic D, Djokovic N et al (2020) In silico methods for design of kinase inhibitors as anticancer drugs. Front Chem 7. https://doi.org/10.3389/fchem.2019.00873

Gillet VJ (2004) Designing combinatorial libraries optimized on multiple objectives, pp 335–354

Hassan Baig M, Ahmad K, Roy S et al (2016) Computer aided drug design: success and limitations. Curr Pharm Des 22:572–581. https://doi.org/10.2174/1381612822666151125000550

Hiss JA, Hartenfeller M, Schneider G (2010) Concepts and applications of natural computing techniques in de novo drug and peptide design. Curr Pharm Des 16:1656–1665. https://doi.org/10.2174/138161210791164009

Huang G, Li J, Wang P, Li W (2018) A review of computational drug repositioning approaches. Comb Chem High Throughput Screen 20:831–838. https://doi.org/10.2174/1386207321666171221112835

Hutter M (2009) In Silico prediction of drug properties. Curr Med Chem 16:189–202. https://doi.org/10.2174/092986709787002736

Kania RS (2009) Structure-based design and characterization of axitinib. In: Kinase inhibitor drugs. John Wiley & Sons, Inc., Hoboken, NJ, USA, pp 167–201

Khalifa ME (2021) Design, synthesis and molecular docking study of new purine derivatives as Aurora kinase inhibitors. J Mol Struct 1229:129843. https://doi.org/10.1016/j.molstruc.2020.129843

Khushal A, Mumtaz A, Shadoul WA et al (2022) Synthesis, carbonic anhydrase II/IX/XII inhibition, DFT, and molecular docking studies of hydrazide-sulfonamide hybrids of 4-methylsalicyl- and acyl-substituted hydrazide. Biomed Res Int 2022:1–16. https://doi.org/10.1155/2022/5293349

Kumar S, Rulhania S, Jaswal S, Monga V (2021) Recent advances in the medicinal chemistry of carbonic anhydrase inhibitors. Eur J Med Chem 209:112923. https://doi.org/10.1016/j.ejmech.2020.112923

Mboge MY, Chen Z, Wolff A et al (2018) Selective inhibition of carbonic anhydrase IX over carbonic anhydrase XII in breast cancer cells using benzene sulfonamides: Disconnect between activity and growth inhibition. PLoS ONE 13:e0207417. https://doi.org/10.1371/journal.pone.0207417

McDonald PC, Chia S, Bedard PL et al (2020) A phase 1 study of SLC-0111, a novel inhibitor of carbonic anhydrase IX, in patients with advanced solid tumors. Am J Clin Oncol 43:484–490. https://doi.org/10.1097/COC.0000000000000691

Meadows KL, Hurwitz HI (2012) Anti-VEGF therapies in the clinic. Cold Spring Harb Perspect Med 2:a006577–a006577. https://doi.org/10.1101/cshperspect.a006577

Milite C, Amendola G, Nocentini A et al (2019) Novel 2-substituted-benzimidazole-6-sulfonamides as carbonic anhydrase inhibitors: synthesis, biological evaluation against isoforms I, II, IX and XII and molecular docking studies. J Enzyme Inhib Med Chem 34:1697–1710. https://doi.org/10.1080/14756366.2019.1666836

Parate S, Kumar V, Lee G et al (2021) Marine-derived natural products as ATP-competitive mTOR kinase inhibitors for cancer therapeutics. Pharmaceuticals 14:282. https://doi.org/10.3390/ph14030282

Peerzada MN, Khan P, Ahmad K et al (2018) Synthesis, characterization and biological evaluation of tertiary sulfonamide derivatives of pyridyl-indole based heteroaryl chalcone as potential carbonic anhydrase IX inhibitors and anticancer agents. Eur J Med Chem 155. https://doi.org/10.1016/j.ejmech.2018.05.034

Peerzada MN, Khan P, Khan NS et al (2020a) Identification of morpholine based hydroxylamine analogues: selective inhibitors of MARK4/Par-1d causing cancer cell death through apoptosis. New J Chem 44. https://doi.org/10.1039/d0nj03474f

Peerzada MN, Khan P, Khan NS et al (2020b) Design and development of small-molecule Arylaldoxime/5-Nitroimidazole hybrids as potent inhibitors of MARK4: a promising approach for target-based cancer therapy. ACS Omega 5:22759–22771. https://doi.org/10.1021/acsomega.0c01703

Philoppes JN, Khedr MA, Hassan MHA et al (2020) New pyrazolopyrimidine derivatives with anticancer activity: design, synthesis, PIM-1 inhibition, molecular docking study and molecular dynamics. Bioorg Chem 100:103944. https://doi.org/10.1016/j.bioorg.2020.103944

Pinard MA, Mahon B, McKenna R (2015) Probing the surface of human carbonic anhydrase for clues towards the design of isoform specific inhibitors. Biomed Res Int 2015:1–15. https://doi.org/10.1155/2015/453543

Piotrowska Z, Costa DB, Oxnard GR et al (2018) Activity of the Hsp90 inhibitor luminespib among non-small-cell lung cancers harboring EGFR exon 20 insertions. Ann Oncol 29:2092–2097. https://doi.org/10.1093/annonc/mdy336

Schüller A, Schneider G (2008) Identification of hits and lead structure candidates with limited resources by adaptive optimization. J Chem Inf Model 48:1473–1491. https://doi.org/10.1021/ci8001205

Schwefel H-P (2002) Deep insight from simple models of evolution. Biosystems 64:189–198. https://doi.org/10.1016/S0303-2647(01)00186-1

Sharif Siam MK, Sarker A, Sayeem MMS (2021) In silico drug design and molecular docking studies targeting Akt1 (RAC-alpha serine/threonine-protein kinase) and Akt2 (RAC-beta serine/threonine-protein kinase) proteins and investigation of CYP (cytochrome P450) inhibitors against MAOB (monoamine oxida. J Biomol Struct Dyn 39:6467–6479. https://doi.org/10.1080/07391102.2020.1802335

Sheng Z, Sun Y, Yin Z et al (2017) Advances in computational approaches in identifying synergistic drug combinations. Brief Bioinform. https://doi.org/10.1093/bib/bbx047

Sun J, Lv X-H, Qiu H-Y et al (2013) Synthesis, biological evaluation and molecular docking studies of pyrazole derivatives coupling with a thiourea moiety as novel CDKs inhibitors. Eur J Med Chem 68:1–9. https://doi.org/10.1016/j.ejmech.2013.07.003

Swamy PMG, Abbas N, Dhiwar PS et al (2021) Discovery of potential Aurora-A kinase inhibitors by 3D QSAR pharmacophore modeling, virtual screening, docking, and MD simulation studies. J Biomol Struct Dyn 1–22. https://doi.org/10.1080/07391102.2021.2004236

Trott O, Olson AJ (2010) AutoDock Vina: improving the speed and accuracy of docking with a new scoring function, efficient optimization, and multithreading. J Comput Chem 31:455–461. https://doi.org/10.1002/jcc.21334

Usha T, Shanmugarajan D, Goyal AK et al (2018) Recent updates on computer-aided drug discovery: time for a paradigm shift. Curr Top Med Chem 17:3296–3307. https://doi.org/10.2174/1568026618666180101163651

Xu C, Liu Y, Zhao G (2022) The Development of 3-substituted Indolin-2-one Derivatives as Kinase Inhibitors for Cancer Therapy. Curr Med Chem 29:1891–1919. https://doi.org/10.2174/0929867328666210831142311

Yamali C, Sakagami H, Uesawa Y et al (2021) Comprehensive study on potent and selective carbonic anhydrase inhibitors: Synthesis, bioactivities and molecular modelling studies of 4-(3-(2-arylidenehydrazine-1-carbonyl)-5-(thiophen-2-yl)-1H-pyrazole-1-yl) benzenesulfonamides. Eur J Med Chem 217:113351. https://doi.org/10.1016/j.ejmech.2021.113351

Zhong S, Hou Y, Zhang Z et al (2022) Identification of novel natural inhibitors targeting AKT Serine/Threonine Kinase 1 (AKT1) by computational study. Bioengineered 13:12003–12020. https://doi.org/10.1080/21655979.2021.2011631

The manufacturer's authorised representative in the EU is Springer Nature Customer Service Centre GmbH, Europaplatz 3, 69115 Heidelberg, Germany. If you have any concerns regarding our products, please contact ProductSafety@springernature.com

Printed and bound by CPI Group (UK) Ltd, Croydon, CR0 4YY
30/09/2026
02225242-0013